Water Projects and Technologies in Asia

This book is a collection of highly refined articles on historical water projects and traditional water technologies of international interest in the Asian regions, addressing information on past water projects (mostly before the 20th century) that are technically and culturally of interest and educationally valuable.

This book explores historical water projects in these regions, presenting technologies used at the time, including calculation and forecasting methods, measurement, material, labor, methodologies, and even water culture. It is expected that the old Asian wisdom of "reviewing the old and learning the new" would be realized to a certain extent in modern planning and practice of water projects.

This book will enable the reader to understand historical water projects and technologies in the Asian regions. It can be used as a one-stop resource to source notable Asian water projects and their relevance to modern-day technology. In this regard, this book is expected to be of interest to a variety of audiences, including the corresponding Asian regions and other international audiences interested in Asian water history from an engineering perspective.

IAHR Books

Series Editor: Robert Ettema

Department of Civil and Environmental Engineering, Colorado State University, Fort Collins, USA

The International Association for Hydro-Environment Engineering and Research (IAHR), founded in 1935, is a worldwide, independent organization of engineers and water specialists working in fields related to hydraulics and its practical application. Activities range from river and maritime hydraulics to water resources development and ecohydraulics, to ice engineering, hydroinformatics, and continuing education and training. IAHR stimulates and promotes both research and its application and, by doing so, strives to contribute to sustainable development, the optimization of world water resources management, and industrial flow processes. IAHR accomplishes its goals by a wide variety of member activities, including the establishment of technical committees, working groups, congresses, specialty conferences, workshops, and short courses; the commissioning and publication of journals, monographs and edited conference proceedings; involvement in international programs, such as the UNESCO, WMO, IDNDR, GWP, ICSU, and The World Water Forum; and by cooperation with other water-related (inter)national organizations.

www.iahr.org

IAHR AIRH

Supported by
Spain Water
and IWHR, China

Water Projects and Technologies in Asia
Historical Perspectives
Edited by Hyoseop Woo, Hitoshi Tanaka, Gregory De Costa, and Juan Lu

For more information about this series, please visit: www.routledge.com/IAHR-Book/book-series/IAHRMON

IAHR members benefit from a 30% discount on all books published under this series.

Water Projects and Technologies in Asia

Historical Perspectives

Edited by

Hyoseop Woo
Sejong University, Korea

Hitoshi Tanaka
Tohoku University, Japan

Gregory De Costa
Open Polytechnic of New Zealand, New Zealand

Juan Lu
Water History Department, IWHR, China

CRC Press
Taylor & Francis Group
Boca Raton London New York Leiden

CRC Press is an imprint of the
Taylor & Francis Group, an **informa** business
A BALKEMA BOOK

First published 2023
by CRC Press/Balkema
4 Park Square, Milton Park, Abingdon, Oxon, OX14 4RN
e-mail: enquiries@taylorandfrancis.com
www.routledge.com – www.taylorandfrancis.com

CRC Press/Balkema is an imprint of the Taylor & Francis Group, an informa business

Library of Congress Cataloging-in-Publication Data
Names: Woo, Hyoseop, editor. | Tanaka, Hitoshi, 1956- editor. | De Costa, Gregory, editor. | Lu, Juan (Water historian), editor.
Title: Water projects and technologies in Asia : historical perspectives / edited by Hyoseop Woo, Sejong University, Korea, Hitoshi Tanaka, Tohoku University, Japan, Gregory De Costa, Open Polytechnic of New Zealand, New Zealand, Juan Lu, Water History Department, IWHR, China.
Description: Boca Raton : CRC Press, 2023. | Series: IAHR book series | Includes bibliographical references and indexes.
Identifiers: LCCN 2022045553 (print) | LCCN 2022045554 (ebook) | ISBN 9781032120386 (hbk) | ISBN 9781032120423 (pbk) | ISBN 9781003222736 (ebk)
Subjects: LCSH: Hydraulic engineering—Asia—History. | Water-supply engineering—Asia—History. | Water and civilization—Asia. | Flood control—History—Asia. | Irrigation—History—Asia.
Classification: LCC TC99. W38 2023 (print) | LCC TC99 (ebook) | DDC 627.095—dc23/eng/20221012
LC record available at https://lccn.loc.gov/2022045553
LC ebook record available at https://lccn.loc.gov/2022045554

ISBN: 9781032120386 (hbk)
ISBN: 9781032120423 (pbk)
ISBN: 9781003222736 (ebk)

DOI: 10.1201/9781003222736

Typeset in Times New Roman
by codeMantra

About the IAHR Book Series

An important function of any large international organization representing the research, educational, and practical components of its wide and varied membership is to disseminate the best elements of its discipline through learned works, specialized research publications, and state-of-the-art reviews. The IAHR is particularly well-served in this regard by its flagship journals, and by the extensive and wide body of substantive historical and reflective books that have been published through its auspices over the years. The IAHR Book Series is an initiative of IAHR, in partnership with CRC Press/Balkema—Taylor & Francis Group, aimed at presenting state-of-the-art themes relating to all areas of hydro-environment engineering and research.

The Book Series will assist researchers and professionals working in research and practice by bridging the knowledge gap and by improving knowledge transfer among groups involved in research, education, and development. The series includes design manuals and monographs. The design manuals, usually prepared by multiple authors, guide the application of theory and research findings to engineering practice, and the monographs give state-of-the-art coverage of various significant topics in water engineering. The first and highly successful IAHR book was *Turbulence Models and their Application in Hydraulics* by W. Rodi, published in 1984 by Balkema.

Turbulence in Open Channel Flows by I. Nezu and H. Nakagawa, also published by Balkema (in 1993), had an important impact on the field and, during the period 2000–2010, further authoritative texts (published directly by IAHR) included *Fluvial Hydraulics* by S. Yalin and A. Da Silva and *Hydraulicians in Europe* by W Hager. All of these publications continue to strengthen the reach of IAHR and to serve as important intellectual reference points for the Association.

Since 2011, the Book Series has once again been a partnership between CRC Press/ Balkema—Taylor & Francis Group and the Technical Committees of the IAHR. The present book is an exciting further contribution to IAHR's Book Series, substantially aiding water engineering research, education, and practice, and showcasing the expertise the IAHR fosters.

IAHR Book Series Editor
Robert Ettema

Contents

Editors

Hyoseop Woo is currently an (Industry-cooperation) Professor at Sejong University, Seoul, Korea. Research interests are river hydraulics, including sedimentation engineering, ecohydraulics, and Nature-based Solutions for water. He was a co-chairs of the special sessions of "Historical water projects and traditional water technologies in the Asian region" in the three IAHR–APD Congresses from 2016 to 2021. Presently working as a Vice President of IAHR.

Hitoshi Tanaka is presently in the position of President-appointed Extraordinary Professor at Tohoku University, Sendai, Japan. Served as a chairman of IAHR–APD from 2011 to 2014 and a council member of IAHR from 2013 to 2017. Research interests are fluid mechanics, such as turbulent wave boundary layers, related sediment movement, and resulting morphodynamics in coastal and estuarine environments. Study sites are not limited to Japan, but cover other regions worldwide, such as Vietnam, Thailand, Korea, Indonesia, Oman, Bolivia, etc.

Gregory De Costa is presently in the position of Principal Academic Engineering, Open Polytechnic of New Zealand, Chartered Engineer, and Fellow of Engineering New Zealand. Research interests are climate change, salinity intrusion, water projects in New Zealand, Sri Lanka and other regions, ancient technologies, and water resources management. Presently working as Chair of IAHR–Asia Pacific Division since 2019.

Juan Lu is currently a Professor at the China Institute of Water Resources and Hydropower Research, Beijing, China. Research interests are water history and culture, flood control, and drought reduction. Presently working as Director of the Research Center of Flood Control and Drought Disaster Reduction, Ministry of Water Resources, China.

Contributors

Ryosuke Akoh
Graduate School of Environmental and Life Science
Okayama University
Okayama, Japan

Yulduz A. Ergasheva
Department of Uzbekistan History
Karshi Engineering Economics Institute
Karshi, Uzbekistan

Jingdong Cai
China Institute of Water Resources and Hydropower Research
Beijing, China

Chun Kiat Chang
River Engineering and Urban Drainage Research Centre (REDAC)
Universiti Sains Malaysia
Nibong Tebal, Penang, Malaysia.

Jin-yong Choi
Landscape Architecture and Rural Systems Engineering Department, College of Agriculture and Life Sciences
Seoul National University
Seoul, Korea

Gregory S. De Costa
Engineering
Open Polytechnic of New Zealand
New Zealand

Jun Deng
China Institute of Water Resources and Hydropower Research
Beijing, China
Research Centre on Flood and Drought Disaster Reduction of Ministry of Water Resource
Beijing, China

Sirapee Ditthabumrung
Panya Consultant
Thailand

Keping Fu
River and Lake Management Center of Pujiang County
Pujiang, Zhejiang Province, China

Hanbin Gu
Institute of Ocean Engineering
Ningbo University
Ningbo, China

Jianzhao Guan
China Institute of Water Resources and Hydropower Research
Beijing, China

Shushu Guo
Development Research Center of the Ministry of Water Resources of People's Republic of China
Beijing, China

H. W. Harindra
Department of Civil Engineering
The Open University of Sri Lanka
Nawala, Sri Lanka

Kiyoshi Hashimoto
Department of Construction
Miyagi Prefecture
Sendai, Japan

Michiko Hayashi
Faculty of Environmental and Urban Engineering
Kansai University
Suita, Japan

Yoshiya Igarashi
Graduate School of Science and Engineering
Saitama University
Saitama, Japan

Toshikazu Ikemoto
Faculty of Geoscience and Civil Engineering
Kanazawa University
NPO for Wise Learning on Tatsumi Aqueduct
Kanazawa, Japan

Taisuke Ishigaki
Faculty of Environmental and Urban Engineering
Kansai University
Suita, Japan

Tadaharu Ishikawa
Tokyo Institute of Technology
Tokyo, Japan

Noor Effarizan Ismail
Irrigation and Drainage Services Division
Muda Agricultural Development Authority (MADA), MADA Headquarters
Ampang Jajar, Alor Setar, Kedah Darul Aman, Malaysia

Cheolhee Jang
Department of Hydro Science and Engineering Research
Korea Institute of Civil Engineering and Building Technology
Goyang-si, Korea

J. M. Jayasundara
Department of Civil Engineering
The Open University of Sri Lanka
Nawala, Sri Lanka

Un Ji
Department of Hydro Science and Engineering Research
Korea Institute of Civil Engineering and Building Technology
Goyang-si, Korea

Xiaoming Jiang
China Institute of Water Resources and Hydropower Research
Beijing, China
Research Centre on Flood and Drought Disaster Reduction of Ministry of Water Resource
Beijing, China

Wataru Kawahara
Saga Prefecture
Saga, Japan

Ryuji Kawanaka
Hydro Technology Institute Co. Ltd.
Osaka, Japan

So Kazama
Department of Civil and Environmental Engineering
Tohoku University
Sendai, Japan

Hyeonjun Kim
Korea Institute of Civil Engineering and Building Technology
Goyang-si, Korea

Masaru Kitaura
Kanazawa University
NPO for Wise Learning on Tatsumi Aqueduct
Kanazawa, Japan

Koichi Koyamada
Osaka Prefectural Sayama-ike Museum
Japan

Djoko Legono
Department of Civil and Environmental Engineering
Universitas Gadjah Mada
Indonesia

Yunpengi Li
Key Laboratory of Water Heritage Protection
China Institute of Water Resources and Hydropower Research
Beijing, China

Jiangang Liu
China Institute of Water Resources and Hydropower Research
Beijing, China
Research Centre on Flood and Drought Disaster Reduction of Ministry of Water Resource
Beijing, China

Juan Lu
Research Centre on Flood and Drought Disaster Reduction
China Institute of Water Resources and Hydropower Research
Beijing, China

Nik Kun Nik Man
Engineering Division
Muda Agricultural Development Authority (MADA), MADA Headquarters
Ampang Jajar, Alor Setar, Kedah Darul Aman, Malaysia

Vivek L. Manekar
Department of Civil Engineering
SVNIT, Sardar Vallabhbhai National Institute of Technology
Surat, Gujarat, India

Alisher N. Khazratov
Department of Hydraulics and Hydraulic Structures
Karshi Engineering Economics Institute
Karshi, Uzbekistan

Fatchan Nurrochmad
Department of Civil and Environmental Engineering
Universitas Gadjah Mada
Indonesia

Pujianiki N. Nyoman
Department of Civil Engineering
Udayana University
Bali, Indonesia

Koichiro Ohgushi
Faculty of Science and Engineering
Saga University
Saga, Japan

G. N. Paranavithana
Department of Civil Engineering
The Open University of Sri Lanka
Nawala, Sri Lanka

Jing Peng
China Institute of Water Resources and Hydropower Research
Beijing, China

Ruwan Rajapakse
Consultant
New York, USA

R. S. Ranasinghe
Department of Civil Engineering
The Open University of Sri Lanka
Nawala, Sri Lanka

W. D. Ranasinghe
Department of Civil Engineering
The Open University of Sri Lanka
Nawala, Sri Lanka

Hiroshi Senoo
TOKEN C. E. E. Consultants Co., Ltd.
Tokyo, Japan

Tetsuya Sumi
Disaster Prevention Research Institute
Kyoto University
Kyoto, Japan

Nobuyuki Tamai
The University of Tokyo
Tokyo, Japan
NPO for Wise Learning on Tatsumi Aqueduct
Kanazawa, Japan

Xuming Tan
China Institute of Water Resources and Hydropower Research
Beijing, China

Hitoshi Tanaka
Institute of Liberal Arts and Sciences
Tohoku University
Sendai, Japan

Norio Tanaka
Graduate School of Science and Engineering
Saitama University
Saitama, Japan

Ritica Thakur
Department of Civil Engineering
SVNIT, Sardar Vallabhbhai National Institute of Technology
Surat, Gujarat, India

Phruetthiphong Thatanchuleekun
Royal Irrigation Department
Thailand

Nguyen Xuan Tinh
Department of Civil Engineering
Tohoku University
Sendai, Japan

Haruhiko Todo
Building Repairs Section, Civil Engineering Affairs Department
Kanazawa City
Kanazawa, Japan

Radianta Triatmadja
Department of Civil and Environmental Engineering
Universitas Gadjah Mada
Indonesia

Chengcheng Wang
Institute of Ocean Engineering
Ningbo University
Ningbo, China

Sutat Weesakul
Hydroinformatics Institute
Thailand

Uruya Weesakul
Thammasat University
Thailand

Hyoseop Woo
Department of Civil and Environmental Engineering
Sejong University
Seoul, Korea

Tibing Xu
Institute of Ocean Engineering
Ningbo University
Ningbo, China

Zhichao Yin
Institute of Ocean Engineering
Ningbo University
Ningbo, China

Cheolsang Yoo
School of Civil, Environmental and Architectural Engineering
Korea University
Seoul, Korea

Nor Azazi Zakaria
River Engineering and Urban Drainage Research Centre (REDAC)
Universiti Sains Malaysia
NibongTebal, Penang, Malaysia

Bo Zhou
China Institute of Water Resources and Hydropower Research
Beijing, China
Research Centre on Flood and Drought Disaster Reduction of Ministry of Water Resource
Beijing, China

Changhai Zhou
Zhuji Water Conservancy and Hydropower Bureau
Zhuji, Zhejiang Province, China

Changrong Zhou
Zhuji Water Conservancy and Hydropower Bureau
Zhuji, Zhejiang Province, China

Foreword

On behalf of the International Association for Hydro-environment Engineering and Research (IAHR), I would like to congratulate the Editors on the successful completion of this unique IAHR book on the history of hydraulics and its relationship to social and economic development in the Asia and Pacific region.

Over the past three decades, population growth, urbanization, and economic development of many Asian countries led to challenging problems in water and environmental engineering. We have also seen a corresponding rise in research, innovation, and education activity in the hydro-environment engineering sector. The IAHR Asia and Pacific Division (APD) has been a growth sector within IAHR, and contributed increasingly to the association's membership, intellectual strength, and vitality. The launch of the *Journal of Hydro-environment Research* (JHER) in 2007 was an example of a successful collaboration between IAHR, IAHR–APD, and the Korea Water Resources Association. JHER is the venue of publication of some of the best papers presented at the IAHR–APD Biennial Congresses and has become a signature journal that underlies the strength of our 86-year-old organization.

A key philosophy shared by many experts involved with IAHR–APD, and JHER is the publication of noteworthy historic Asia water projects to share water engineering knowledge and technology that has stood the test of time. Unlike traditional IAHR monographs, which are mostly concerned with the recent advances in a thematic research area of practical engineering, this book is about historic water projects and technologies in Asia. It is always felt that many ancient water projects in Asia have distinctly different features and characteristics from the western civilization that are of historical, cultural, and scientific interest. It is also believed that some projects may serve as good teaching material if documented properly, especially if they can be given a fresh perspective using the modern science of water, e.g., CFD modeling with in-depth theoretical interpretation. Also, reviving the past technologies used for the various water projects in Asia is equivalent, in many cases, to practicing modern-day's ecological engineering in water, which implies "Gaining new knowledge by reviewing old ones" in Asian philosophy.

For the reasons above, this book is a welcome and refreshing addition to the IAHR repertoire of books. The material is drawn from invited papers selected from special sessions convened during three past IAHR–APD Congresses (in 2016, 2018, and 2020), with representative projects from countries including China, Japan, Korea, India, Sri Lanka, and the Southeast Asia region. I would like to thank the Editors of this volume, especially the Convener and Editor-in-Chief, Dr. Hyoseop Woo of South Korea, and

Co-Editors Prof. Hitoshi Tanaka of Japan, Dr. Gregory De Costa of New Zealand, and Prof Juan Lu of China for the careful paper selection, coordination, and editing that culminated in this quality work.

Professor Joseph Hun-wei Lee
IAHR President
President, Macau University of Science and Technology

Preface

Historians tell us that the Indus River in South Asia and the Yellow River in East Asia were the birthplaces of the Great Ancient Civilizations, leaving aside the Fertile Crescent. This tells us that the Asian region and water have played a major role in world history. We also know that the staple food has been rice in vast areas of Asia, and paddy cultivation has been very prominent. The Asian region has been associated historically with water and rivers.

This book presents a brief history of the relationship between human society and water in Asia from an engineering aspect. At this moment, the authors focus especially on rivers, as they embrace only a tiny portion of water on the earth; 1.2/1,000,000 of the total amount of water found on the earth's surface and the ground, but have profoundly affected human life and society in Asia. This book's main interests are the relationship between human society and water. i.e., irrigation for paddy fields, flood management, inland navigation, and water culture.

In detail, this book explores historical water projects in this region to present the technologies used at the time, including the calculation and forecasting methods, measurement, material, labor methodologies, and even water culture. It intends to rediscover, recognize, and disseminate throughout the world, the historical water projects and traditional water technologies of international importance or a particular interest in the Asian region. The Asian region includes the Indian subcontinent, Southeast Asia, and East Asia, excluding Southwest Asia. The timespan considered in the book is from the ancient civilizations, until modern water technologies were implemented, say before the large-dam era started in the late 19th century.

Many articles in this book are based largely on the past three IAHR–APD Congress special sessions in Colombo 2016, Yogyakarta 2018, and Sapporo 2021. The co-editors also invited authors from different regions in Asia as lead writers covering particular interests of those regions, such as irrigation, flood management, navigation, and relevant technologies. In addition, the co-editors wrote one article each to cover information gaps among the selected and invited articles. Finally, the Editor-in-Chief wrote an overview article, "A historical review on the relationship between human society and water in Asia from the engineering aspect."

This book presents multiple articles from China, Japan, Korea, India, and Sri Lanka, and one article from each Southeast Asian country, including Cambodia, Thailand, Malaysia, and Indonesia. In addition, we invite one article from Uzbekistan in Central Asia. We, the co-editors, hope this book will enable the reader to understand historical water technologies in the Asian region and be used as a one-stop resource to source world notable water projects in the region and their relevance to modern-day technology.

Editor-in-Chief: Hyoseop Woo, Professor, Sejong University, Korea
Co-editors: Hitoshi Tanaka, Professor, Tohoku University, Japan
Gregory De Costa, Principal Academic, Open-polytechnic University, New Zealand
Juan Lu, Director of Water History Department, International Water Resources and Hydropower Research, China

Congratulatory Remarks

The earth is often called the Water Planet. Abundant water is one of the distinct characteristics of earth compared to other planets, producing many species and lives and supporting the ecosystem. As one of the lives on the earth, human beings have achieved remarkable development by utilizing water actively in a manner different from other species.

Meanwhile, in the IPCC (Intergovernmental Panel on Climate Change) Fifth Assessment Report, there is no doubt that the global climate system is warming. Extreme precipitation and prolonged drought will likely be stronger and more frequent in most regions by the end of the 21st century. In recent years, flood damages, for example, have become more severe and frequent, such as those in Thailand in 2011, Japan in 2019, China and Korea in 2020, etc. For this reason, the entire society needs to prepare for extreme floods and drought and rebuild a water-conscious society by implementing and combining hard and soft countermeasures.

Under such circumstances, looking back on the history of water projects and water technologies in Asia will be valuable in providing a guideline for the future direction of water and river management. It is called in Asian philosophy, "The old comes alive to find out what's new." Especially the Asian region, which accounts for 60% of the world population, has a far-reaching history of water technologies, not to mention the Great Ancient Civilizations. Understanding the history of water project development in Asia and absorbing the wisdom and experience of the ancient sages will help conform to the law of natural evolution and explore the sustainable development model of harmony between man and water. In that sense, it is significant to introduce a book containing various articles on the historical water projects and technologies in the Asian region.

The Advisory Committee would like to emphasize the importance of understanding localities and global features. Under the wide variety of climate and cultural backgrounds of the Asian region, we need to pay attention to the long history of severe natural catastrophes. Achievements obtained in science and technology on hydro-environmental heritages in the Asian region clarify historical wisdom on water and human societies in the region and give us light and hints towards the future.

We believe this book would benefit researchers, engineers, and water engineering and science students, not only in the Asian region but also in other parts of the world. We congratulate the authors and the editors on publishing this valuable IAHR book.

Professor Xiaotao Chen
China Institute of Water Resources and Hydropower Research (IWHR), China

Professor Subhasish Dey
Indian Institute of Technology Kharagpur, India

Professor Emeritus Nobuyuki Tamai
University of Tokyo, Japan

The Advisory Committee for the book of *Water Projects and Technologies in Asia: Historical Perspectives*

Chapter 1

A historical review of the relationship between human society and water in Asia – an engineering perspective

An introduction

Hyoseop Woo
Sejong University

CONTENTS

1.1 INTRODUCTION

In an article titled "Asia's sacred, wicked waters," Dipak Gyawali wrote in 2015 that Asia encompasses an enormous diversity: from population giants like China and India to sparsely populated small countries; and from the tropical deltas of the Mekong River to the hot deserts of the lower Indus or the cold plateaus of Tibet and Mongolia.

DOI: 10.1201/9781003222736-1

The writer mentions additional features marking the diversity of Asia: developed regions (like Japan, South Korea, and East China); eagerly developing regions; mighty rivers flowing into oceans after a long journey of thousands of kilometers on the continent (Yangtze River and the Ganges River); to rivers, like the Amu Darya, wandering over the continent and flowing into landlocked lakes and seas. Geographically, this article covers most regions of the Asian continent, ranging from the Japanese archipelago to the east, Central Asia and the Indian Subcontinent to the west, North China to the north, and the Indonesian archipelago to the south. Not covered are the Middle East, Near East, and Siberia.

This article presents a brief history of the relationship between human society and water in Asia from an engineering aspect. At this moment, the authors focus especially on rivers, as they embrace only a tiny portion of water on the earth; 1.2/1,000,000 of the total amount of water found on the earth's surface and the ground, but have profoundly affected human life and society in Asia. In addition, this article covers the period ranging from the birth of human civilization, about 3,500 years BC, to the end of the 19th century, at the height of the Second Industrial Revolution.

A short, scientific explanation is offered of water distribution on earth: ocean and sea, icecaps and glaciers, groundwater and atmosphere, and lakes and rivers. Then, this article briefly reviews the historical changes in the relationship between human society and water use (irrigation, flood management, navigation, fishery, esthetics, and religions). This does not cover municipal water supply and distribution to simplify the review, which came to the fore in most parts of Asia in the past century.

Four approximate historical eras are covered; (1) four ancient civilizations, (2) the beginning of the Common Era, (3) AD 1,000 years, and (4) the era of the Western (European) power encroaching on the East (Asia). This division of eras and the following brief explanations of human activities and water in each era are based on Woo's presentation (2005).

Lastly, in alphabetical order, this article describes a brief history of paddy-field irrigation activities in Asia, including China, Japan, Korea, South Asia, and Southeast Asia. The book extensively uses this regional division. Irrigation has been one of the main human activities related to water and river in Asia and is widely discussed throughout the book.

1.2 BRIEF REVIEW OF THE HISTORICAL RELATIONSHIP BETWEEN HUMAN SOCIETY AND WATER

1.2.1 Water and rivers on the earth

Water is the foundation of life. No living creatures on the earth can survive without water. Humans have found water relatively easily from streams, rivers, and lakes. Human history is linked to these sources of water. For example, water generally makes up more than 70% of the human body weight.

About 97% of all the water on the earth is seawater, and only 3% is freshwater in places other than oceans and seas. Among this tiny amount of fresh water, 99% is found in groundwater, ice caps, and glaciers. We can find 1% of the remaining water from lakes, atmosphere, soil, and rivers. Among them, only 0.4% of fresh water is found in rivers. Thus, as mentioned above, freshwater accounts for merely 1.2/1,000,000 or 1.2 ppm of the total water on the earth!

River water, such a tiny portion of water on the earth, has had a profound relationship with human society for more than 5,000 years. History has shown this dynamic relationship through political, cultural, industrial, social, and ecological perspectives in national and transnational settings. As integral sources of food and water, local and international transportation, recreation, and esthetic beauty, rivers have dictated where cities have risen and, in times of flooding, drought, and war, where they have fallen (Christof and Zellar, 2008). Civilizations sought to control rivers by channeling and diverting them for irrigation, taming them in flood control and canal systems, and damming them for power generation.

How have these actions been done in the Asian region? This article starts from the cradle of human civilization about 5,000 years ago and jumps to the beginning of the Common Era, jumps again to AD 1000 (the 11th century), and finally to the onset of the 20th century. This division of eras seems rather arbitrary. However, these eras approximately align in world history, with the beginning of the Common Era when two civilized societies coexisted in the West and East—The Roman Empire (27 BC–AD 476) and the Han dynasty (202 BC–AD 220). The era of AD 1000 is also associated with several civilized societies in the East, say, the Islam empires (caliphates) in the Middle East, various regional kingdoms in the Indian Subcontinent, and the Song dynasty (AD 960–1279) in China. However, it is called the "Medieval Age" or even the "Dark Age" in a less formal term in Europe. This division was inspired by "The World in the Year 1000" (Heitzman and Schenkluhn, 2004). The last division is the end of the 19th century or the beginning of the 20th century. It was the time when the Second Industrial Revolution started the engineered exploitation of water resources, as evident in the construction of many large dams.

1.2.2 The four ancient civilizations around the river valleys

In *A Study of History*, British historian Arnold Toynbee posited that the history of civilizations was centrally driven by a dynamic process of responses to environmental challenges. He emphasized that difficult challenges provoked exceptional, civilizing responses in ascendant societies, while inadequate responses contributed to stagnancy, subordination, and collapse in declining ones (Solomon, 2011). Water or river was a major environmental challenge in ancient times.

A river valley civilization is an agricultural nation or civilization situated beside and drawing sustenance from a river. A river gives the inhabitants a reliable water source for drinking and agriculture. Additional benefits include fishing, fertile soil due to annual flooding, ease of transportation, and a natural military defense barrier. A "civilization" means a society with large, permanent settlements featuring urban development, social stratification, specialization of labor, centralized organization, and written or other formal means of communication. As shown in Figure 1.1, the first great civilizations, such as those in Mesopotamia, Indus Valley, ancient Egypt, and later China, grew up in river valleys. The Mesopotamian civilization, the earliest civilization among the four, flourished from about 4000 to 3100 BC in Mesopotamia, meaning the land between the rivers, i.e., the Youprites and Tigris Rivers. The Nile Valley in Egypt had been home to agricultural settlements as early as 5500 BC, but

Figure 1.1 The four ancient civilizations in the Asian continent and Egypt.

the growth of Ancient Egypt as a civilization began around 3100 BC. A third civilization grew up along the Indus River around 3300 BC in parts of what is now Pakistan. The fourth civilization emerged around 1700 BC along the Yellow River in China.

Historians say civilizations tended to develop in river valleys (Mindsparks, 2007). The most obvious is access to a reliable water source, mostly for agriculture. Plentiful water and soil enrichment due to annual floods made it possible to grow excess crops beyond the need to sustain an agricultural village. This enabled some community members to engage in nonagricultural activities such as the construction of buildings and cities, metalworking, trade, and social organization. Besides, boats on the river provided an easy and efficient way to transport people and goods, encouraging trade development and centralized control of outlying areas. Last but not least, a large river also played as an effective military defense barrier against distant enemies.

Here, it is worthwhile mentioning the term "hydraulic civilization." This term is, according to the historian Karl A. Wittfogel (1957), any culture having an agricultural system that is dependent upon large-scale government-managed waterworks—productive (for irrigation) and protective (for flood control). Wittfogel asserted that such civilizations, mostly found in the East (the Near, Middle, and Far East), were quite different from those of the West. He believed that wherever irrigation required substantial and centralized control, government representatives monopolized political power and dominated the economy, resulting in an absolutist managerial state. The bureaucratic network directed the forced labor for irrigation projects. Wittfogel listed ancient Egypt, Mesopotamia, and China among these hydraulic civilizations. However, this theory has been disputed by other historians saying that irrigation is not the most important factor in social development in the East, but it may help consolidate political control. Aside from debates regarding that theory, the theory itself strongly indicates that water has been crucial for the birth and sustainability of civilization from the ancient civilization period.

1.2.3 Rivers and human activities around the beginning of the Common Era

When the Common Era began about 2,000 years ago, there were numerous civilizations in the world: The Roman Empire in the West and Han dynasty in the East, and a few more, though less prominent, such as the Parthian Empires (modern-day Iran) and Kushan Empire in the Indian Subcontinent.

In the period of the Roman Empire, these civilizations focused on the power of water flowing on the steep hills of the Mediterranean terrains, such as water mills for grinding grains and hydraulic jets for hydraulic mining (Viollet, 2007). However, the most symbolic, water-related infrastructure was probably the extensive network of aqueducts that enabled Rome to access, convey, and manage prodigious supplies of wholesome freshwater for drinking, bathing, cleaning, and sanitation on a scale exceeding anything realized before in history. Without such gigantic and efficient infrastructure, Rome as a large metropolis would not have been possible. Another example of human activities involving water is the use of rivers as a natural military defense barrier. A good example is the Rhine River northwest of the Roman Empire and the Danube River in the northeast; Rome had 15 regions along these two rivers.

On the other hand, the relationship between human activities and water in the East (for example, in China) focused mostly on irrigation and flood control, as Wittfogel emphasized in his book *Hydraulic Civilizations*. According to a Chinese legend, before the Shang dynasty (c. 1600–1046 BC), the old civilization near the Yellow River, the traditional founding father of China's Yellow River civilization was Yu the Great. As a river engineer, Yu rose to power on the merits of his accomplishment as the tamer of the great floods that ravaged settled life in the Yellow River basin before recorded history. By having "mastered the waters and caused them to flow in great channels," he made the world habitable for human society. He founded the Bronze Age Xia dynasty from about 2200 to 1750 BC and became revered as the Lord of the Harvest in association with the river's early irrigation works.

A good example of the extensive irrigation and flood control projects before the Common Era, yet still being used, is the Dujiangyan project, completed in 256 BC, during the Qin dynasty (Warring States period to 206 BC), just before the Han dynasty. Planned and implemented by Li Bang—an avatar of Yu the Great, over the Min River, a tributary of the Yangtze River, 57 km southwest of modern-day Chengdu in Sichuan Province, enabled multipurpose agricultural irrigation and floodwater diversion using a new method of channeling, and dividing the water and sediment rather than simply damming it. The water management scheme still irrigates over 5,300 km^2 of land (Zhang and Hu, 2006). An article in this book introduces this project in detail (Chapter 4).

The first known canal to cross a marked watershed between river basins, probably the first in the world, was the Ling Canal or "Magic Transport Canal," constructed in China in 219 BC in the Qin dynasty, which connects the Yangtze River system with the Pearl River system in southwest China (UNESCO, 2022). It was the world's first transport contour canal, 36 km long, dug by following the natural topography of the surrounding landscape to avert complex tunneling and water-level management problems.

During the four centuries of the Han rule, from 206 BC to AD 220, China's powerful centralized state and high civilization flourished as one of the two greatest on earth. Historians frequently have noted the many historical parallels between the Han and Roman empires, including Han's irrigation and navigation projects and the Roman aqueducts as a water-related civilization.

The outstanding, transformational event that catapulted Chinese civilization above all its contemporaries and marked one of water history's turning points was the Grand Canal (Solomon, 2011), connecting the water-abundant Yangtze River in the south and the water-scarce Yellow River in the north. The Grand Canal project was completed in the early 7th century when China was unified again by the Sui dynasty (AD 581–618). Parts of the navigation route are still being used today. An article in this book also discusses this project in detail (Chapter 6).

1.2.4 Rivers and human activities around AD 1000

How was the world during the year of AD 1000? How was it, especially for the relationship between human activities and water? An impression can be obtained by glancing at a few examples in the book *The World in the Year 1000*, edited by German scholars Brüggemeier and Schenkluhn (2000). They explained the agricultural structure, urban setting, trade, and transportation of that era with the division of geographic civilization of East Asia, Indo-Southeast Asia, Islam, and Europe–Byzantine.

During this era, the relationship between human activities and water was especially evident in the East Asian civilization, including rice-cultivating and inland navigation technologies. For example, the Song dynasty in China, which lasted from AD 960 to 1279, first adopted the transplantation of rice paddies to enhance the harvest yield. Also, this dynasty advanced irrigation and drainage technologies for the reclamation of riparian lands. Another example is the floating paddy in the river. Such paddies were formed with the wooden matrix frame with bush, rice hays, twigs, clays, and water plants filling the frame. This rice farming technology had remarkable advantages compared to the traditional farming method of planting on the ground, with no irrigation and drainage and flood hazard problems, since the frame itself moves up and down according to water level.

Another example of the splendor of that period is apparent from the extensive use of inland navigation. During the Song dynasty, inland navigation was concentrated, especially in the south of China, around the nowadays Huanan province. According to contemporary documents, about 3,000 boats were navigating the rivers and canals, and 70,000 workers could transport 100,000 tons of goods.

What happened in Europe around AD 1000 in terms of humans and water? Historians call this era the Medieval Age (from the 5th to 15th century) or, pejoratively, the Dark Age. A well-written and informative book by Steven Solomon (2011) only stresses this period's water wheels and oceanic voyages. Solomon's second topic concerns the sea routes, not related to fresh water. Europeans used the rivers connecting the seas or inland seas as navigation routes for trade. Rivers were also used for sea-born invaders' routes infiltrating the inlands of the European continent. During this period, agricultural technologies related to water were not improved in Europe, since they cultivated wheat, barley, and other dryland crops, which did not need much water for cultivation, unlike their Asian counterparts.

1.2.5 Industrial revolution and end of Asian leads over European counterpart

In general, historians say that European powers started overtaking their Asian counterparts, including Saracen, Mugul, and Chinese empires, in the late 15th and the early 16th centuries, nominally about the time Christopher Columbus tried to discover an oceanic route going to India from the west, and Ferdinand Magellan led the 1519 Spanish expedition to the East Indies across the Pacific Ocean to open a maritime trade route. As a result of the great voyages, which had opened a way across the Atlantic, a way around the Cape of Good Hope, and a way around Cape Horn, Western traders and missionaries had begun to reach the coast of China by sea before the end of the seventeenth century. Asian historians say this historical trend is "Western power encroaching East." To be kept in mind, however, is that in the early 15th century, Zheng He, a Chinese explorer, diplomat, fleet, and admiral during China's early Ming dynasty (from AD 1368 to 1644), commended four expeditionary voyages to Southeast Asia, the Indian Subcontinent, Western Asia, and East Africa. His larger ships carried hundreds of sailors on four decks and were almost twice as long as any wooden ship ever recorded.

Andre Gunter Frank, an economic historian, argued in his book *Reorient* that, at least until the early 19th century (say AD 1800), the Asian civilizations, especially in terms of economic power, were ahead of their European counterparts. He questioned many assumptions about Europe as the birthplace of capitalism, industry, and modernity. He asserted that Asia did it bigger, better, and much earlier than Europe. As shown in Figure 1.2, since the 11th century, populations in Asia and China as individual states always far exceeded Europe's. During the 11th–19th centuries, it was comparable only to that of India alone. With highly labor-intensive industries, such as rice cultivation and the cotton industry, China exceeded Europe until the onset of the 19th century in GNP per capita with USD 228 (constant price in 1960) vs. USD 150–200 (Frank, 1998). The population in China at that period was about 345 million

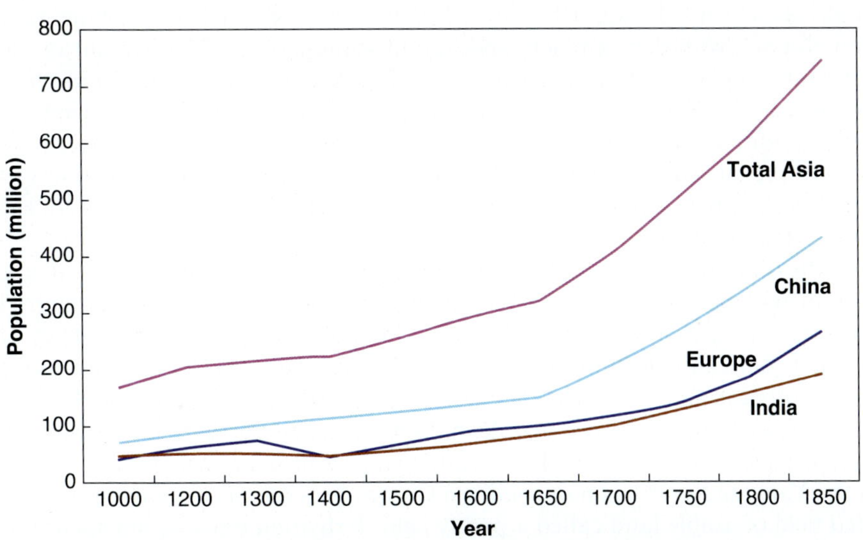

Figure 1.2 Population trends in different regions. (Frank, 1998.)

while that of Europe was only 188 million, which means the GNP of China was 2.4 times higher than that of Europe.

Economic power, however, had shifted from Asia to Europe since the onset of the 19th century when the Industrial Revolution (starting from around the 1760s) accelerated with the invention and spread of efficient steam engines and steam-driven pumps, and transportation materialized in European societies. With the invention of Portland cement in the mid-19th century and the spread of electricity generation in the late 19th century, water and rivers were not only considered merely irrigation sources or navigation routes but also the source of electric power transfigured from the potential energy of water. The dam era began in the late 19th century. Europe began to overwhelm Asia in every aspect of society and nation, including human activities with water and rivers.

1.2.6 Summary of the relationship between rivers and human activities

Since the cradle of civilization in Mesopotamia and Egypt 5,500 years ago, human societies have had intimate but sometimes distant interactions with water, especially rivers, in several distinct ways: irrigation, flood management, and inland navigation. The degree to which human societies have interacted with water in these ways is summarized in Table 1.1. As shown in this table, relationships between human societies and water in Asia were intimate, as expressed as "active," while they have been less active in Europe than in Asia. Irrigation and flood-management activities were active in Asia throughout the era, while they were less active in Europe. Inland navigation was active in both regions. Especially in China, since the Qin dynasty in the 3rd and 4th centuries BC, numerous canals were constructed, including the Ling Canal and the Grand Canal, to connect existing river channels on the relatively level plains with the web-like network of natural river channels, especially in the eastern and southern parts of China.

Two factors may be identified as causing the above results: the prevailing climate and cultivated crops. In southern Europe, they were affected mostly by the Mediterranean climate, with dry summers and humid winters in the Mediterranean region. In contrast, they were affected by the westerly oceanic climate with mild temperatures and seasonally relatively uniform precipitation in western Europe. Under these climatic conditions, they cultivated wheat and barley—the staple crops from the beginning of the ancient civilizations, depending less on the artificial irrigation system. Under this climatic condition, flood risks were less common, and floods were less severe than those in Asia.

On the other hand, in Asia, the prevailing climate is the monsoon climate (Asian monsoon), with hot summer rain and relatively dry seasons other than summer. These climate features are called southwest monsoon and northeast monsoon. They affect the Indian Subcontinent and Sri Lanka in the summer. The East Asian monsoon affects Southeast and East Asia, including Eastern China, Japan, and Korea. Because of these climate features, people in Asia (except in northern and western parts of China) have cultivated rice. This semiaquatic crop is known to have originated from the Yangtze River basin in southern China. This crop is usually cultivated in a flooded field of arable land called a paddy field. Irrigation must be indispensable for such an aquaphilic plant.

Table 1.1 A summary of the degree of activities of Asia and Europe by different water functions

Old Civilizations	*Irrigation*		*Inland Navigation*		*Flood Management*		*Note*
	Active		*Active (Except China Civilization)*		*Active*		
Regions	*Asia*	*Europe*	*Asia*	*Europe*	*Asia*	*Europe*	
Around 1st century	Active	Less	Active	Less	Active	Less	
Around 11th century	Active	Less	Active	Active	Active	Less	a
Before 20th century	Active	Less	Active	Active	Active	Less	b

[a] Inland navigation in Europe was active only in natural waterways.
[b] Inland navigation in Europe was active both in natural waterways and canals.

Another characteristic of the relationship between human society and water in Asia is managing excessive water (i.e., floods). Statistics[1] of the historically deadliest floods (excluding coastal storm surges) that occurred in Europe and Asia for the last 500 years, causing more than 1,000 fatalities, show approximately 41 flood cases in Asia, while about seven cases in Europe (mostly in Italy). Naturally, human society on flood management has been more active in Asia than in Europe. This is another typical example of how Asia differed from Europe, as shown in Table 1.1.

From now on, among the three topics in the engineering aspect of the relationship between human society and water, this article focuses on the activities of human society in the irrigation of paddy fields in Asia. This topic is one of the prevailing focal points in this book, where most regions (countries) have at least one article on this topic. The next chapter is based largely on the book's articles introducing paddy-field irrigations.

1.3 BRIEF REVIEW OF THE HISTORY OF PADDY-FIELD IRRIGATION ACTIVITIES IN ASIA

Among the various human activities on water and river over 5,000 years, the most abundant artifacts and records are related to the irrigation work to paddy fields in Asia. A paddy field is a flooded field of arable land used for growing rice—the most important staple crop for Asian people.

Despite different theories of the origin of rice cultivation, it is said that Asian rice (Oryza Savita) was first domesticated in the Yangtze River basin in China 13,500–8,200 years ago (Normile, 1997). Since the first rice cultivation, human migration and trade spread the rice to Far East Asia and South China, including Taiwan, from 3500 to 2000 BC. It spread further to Southeast Asia through the Austronesian expansion from 2000 to 500 BC, including numerous large islands between the Pacific and Indian Oceans. Meanwhile, intensive paddy rice farming was introduced into the Korean peninsula from 850 to 550 BC and reached Japan around 300 BC.

On the other hand, inhabitants of the Indian Subcontinent started the early domestication of rice based on the wild species (Oryza Nivara) in the Belgian and Ganges valley region of northern India from 5400 to 4500 BC. Meanwhile, the wetland rice (Japonica) —a variation of Oryza Savita arrived in the Indian Subcontinent around 2000 BC.

While rice is a versatile crop that can be grown in various environments, it is still a tropical plant that needs water and heat. Thus, it is unsurprising that about 90% of the global rice supply is produced in tropical and subtropical nations with high rainfalls, such as Vietnam, Thailand, Indonesia, India, and South China. Meanwhile, regions under moderate climate zones, such as the eastern part of China, Korea, and Japan, have had a long rice cultivation history due to the monsoon climate, which provides abundant rainfall during the rice-growing season.

Rice crop water-intensive characteristics mean most farmers must flood their paddy fields according to the crop's growth cycle. Thus, most farmers must rely on a proper irrigation system, including reservoirs (dams), canals, water wheels, and shadoofs.

1.3.1 Some historical irrigation projects in China

Irrigation in China has a history of more than 3,500 years. The evolution of civilization is closely related to and has affected the origin, development or decline, and revival of irrigation. There are many irrigation projects in China with more than several centuries or even millennia of history. Irrigation practices are also subject to the natural environment, technology, and scale of different regions, due to the different geographical latitudes, topographies, and climate conditions in different regions of China. The terrain and climate conditions directly impacted the ancient irrigation projects and various natural and geographical conditions. The origin and development of irrigation engineering are sometimes of historical significance, representing the regions' evolution and civilization (Chapter 3).

Part I of this book introduces several historical irrigation projects in China, including the famous Dujiangyan irrigation system (Chapter 4). This historical water project is mentioned above.

Fast forward to 1,000 years, this book introduces Tuoshan Weir—an ancient hydraulic project constructed to regulate and control water in the estuary in ancient Ningbo, a modern coastal city in Zhejiang Province. The project was built in AD 833 during the Tang dynasty (618–907). The weir played an important role in water distribution for the western part of Yin-Feng plain and ancient Ningbo. This ancient hydraulic project for estuarial river regulation could be considered a typical coastal reservoir applied to restore freshwater, prevent saline water intrusion, protect against flooding, and irrigate local farmland (Chapter 7).

This book introduces an old groundwater irrigation project called Shadoof-well irrigation—one of the oldest water-lifting irrigation systems (Chapter 5). The system involves a crane-like tool with a lever mechanism that has been used in irrigation, since around 3000 BC, by the Mesopotamians. This irrigation system can be traced back to the 12th century AD, during the era of the Song dynasty. The Shadoof-well irrigation suggests that people in that period understood the mechanism of the groundwater cycle, and they built a dam in the brook to increase the seepage supply of groundwater.

1.3.2 The oldest irrigation project in Japan

The basic features of irrigation systems in western Japan were formed during the Kofun and Nara periods (AD 300–800) (Tabayashi, 1987), based largely on ponds. On the other hand, the main irrigation systems in eastern Japan were mostly built during the Sengoku and Edo periods AD (1450–1867), largely based on streams. Especially during the Edo period (AD 1603–1867), a hierarchical structure of water management cooperatives and water delivery methods was established according to the branch-like network of canals.

An article in Part II introduces the "Sayama-ike"—an early pond-based irrigation system and one of the oldest reservoirs in Japan. The system provided an irrigation network to the Osaka plain in western Japan since the early 7th century. Up to the present, this reservoir's dike has been reinforced and heightened repeatedly to increase the reservoir capacity as the irrigation area has expanded. The irrigation network has a unique distribution system with the Sayama-ike as the first-stage reservoir supplying water to the second and third ones in a cascade. From the 16th to the 17th century, the irrigation area of Sayama-ike became the largest, which amounted to approximately 4,000 ha.

1.3.3 Some historical irrigation projects and technology in Korea

Wetland rice cultivation in Korea has a long history, first spreading from East China about 2,000 years ago. Due to the extreme variations in rainfall in the Korean peninsula, characterized by frequent drought and floods, the history of paddy farming in Korea is well exemplified by its constant struggles to overcome water shortages during the dry season. Rice seeding and transplantation occur before the monsoon starts in June. The reservoir was the streamlined solution for irrigation during the water scarcity season in spring and is well described in this monograph (Chapter 16).

This book introduces a few historical irrigation projects and water-related technologies in Korea. One is the Byeokgolje—an ancient embankment located in Gimje City, Chollabuk-do province (Chapter 16). The historical record shows that Byeokgol-Je, a large-scale reservoir that can irrigate about 10,000 ha of farmland, was built in AD 330 in the Baekje kingdom during the three kingdoms era (AD 57–668), and a part of it remains. The reservoir originally had five floodgates, while, at present, it has a three km long bank, and only the second and fifth floodgates are left (KCID, 2001).

This book also introduces the ancient rainfall gauge—Chugugi invented in 1,441 in the Joseon dynasty (AD 1392–1910). This instrument was used for weather alerts, flood forecasting, and water management (Chapter 19) and is recognized as the world's first instrumental rain gauge. The depth of rainfall collected by Chugugi was measured with a standard ruler. Chugugi were located at the Palace and Meteorological Agency in the capital area and the provincial offices of the Joseon dynasty. The daily rainfall data in Seoul have been collected since 1770, representing one of the world's longest daily precipitation records.

1.3.4 Irrigation works in South Asia

Rice agriculture had spread across the Indian Subcontinent by about 500 BC. The earliest references to irrigation are found in Rigveda. The Veda mentions only well-styled irrigation. The initial spread of rice outside the core zone of central Gangetic Plains is thought to have been limited by climatic constraints, particularly seasonal rainfall levels. Accordingly, the later spread of rice into the dry regions of South India is largely supposed to have relied on irrigation. In South India, rice farming in the Iron Age (about 1000–500 BC) to Early Historic (about 500 BC–AD 500) may not have been supported by irrigated paddy fields. However, it may have relied on heavy seasonal rainfall elsewhere in the Subcontinent (Kingwell-Banham, 2019).

In the central Gangetic Plains, the most widespread irrigation system in India was undertaken in the medieval period by the Sultanate rulers. For example, Firoz Shah Tughlaq (AD 1309–1388) built the most extensive canal irrigation system around the Indo-Gangetic doab and the region west of the river Yamuna in the 14th century. Irrigation systems were continued by the subsequent rulers of northern India, particularly the Mughal rulers, until the early 19 century. From the 19th century, the British built colonial canal networks based on these medieval canal systems, mostly in the Gangetic valley and east of it and later Punjab region, mainly for growing wheat, tea, and opium, but not for rice.

This book does not elaborate on historical rice irrigation projects but does explain a few projects set in Sri Lanka. Rather, this book introduces a brief overview of the irrigation practices in Punjab—northwest of the Indian Subcontinent, done in a relatively recent era (Chapter 20). A special interest is the development of the empirical formulas predicting the equilibrium geometry of irrigation canals called the Regime Theory. They were developed in Punjab in the late 19th century by British engineers working on designing and managing huge networks of irrigation canals.

This book also briefly introduces, from the technological aspect, the history of paddy-field irrigation in Sri Lanka—a part of the Indian Subcontinent. About 2,500 years ago, when paddy cultivation was introduced, and water was the most important resource for securing rice production, they developed various irrigation schemes and technologies, including the Bisokotuwa (Chapter 21). It was a pressure- and volume-control system placed at the sluice gate, considered one of the most advanced irrigation systems that have survived throughout history. There was a special arrangement before Bisokotuwa called the inlet barrel, which extended toward the middle of an irrigation tank by more than 30m. An example of this arrangement was the "Nuwara wewa" (Chapter 22)—an old sluice barrel that extended up to 45m. This arrangement was first constructed in the 1st century BC.

1.3.5 Irrigation works in Southeast Asia

Rice farming spread from South China to Taiwan at about 3500 and 2000 BC, then to the Southeast Asian region, including Vietnam, Cambodia, Thailand, and Malaysia, and eventually to Indonesia at about 500 BC. This book includes various articles introducing historical paddy-field irrigation in those regions, but only in historically recent years.

The prehistory of Vietnam is characterized by attention to water engineering as a foundation for developing paddy rice irrigation. For the last 1,500 years in the feudal age (AD 6–19th century), people constructed water infrastructures (such as dikes and canals) to enlarge fertile land for cultivation and husbandry and cope with the common hydro-meteorological hazards of such as floods and droughts. A relatively recent example of the waterworks for multiple purposes of flood protection, inland reclamation, and sea encroachment was recorded in the years 1827–1830, when Nguyễn Công Trứ (a Vietnamese poet and scholar) ordered the construction of coastal dikes enabling people to establish new hamlets and build two new lands for farming.

In Cambodia, rice cultivation using flooded water was practiced more than 1,000 years ago (Chapter 24). Since then, the colmatage system (a series of comb-shaped narrow canals to the main river course excavated the natural levee) was expanded before and after French rule, which began in 1863. Colmatage involves impounding of silt-laden water to build up low-lying areas. This book only introduces the colmatage system—an authentic multipurpose water and river utilization system, including paddy-field irrigation, flood control, and navigation.

Thailand was an agricultural country where irrigation was a backbone for supplying water to farmers, especially for rice cultivation, and locally transporting people and goods in the past. Modern irrigation work was gradually developed more than 100 years ago. The starting point was the Rangsit area—a floodplain, swamp, and abandoned area. The Rangsit Canal was the first irrigation canal to be dredged and is still functioning today. It has been frequently used as a drainage part in the flood control system for disaster risk reduction in the Chao Phraya River basin. The Rangsit Canal has been proven for multipurpose uses in the past, and the canal will undoubtedly be used as a resilience structure for flood disaster reduction in the future.

Rice farming in Malaysia spread from mainland Southeast Asia with the flow of human culture through the Malay Archipelago about 1500 BC. Before the 19th century, however, dry paddy cultivation was more popular than wet paddy cultivation among the Malays, although wet paddy cultivation was introduced in about 15–16 centuries via Thailand to North Malaya (Cheng, 1969). This book only introduces the Wan Man Saman Canal, 36 km long, completed in 1895, which enabled the transformation of swamp lands along the coastal plain into another vast area for rice cultivation in Kedah, the northwest part of the Malay peninsula (Chapter 26). The irrigation canal enabled Kedah to boost its rice production and enabled Kedah to earn the nickname "Malaysia's rice bowl."

Lastly, rice cultivation in Indonesian history is linked to the development of iron tools and the domestication of wild Asian water buffalo as water buffalo for the cultivation of fields and manure for fertilizer in about 1500 BC. Rice production requires exposure to the sun. Once covered by dense forest, much of the Indonesian landscape had been gradually cleared for permanent fields and settlements as rice cultivation developed over the last 1,500 years (Taylor, 2003). This book introduces Subak—a traditional irrigation system well preserved in Bali. A Subak is a social–religious communal organization of farmers and its irrigation system in Bali. One of the systems, probably more than a thousand years old and UNESCO's World Heritage award, is still functioning with many tunnels built hundreds of years ago (Chapter 27).

1.4 SUMMARY AND CONCLUSIONS

Freshwater, especially rivers, comprises a relatively tiny portion of water. However, freshwater has profoundly affected human society and vice versa. Throughout the 5,500 years of the intimate but sometimes distant relations between human societies and water, human societies have interacted with water in several distinct ways: irrigation, flood management, and inland navigation. Irrigation and flood management activities were active in Asia throughout the era, while they were less active in Europe. Inland navigations were active in Asia, both in natural waterways and artificial canals.

In Europe, however, they were active only in natural waterways until the onset of the 17th century.

Two factors may be identified as causing the above results: the prevailing climate and the cultivated crops (notably rice). The monsoon climate (Asian monsoon) is dominant in Asia, with rainy and hot summers and relatively dry seasons besides summer affecting the Indian Subcontinent. In contrast, the East Asian monsoon affects Southeast and East Asia, including East China, Japan, and Korea. Because of these climate characteristics, inhabitants in Asia have cultivated rice—a semiaquatic plant requiring abundant water during the growing season. Therefore, it is essential for rice cultivation on a proper irrigation system, including reservoirs (dams) and canals, water wheels, and even rain gauges. Also, flood risks have been more common in Asia than in Europe under this climate condition, as recognized in history. To utilize and, at times, overcome these climate conditions, people built and managed many historical and grandiose water projects and invented and used localized but sophisticated water-related technologies.

Some of those water projects and technologies include the following examples: the Dujiangyan irrigation system and the Ling Canal project in China in the 3rd century BC; the "Sayama-ike," the reservoir-based irrigation system in Japan in the 4th–7th centuries; the Chugugi—the oldest rainfall gauge in the world, invented in 1441 and since then had been used for about 400 years in Korea; the extensive canal irrigation system around the Indo-Gangetic doab and the region west of the river Yamuna in the 14th century, leaving aside the sophisticated municipal water supply and sewerage system in the old civilization of Harappa and Mohenjo-Daro along the Indus River valley; the unique irrigation technology of the Bisokotuwa in Sri Lanka; and numerous irrigation schemes that had been manipulated to fit the local cultural geography in Southeast Asia, a typical example of which is the Subak—a social–religious communal organization of farmers and its irrigation system in Bali.

ACKNOWLEDGMENTS

This article was reviewed for English writing by Ms. Chorong Kim in KICT, Korea, and by Dr. Seokyun Woo, Northwestern University, USA, both for English writing and contents, and finally, by Prof. Robert Ettema in Colorado State University, USA. Their careful reviews and valuable suggestions are greatly appreciated.

REFERENCES

Brüggemeier, F.-J. and Schenkluhn, W. (2000). *Die Welt imJahr 1000.* Herder, Freiburg (in German).

Cheng, S. H. (1969). The rice industry of Malaya: A historical survey. *Journal of the Malaysian Branch of the Royal Asiatic Society*, 42(2), 130–144.

Christof, M. and Zeller. T. (2008). *Rivers in history: Perspectives on waterways in Europe and North America.* University of Pittsburgh Press.

Frank, A. G. (1998). *ReOrient: Global economy in the Asian age.* University of California Press, Berkeley.

Gyawali, D. (2015). Asia's sacred, wicked waters. Global Asia. *A Journal of the East Asia Foundation*, 10(1), 8–13.

Heitzman, J. and Schenkluhn, W. (2004). *The world in the year 1000*. A collection of papers at a conference organized in April 2000 by F. J. Brüggemeier and W. Schenkluhn. University Press of America, Lanham.
KCID (Korean National Committee on Irrigation and Drainage). (2001). *History of irrigation in Korea*. Seorabul Data Co., Korea.
Kingwell-Banham, E. (2019). Dry, rained or irrigated? Reevaluating the role and development of rice agriculture in Iron Age - early historic south India using archaeo-botanical approaches. *Archaeological and Anthropological Sciences*, 11, 6485–6500.
Mindsparks. (2007). *Rivers and civilization: what's the link?* p. 8.
Normile, D. (1997). Yangtze seen as earliest rice site. *Science*, 275(5298), 309–310.
Solomon, S. (2011). *Water: the epic struggle for wealth, power, and civilization*. Harper Collins, New York.
Tabayashi, A. (1987). Irrigation system in Japan. *Geophysical Review of Japan*, 60-series B(1), 41–65.
Taylor, J. G. (2003). *Indonesia: peoples and histories*.Yale University Press, New Haven and London, pp. 8–9.
UNESCO. (2022). https://whc.unesco.org/en/tentativelists/5814/. Accessed June 30, 2022.
Viollet, P.-L. (2007). *Water engineering in ancient civilizations, 5,000 years of history (IAHR Monograph)*. CRC Press, Boca Raton, FL.
Wittfogel, K. A. (1957). *Oriental despotism; a comparative study of total power*. Random House, New York.
Woo, H. (2005). River and human activities: how has the interaction between rivers and humans changed since civilization? Public Lecture, *AOGS 2nd Annual General Meeting*, Singapore, June.
Zhang, K. and Hu, C. (2006). *World Heritage in China*. The Press of South China University of Technology, Guangzhou, pp. 95–103.

NOTE

1 https://en.wikipedia.org/wiki/List_of_deadliest_floods. May 1, 2022 accessed.

Part I

Historical water projects and traditional water technologies in China

Stone Portraits of Yu the Great made in the Eastern Han Dynasty—the first person to control the Great Flood of China and the first emperor of the Xia Dynasty at about 2100 BC (collections in Jiaxiang Han Dynasty Stone Portraits Museum, Shandong, China).

China is an ancient agricultural country with more than 4,000 years of history. River control and water diversion are the most important works for all the dynasties of China. A large number of water projects for irrigation and water diversion have been built, and some of them are still being used today. In this Part, typical Chinese historical

DOI: 10.1201/9781003222736-2

water projects and traditional water technologies were introduced, including the article "Sustainability of Chinese civilization and historical irrigation projects" looking at the influence of traditional water technologies on sustainable Chinese civilization and the types, location, and development of water projects in China, and the article "An analytical literature review of Chinese water culture," analyzing classifying the Chinese water culture in different regions. Also, four typical historical water projects were introduced, including the Beijing–Hangzhou Grand Canal—the earliest and the longest canal in the world; Dujiangyan irrigation systems—the oldest water diversion project in China; Tuoshan Weir—an ancient estuarial river regulation project; and Zhuji Jiegao irrigation system—shadoof–well irrigation in southeastern China.

Chapter 2

The Chinese water culture

An analytical literature review

Juan Lu
China Institute of Water Resources and Hydropower Research

CONTENTS

2.1 INTRODUCTION

Culture has played a pivotal role throughout human history. The path from barbarism to civilization turns humans from natural beings to social ones. It has generated people's diverse personalities, temperaments, and sentiments, such as joy and pain, happiness and sorrow, and nobleness and vulgarity; it has established various values and outlooks on life. People are shaped by culture and, in turn, promote continuous cultural development and progress. Generally speaking, culture is created by people, and there is no culture without people. Culture comes from history and exists in every corner of the earth where human beings exist.

Water culture is an important form of culture. In China, "water culture" was formally proposed in 1988. Over the past 30 years, water culture has attracted widespread

DOI: 10.1201/9781003222736-3

attention from all walks of life. Experts and scholars have discussed its definition and connotations from multiple scopes. The exploration, inheritance, and implementation of water culture have been widely practiced at various places, and governments have increasingly valued it at all levels. Water culture also presents different characteristics in different regions. Like all other advanced cultures, advanced water culture can lead to positive trends, educate people, serve society, and promote development. Therefore, the systematic definition of the connotation of water culture and the establishment of a theoretical system are of great significance for guiding the practice of contemporary water conservancy.

2.2 SUMMARY OF CHINESE WATER CULTURE RESEARCH

Cultures originate from the accumulation of history, and water culture derives from the history of the development and utilization of water resources. Since human beings cannot live without water, the history of leveraging water resources has always been synchronized with the history of human development. The development and utilization of water in China began with the legend of Da Yu, thrived in the Spring and Autumn and the Warring States periods, and developed in various degrees throughout different dynasties but generally served the ruling class. Only in the 70 years since the founding of the People's Republic of China did the purpose of water resources development shift to serving ordinary people.

The Water Conservancy History Research Group of the China Institute of Water Resources and Hydropower Research has been the leading force in water culture and conservancy history research. In the 1980s, supported by culture-preserving activists from all sectors, the Research Group identified a series of renowned water culture heritage, including Dujiangyan Irrigation System, Zhengguoqu Canal, Lingqu Canal, Quebei Irrigation System, Tashanyan Wier, and Tongjiyan Wier. It also promoted the conservation of these sites and used the research results on water culture to guide urban planning and construction.

2.2.1 The concept of water culture

Cultural studies began to thrive in China in the 1980s when nearly all academic fields inspected their own cultures. Water conservation has a history of 5,000 years, so water culture research also came into being.

The term "water culture" was first mentioned at the Huaihe River Basin Publicity Conference on October 25, 1988. As Mr. Li Zongxin, the then director of the Publicity and Education Department of the Huaihe River Water Conservancy Commission, proposed in his speech entitled Strengthening the Publicity of Huaihe River Management and Promoting the Development of Huaihe River Management,

> Some have proposed the study of water culture, that is, to explore the development history and mutual relationship of water-related activities, water policies, and water conservancy; the connection of water culture, human civilization, and social development; and the shared beliefs and values of the water conservancy

sector. We think this kind of research will be highly relevant and should become an important part of our publicizing efforts.

(Li, 2007)

According to the literature search on CNKI, an article on water culture first appeared in May 1989, titled "Research on water culture," published in the fourth issue of the *Zhihuai* magazine by Li Zongxin. He noted that "water conservancy should have a culture with its characteristics to become a well-established industry. Our proposal to research water culture is well-grounded." "What is water culture? What should it include? What are the approaches and methods for studying water culture? Further exploration is needed in such issues." Meanwhile, he also gave a preliminary explanation of the concept of water culture:

> While it is not yet possible to give an accurate definition to the term "water culture", in a general sense, we can take water culture as a collection of values, moral standards, and practices that are shared and observed by people engaging in water-related activities. In other words, it is the sum of the cohesion, sense of belonging, honor, and other spiritual forces shared by people engaged in water-related activities.
>
> (Li, 1989)

In October 1989, Wu Zongyue, the then deputy director of the Publicity and Education Department of the Huaihe River Water Conservancy Commission, published an article titled "Talking about water culture" in the fifth issue of the "*Water Conservancy World*" (now renamed as "*Hydro Science and Cold Zone Engineering*") (Wu, 1989). On November 5 of the same year, the Huaihe reporter station of "China Water Resources News" and the editorial department of the *Zhihuai* magazine jointly issued the "Proposal for Holding a Water Culture Seminar" and the "Water Culture Discussion Reference Outline." The proposal noted that

> water is the source of life and culture, and civilization of human society. The relationship between people and water greatly affects the formation and development of people's thoughts and ideologies, which eventually form values, social beliefs, and lifestyles with characteristics related to the water conservancy industry.
>
> ...
>
> Water culture is a rich culture with a long history and diverse connotations. However, we still lack a thorough understanding and systematic research on it, and some even ignore its existence.

Based on the above understanding, four suggestions were put forward: (1) All interested in water culture research should act immediately to conduct research and discussion on water culture. (2) The current recommendations for research and discussion topics include the concept, content, significance, and research methods. Specifically, it may include the influence of water on society and politics, social and economic development, military, literature and art, people's mentality, customs and morals, science and technology, religion, myths and legends, folk tales, and folklore that are related to water culture. (3) To encourage communication and exchange of study results, a meeting is to be held in the first half of 1990 to discuss the establishment of a research committee. (4) For those interested in the initiative, research papers or outlines may be mailed to Li Zongxin and Wu Zongyue of the Huaihe Water Conservancy Commission before the end of 1989.

Since then, from 1990 to 1999, several publications led by journals, including the *Zhihuai*, the *Water Conservancy World*, and the *China Water Resources*, and many experts began to discuss the concept of water culture and related extensions. From 2000 to 2019, more experts participated in the discussion. "Water culture" gradually gained broader recognition in the academic circle, and the number of related articles is increasing. Many interpretations of the concept of water culture were proposed, and debates were carried out. Relevant works were published one after another.

2.2.2 The interpretation of water culture by various scholars

Researchers have probed into the meaning of water culture from various perspectives. Some define water culture as "common cultural concepts, traditional habits, values, ethics, life beliefs, and enterprising goals" (Xing, 1990), while others suggest that it is a "spiritual civilization." Some propose the idea of a "human–water relationship" (Fan, 1990). Some consider it to be a "water conservancy culture" (Feng, 1994). For some studies, it is "the sum of various cultural phenomena with water as the carrier" (Ran, 2000). Others may propose that water culture can be defined from a "broad and narrow sense" (Li, 1998). Some works define it as "water or water culture conservancy spirit" or analyze it from the aspects of "social culture, in culture, institutional, and material culture" or look as a collective representation of "thinking, lifestyle, and behavior" (Wang, 2000). Some focus on "water's role in the development of human society, industry, and humanities and science" (Chen, 2003). Some believe it consists of "spirit and history, and their various changes that have occurred, and their evolution mechanism" (Yuan, 2005), while others identify it as including "water conservancy industry cultural system and water ecological, cultural system" (Yang, 2005). Some believe that it is the "knowledge and experience, concepts, management method, social norms, legislations, and social behaviors associated with water, as well as the cultural result of water management and the transformation of the water environment" (Zhou, 2008). In contrast, others define it as "the sum of material wealth and spiritual wealth" (Meng and Yu, 2008). In short, researchers presented a full spectrum of vibrant ideas on the subject of water culture.

2.3 ANALYSIS OF THE CONNOTATION OF WATER CULTURE

2.3.1 Definition of water culture

To clarify the connotation and extension of water culture, this chapter first analyzes the connotation and extension of culture as the basis of establishing a water culture theoretical system.

There are three levels of culture: spiritual culture, institutional culture, and material culture. Spiritual culture is the core of culture, the basis of material culture, and what differentiates the cultures.

Different taxonomies apply to the classification of cultures. Hierarchical types include ideological culture, behavior norm culture and material form culture, and spiritual, institutional, and material culture. From an ethnic perspective, there are Han and minority cultures. Also, industry, enterprise, and religious cultures are shaped by different social groups.

By geological locations, cultures may be categorized according to nations, regions, or watersheds.

The cultural structure consists of spiritual, institutional, and material levels. The constituent elements of culture include people or groups, time, and space.

Water culture and general culture share the same core. The connotation and extension of culture apply to water culture, but water culture contains the dimension of water, i.e., specific people or groups + water (human–water relationship). Therefore, the constituent elements of water culture are people or groups, water (human–water relationship), time, and space. The structure of water culture consists of three levels: ideology, behavior norms, and material form.

As discussed above, water culture can be defined as deriving from the human–water relationship of a specific people group in a specific period and/or region and acting on the human–water relationship of the descendants of this region or group. A specific period can be a certain dynasty, age, or time; a specific region can be a country, province, city, county, township, village, region, watershed, etc.; a specific group can be an industry, enterprise, community, etc.; and a specific group can be a family, nationality, ethnic group, etc.

Water culture is attached to people and social groups; the human–water relationship is the connotation of water culture. No water culture can exist without human–water relationships; water culture naturally exists where there is a human–water relationship. The concept of water culture can be summarized as the common ideology or behavior of people or groups in a specific period and region to understand, treat, utilize, control, and manage water. The behavior includes water-related lifestyle (living habits and mindset) and production mode (production habits) of people or groups, with those generated from water management (i.e., water conservancy culture) as the main body of water culture. The water cultural phenomena or achievements with significant historical, artistic, and scientific value can be named water cultural heritage. The ideological ones among them can be named intangible water cultural heritage and material ones material water cultural heritage.

The human–water relationship is complicated and delicate. People choose to live near water for water utilization, but some decide to live on hills to avoid flooding water. The eternal theme of the human–water relationship is boosting water's benefits while reducing its harm. In terms of utilization, water serves our daily life, irrigation, transportation, landscape, and power generation through the construction of various water conservancy projects. However, water in nature is hard to predict, and that causes drought and water shortage, floods, storm surges, and water and soil loss, bringing severe disasters, such as death, food shortage, water transportation interruption, environmental deterioration, etc., to human beings.

During its thousands of years of water management, China has formed a complete scientific and technological system consisting of ideological, theoretical, planning, engineering, and technical approaches to promote the benefits and eliminate the harm of water. Measures to prevent drought, dampness, flood, and waterlogging are

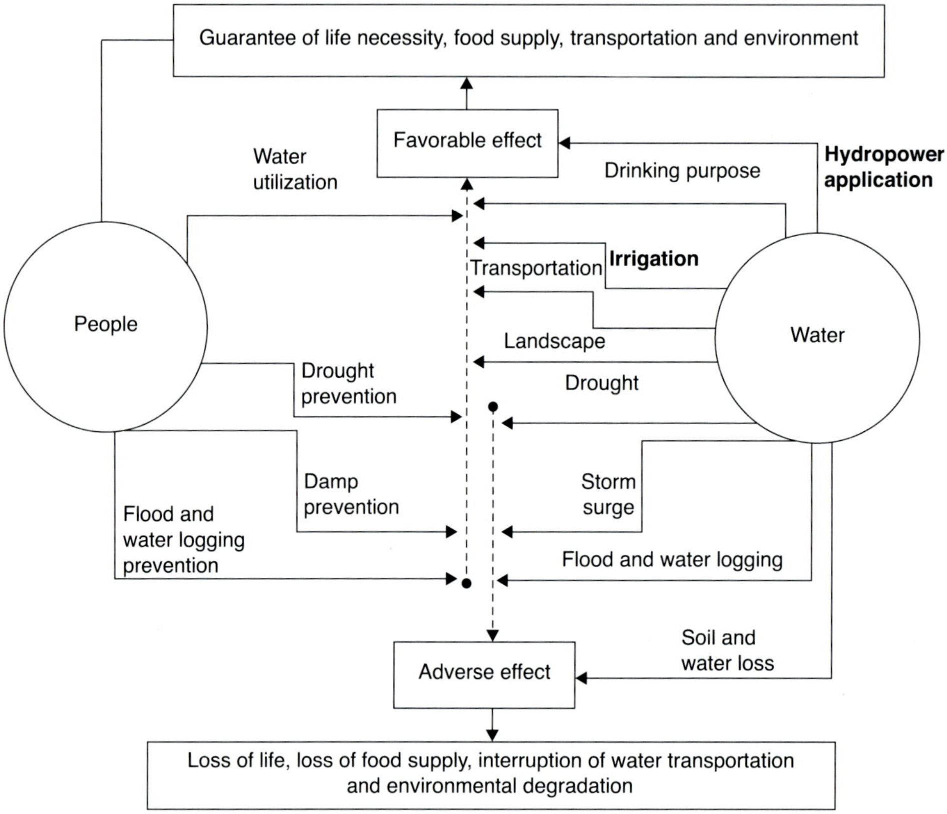

Figure 2.1 Human–water relationship.

required. Nowadays, China advocates water conservation, rational allocation of water resources, outlet construction for floods, water quality guarantee, and construction of a beautiful water environment, thus promoting the harmonious human–water relationship (as shown in Figure 2.1).

2.3.2 Levels of water culture

Water culture can be divided into ideological water culture, behavior norm water culture, and material form water culture.

The structure of water culture consists of three levels from core to extension: spiritual water culture, institutional water culture, and material water culture (as shown in Figure 2.2). Water culture can be divided into ideological water culture (spiritual water culture), behavior norm water culture (institutional water culture), and material form water culture (material water culture). Constituent elements of water culture include people or groups, water (human–water relationship), time, and space.

The first level, ideological water culture, also known as spiritual water culture and conceptual water culture, includes three categories: (1) water culture of pure consciousness, such as psychology, mentality, concept, morality, ethics, beliefs, values,

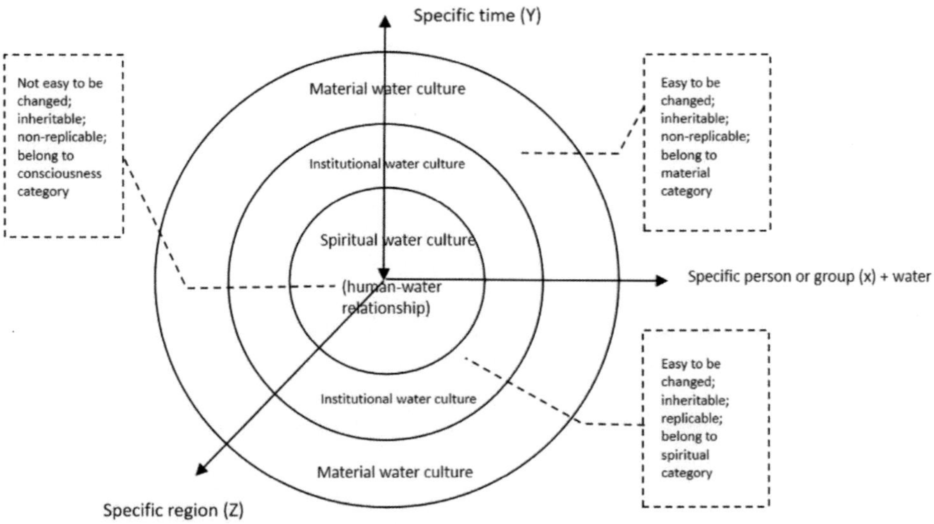

Figure 2.2 Schematic diagram of the structure and constituent elements of water culture.

and cognitive methods of water-related operations; (2) water culture of literature and art, such as esthetic theories, music, poetry, literature, and painting related to water; and (3) water culture of science to control water, such as water-management philosophy, thoughts, theories, and technology.

The second level, behavior norm water culture, also known as institutional water culture, includes two categories: (1) water-related folk rules and regulations, such as customs and religious ceremonies; and (2) laws and regulations for controlling water, production management regulations, and so on.

The third level, material form water culture, also known as material water culture, includes two categories: (1) water-related works, such as paintings and calligraphy related to water, documents related to water management, and so on; (2) water-management buildings and appliances, such as dams, gates, waterwheels, etc.

Ideological water culture is the core of water culture—the foundation of the material form of water culture. It is what differentiates a water culture from another. The material form of water culture is the external expression of ideological water culture. Water cultures formed by the restriction of the behavior norms, i.e., behavior norms become a kind of culture, are called behavior norms water culture. It falls between the ideological and material form levels in the structure. Water culture's ideological and behavioral norms with certain historical, artistic, and scientific values can be called intangible water cultural heritage. The material form water culture with certain historical, artistic, and scientific value can be called material water cultural heritage.

Though hard to be changed once formed, ideological water culture can evolve throughout time. Behavior norm water culture evolves in accordance with the needs of a country. The material form of water culture is not likely to demonstrate great changes over time. Ideological water culture is non-replicable but inheritable; behavior norm water culture is both replicable and inheritable, and material form water culture is non-replicable but inheritable.

2.3.3 Classification of water culture

Categorized by ethnic groups, Han and minority water cultures are distinguished, and by social groups are industry, enterprise, and religious water cultures.

- **By ethnic groups:** There are 56 ethnic groups in China, each with its own water beliefs, technologies, and customs. Therefore, nationality-based water culture can be divided into 56 categories. Examples are as follows.

  In terms of water belief, the folk proverb of Dai goes: "A ditch is dug before a cropland is planted," "to build a village, there must be forests and mountains, and to build a tribe, there must be ditches and rivers," "trees are beautiful because of leaves, and the ground is fertile because of water." Water is indispensable and significant in Dai people's daily production and life. Dai people believe that there is a god of water in the well who grants humans the wellspring. The god and the people share clean and sweet water. The well is called "Namo" in the Dai language. "He who digs a well while he is alive will go to heaven after he dies," and "digging well and building a pavilion show the kindness of a man" are Die proverbs about the wellspring god. When the Dai people built villages and dug wells, they were used to build a simple pavilion, which looks like a Buddhist pagoda or a stilt Dai folk house from a distance, and is brightly decorated above the well to protect it and worship the wellspring god. There are also some proverbs related to respecting, loving, and cherishing water engraved on the inner walls of the pavilions.

  The well is considered an essential and sacred space in Dai villages. Dai people believe that digging well is an act of accumulating merits and virtues. People not only withdraw water from wells for daily production and domestic uses but also use the clean water from wells for various sacrificial and Buddhist activities. Dai villages must worship water wells and water gods. Villagers spontaneously organize well-washing every year, which is also a good deed. Women are not allowed to take a bath in the water wells, and people are strictly prohibited from defecating in the wells.

  Another example is the Yi people. For their traditional livelihood based on "migrating by water," the Yi people have bred abundant water myths since ancient times and have correspondingly formed various mysterious water worship customs. Creation and extinction are the two main motifs of the ancient water myths and rituals of the Yi people, both running through the cycle of birth, destruction, and regeneration of human and the Yi ancestors as the cultural subjects, integrating into the origin and core of the traditional water culture of the Yi group. The Yi proverb goes: "People are born from water" and "people come from water." The Yi people worship the dragon not only because the dragon god takes charge of rainfall and farming but also, more importantly, they believe that "the ancestors came out of the water," so they were "the dragon."

  In terms of water conservancy technology, ethnic minorities have mastered the skills and techniques of digging wells (springs), canals, dams and ponds, water withdrawal, water conveyance, and utilization by waterwheels, water-power rollers, and water mills for production and daily uses. For example, the Hani people's God Forest (Angma) is generally located in the dense forest above the village. During the village building ceremony, the Hani people dig springs in the dense forest to get water for the villagers' production and daily uses. Zhuang people generally live in bamboo forests and use local materials, applying the bamboo tube water distribution technique to solve the problem of long-distance water conveyance.

In terms of water-related customs, the Yi people in Yuxi, Eshan, Xinping, Yuanjiang, Shiping, and other places in south-central Yunnan, for example, hold a folk Migaha Festival on the first day in February of the lunar calendar every year. This is a large-scale worship ceremony held in each Yi village, worshipping God Miga—a high-level god in charge of villages, forests, farming, and fertility. The symbol of God Miga is the banyan tree, which is locally known as "the Evergreen." Under the leadership of Bimo, elders of a Yi village choose a tall, straight, and flourishing banyan tree as the "Miga sacred tree" at the village's water source on an auspicious day. Cutting, climbing, sawing, or shoveling the sacred tree and surrounding trees are strictly prohibited. The whole woodland with the sacred tree at its center is called the "Miga God Forest." No one is allowed to enter on ordinary days, and violators will be punished. Therefore, there is always a lush, dense forest around the Yi villages in south-central Yunnan. The forest is not only the space and carrier for Migaha worship but also the area where its water source and geomantic forest are located. The forest facilitates atmospheric circulation and conserves water, thus regulating the natural ecosystem of the whole village and demonstrating great significance to the well-being of the whole village.

The Zhuang people, who live on rice and fish, rely more on water for production and daily life. They not only invented and created advanced water utilization and management technologies but also formulated strict water laws and regulations (Huang, 2014).

- **By social groups**: Different social group includes industry, enterprise, and religious water culture forms inside different social groups. For instance, the industry (enterprise) water culture refers to the scientific and sustainable development industry norms jointly observed by enterprises and employees in the water sector, which can ensure the harmonious coexistence of people and water. It has the functions of cohesion, guidance, restraint, incentive, coordination, education, maintenance, optimization, and reputation enhancement.

  For example, as a subordinate institution of the Haihe Water Resources Commission of the Ministry of Water Resources of China, the primary function of the Zhang River Upstream Administration is to coordinate the contradiction of water utilization between the upstream and downstream areas, the left and right banks of the Zhang River, and to implement unified management of the 108 km water dispute-prone river section in the border area of Shanxi, Hebei, and Henan Provinces, to maintain the sustainable and stable water order in this region. The Zhang River Upstream Administration, through unified planning, management, dispatching, and management of the river sections, manages directly and comprehensively adoption of administrative, economic, legal, and engineering measures for solving trans-provincial boundary river water disputes that have facilitated the formation of a harmonious water relationship between adjacent provinces, cities, and counties in the upstream area of the Zhang River, continuously maintained the stable status of the Zhang River, and boosted the sustainable socioeconomic development of the region.

2.4 DIVISION OF WATER CULTURE AREAS IN CHINA

Across China's vast territory, significant temperature difference exists between the north and the south and remarkable humidity difference between the east and the west. The complexity in topography, soil, climate, and hydrological environment contribute to various regional characteristics of Chinese culture, with mountains, hills, plateaus, plains, basins, rivers, and lakes crisscrossing nationwide. From the southeast to the northwest, natural vegetation presents three zones: forest, grassland, and desert.

The Chinese culture comprises agricultural culture, which serves as its pillar, and the colorful nomadic culture. Therefore, based on China's geography, Chinese cultural geographers divide the country into different cultural areas with different classes.

Class 1 contains the agricultural and nomadic cultural areas. The vast areas, namely the agricultural and cultural area in the east and the nomadic cultural area in the west, can be subdivided further. The great eastern agricultural and cultural area, located in the monsoon zone with abundant water and heat, flat terrain, and fertile soil, has been the center of Chinese culture since the ancient times. Spanning tropical, subtropical, warm temperate, and subfrigid climatic zones, the area has an agricultural cultivation varying from north to south and a geographical landscape varying in styles and features from place to place. Residents' food, clothing, housing, transportation, customs, artistic style, taste, and interests all feature their flavors.

Therefore, the great eastern agricultural and cultural area and the great western nomadic cultural area can be subdivided according to their historical and cultural traditions and current situations. (1) The great eastern agricultural area is subdivided into 13 regions: the northeast cultural region, the Yanzhao cultural region, the Loess Plateau cultural region, the Central Plains cultural region, the Qilu cultural region, the Huai River Basin cultural region, the Bashu cultural region, the Jingxiang cultural region, the Poyang cultural region, the Wuyue cultural region, the Fujian–Taiwan cultural region, the Lingnan cultural region, and the Yunnan–Guizhou Plateau cultural region. The first 12 regions are all dominated by the Han culture, while the last one is highly complicated in ethnic composition. (2) Although the great western nomadic cultural area is not as complicated as the agricultural one, there are still great differences. Based on its general characteristics, the area can be subdivided into the Inner Mongolia cultural region, the northern Xinjiang cultural region, the southern Xinjiang cultural region, and the Qinghai–Tibet Plateau cultural region.

In summary, the Chinese cultural geography can be divided into 17 secondary regions, i.e., the northeast cultural region, the Yanzhao cultural region, the Loess Plateau cultural region, the Central Plains cultural region, the Qilu cultural region, the Huai River Basin cultural region, the Bashu cultural region, the Jingxiang cultural region, the Poyang cultural region, the Wuyue cultural region, the Fujian–Taiwan cultural region, the Lingnan cultural region, the Yunnan-Guizhou Plateau cultural region, the Inner Mongolia cultural region, the northern Xinjiang cultural region, the southern Xinjiang cultural region, and the Qinghai–Tibet Plateau cultural region.

The author finds that the above Chinese cultural geographical regions almost overlap with China's major river basins' upper, middle, and lower reaches. Therefore, the Chinese cultural geographical regions can be set to river basins to form 17 water culture areas: (1) the Songhua-Liao Rivers in the northeast cultural region, where less water cultural heritage is seen since its agriculture depends on fine weather and well irrigation, although there are the Songhua River, the Liao River, the Nen River, the Northeast China Plains and old-growth forests, with agriculture, forestry, animal husbandry, and fishery all involved; (2) the Hai River in the Yanzhao cultural region, (3) the upper Yellow River in the Loess Plateau cultural region, (4) the middle Yellow River in the Central Plains cultural region, (5) the lower Yellow River in the Qilu cultural region, (6) the Henan–Anhui–Jiangsu area in the Huai River cultural region, (7) the upper Yangtze River in the Bashu cultural region, (8) the middle Yangtze River in the Jingxiang cultural region, (9) the lower Yangtze River in the Poyang cultural region, (10) the Lake Tai in the Wuyue cultural region, (11) the Minjiang River in the Fujian–Taiwan cultural region, (12) the Pearl River in the Lingnan cultural region, (13) the upper Pearl River in the Yunnan–Guizhou Plateau cultural region, (14) the inward flowing rivers in the Gansu–Inner Mongolia cultural region, (15) the inward flowing rivers in the northern Xinjiang cultural region, (16) the inward flowing rivers in the southern Xinjiang cultural region, and (17) the inward flowing rivers in the Qinghai–Tibet Plateau cultural region. Each water culture area can be further divided based on its tributaries. For example, the basin of the Zhang River, a tributary of the Zhangweinan Canal in the Hai River, can be subdivided into areas containing the Zhang River water culture area (as shown in Figure 2.3).

2.5 FUNCTIONS OF WATER CULTURE

The report of the 19th National Congress of the Communist Party of China stated, "Culture is the soul of a country and a nation. A country with prosperous culture will prosper, and a nation with a strong culture will strive." Water culture is an important

Figure 2.3 Regional characteristics of Chinese culture.

Notice: This map has been provided by the author, and the editors and publisher of this book disclaim any disputes regarding the political boundary shown in the South China Sea.

part of Chinese culture and national spirit. Inheriting, protecting, and carrying forward the excellent Chinese water culture and building a modern water culture innovatively are the needs of national development and the people. Water culture has six functions: recording history, improving cognition, inheriting and spreading cultural heritages, regulating and educating people, uniting people's hearts, and maintaining order.

2.5.1 Recording history

Since the first day, humans had recorded history, passing on experience, knowledge, and ideas from generation to generation through oral languages before the writing system came into being. For example, the legend of Yu (Pinyin: Dà Yǔ, c. 2200–2100 BC, famed for his introduction of flood control) has no exact written documentation. Legends of this ancient king were passed down from mouth to mouth and then documented by Sima Qian—a famous Chinese historian of the Han dynasty who spoke highly of the legend and wrote it down in the *Records of the Grand Historian*. The work is known as the first in Chinese history to present general Chinese history in a series of biographies. The first chapter, the Annal of the Five Emperors, and the second chapter, the annal of the Xia Dynasty, recorded Yu's legend, in which he made significant contributions to flood control and so established the first dynasty in China—the Xia Dynasty. Da Yu's tale is still praised, inspiring Chinese people to forge ahead bravely.

2.5.2 Improving cognition

People can constantly accumulate experience, improve their mindset and cognitive ability, gradually understand water habits, characteristics, utilization, and treatment methods, and keep improving the existing production tools and methods for obtaining, utilizing, and controlling water through water culture. In this way, they can create new materials to form water culture, and further expand and deepen their cognitive ability of better quality and faster speed.

2.5.3 Inheriting and spreading cultural heritages

A cultural phenomenon is a social phenomenon. It will naturally spread through social communication when it comes into being and develops in society. Like any other culture, water culture can be passed down from generation to generation. The ancients left us excellent water cultural heritages, and we can create excellent new water cultures for future generations.

For example, the Dujiangyan Irrigation System, built more than 2,200 years ago, has been renowned and revered by the world throughout the ages. The reason is it applied a scientific, sustainable, and harmonious flood control concept and served the functions of spring irrigation and summer flood diversion by setting the water diversion weir (Pinyin: Shǔi Yú Zǔi, in shape like the fish mouth), the water flows control branch (Pinyin: Bǎo Pín Kǒu, in shape like a long narrow bottleneck), the spillway (Pinyin: Féi Shā Yàn, named for its strong ability to desilt flood), and adopting the annual repair concept of "digging the riverbed into sufficient depth and building the weir with an appropriately low weir top." Those actions complying with the natural law enabled the system to operate normally until today. Ever since then, the significant influence of the Dujiangyan Irrigation System has been visible in projects all over China that took the concept of Dujiangyan Irrigation System as a reference, and many of them have praised themselves as Little Dujiangyan.

2.5.4 Regulating and educating

Culture educates people naturally and imperceptibly. Through social behaviors under the impact of culture, people's way of thinking, behavioral habits, values, and esthetic tastes are changed. Once formed, water culture will inevitably become an organic part of our social environment, which, different from the natural one, influences, shapes, regulates, and educates people. Water culture is conducive to leading people in the right direction of water utilization, control, and management, creating more and better achievements in water culture.

2.5.5 Uniting people's hearts

Containing historical, artistic, scientific and technological, economic, and water conservancy values, water culture embodies ancient Chinese people's intelligence. Culture can educate people in a social group in the same cultural form or mode, thus producing a great sense of identity and strength to resist rivals by creating the same way of thinking, values, and behavioral habits, and closely uniting each other. Lofty, progressive, and excellent water culture can enhance national pride, and strengthen people's confidence and determination in water conservancy construction in the new period.

2.5.6 Maintaining order

For their survival and development and to ensure that society operates and develops in a specific order, a social group will inevitably require its members to abide by particular codes of conduct and moral standards. From certain social norms, members of a social group know right from wrong and distinguish good from evil, with their values and esthetic tastes homogenizing. Therefore, recognition of the achievements of water culture indicates new codes of conduct for water conservancy construction form and specific social order maintained by water culture.

Excellent water culture can lead people toward a harmonious and sustainable development of water and human society by influencing water management. Therefore, water culture research and construction should be paid attention to establish a more harmonious relationship between people and water, and achieve more sustainable socioeconomic development. Nowadays, as it is vigorously advocated to build a solid cultural power and boost cultural self-confidence, deeply digging into Chinese water culture and striving to inherit it are required. That will promote China's water conservancy and support the strategy of enhancing cultural power.

2.6 CONCLUSIONS

The multidimensionality of water culture enables countless unique water culture phenomena and water culture achievements. Whether perished, having become heritage, or still in use, these unique cultural phenomena and legacies are all records of China's history of dealings with water, reflecting the glory of Chinese civilization and national wisdom. Water culture records history, improves cognition, inherits and disseminates cultural legacy, unites people's hearts, and maintains order; it shapes individuals' and groups' attitudes and behaviors toward the water, change their ideological concepts of utilizing water, and affects their water-related habits. Activities guided by excellent water culture, such as scientific water control planning, enhanced water control policies, and proper water control actions, foster a harmonious and sustainable relationship between people and water. Therefore, as today's China vigorously advocates for strengthening cultural soft power and cultural self-confidence, it is of great and far-reaching significance for people to explore Chinese water culture and strive to inherit it. That bolsters China's water conservancy development and supports its strategies to build a culturally strong country.

REFERENCES

Chen, Jie. (2003). Preliminary study on water culture construction. *City Planning Review*, 9: 84–86 (in Chinese).

Fan, Youlin. (1990). On the essence of water culture. *Zhihuai*, 4: 55 (in Chinese).

Feng, Guanghong. (1994). What is water culture. *Hydro Science and Cold Zone Engineering*, 3: 50–51 (in Chinese).

Huang, Longguang. (2014). Introduction to water culture of ethnic minorities. *Journal of Yunnan Normal University*, 3: 147–156 (in Chinese).

Li, Keke. (1998). Thoughts on the study of water conservancy culture. *Journal of Yangtze University* (Social Sciences Edition), 1: 41–43 (in Chinese).

Li, Zongxin. (1989). Research on water culture. *Zhihuai*, 4: 37 (in Chinese).

Li, Zongxin. (2007). Present situation and prospect of water culture research. *Environmental Change and Water Security – Proceedings of the Fifth China Water Forum*. Beijing: China Water and Power Press, 7 (in Chinese).

Meng, Yaming and Yu, Kaining. (2008). On the connotation, research methods and significance of water culture. *Journal of Jiangnan University (Humanities and Social Sciences)*, 4: 63–66 (in Chinese).

Ran, Lianqi. (2000). Two essays on eater culture. *Beijing Water*, 4: 59 (in Chinese).

Wang, Dehua. (2000). On the relationship between water culture and urban planning. *Urban Planning Forum*, 3: 29–36, 79 (in Chinese).

Wu, Zongyue. (1989). Talking about water culture. *Hydro Science and Cold Zone Engineering*, 5: 11 (in Chinese).

Xing, Li. (1990). On the connotation of water culture. *Zhihuai*, 2: 47 (in Chinese).

Yang, Danian. (2005). *Chinese Water Culture*. Beijing: People's Daily Press, 50 (in Chinese).

Yuan, Zhiming. (2005). Theoretical discussion on water culture. *Water Resources Development Research*, 5: 59–61 (in Chinese).

Zhou, Kuiyi. (2008). *Introduction to Chinese Water Culture. Preface: Introduction to Chinese Water Culture*. Zhengzhou: Yellow River Water Conservancy Press (in Chinese).

Chapter 3

Sustainability of Chinese civilization and historical irrigation projects

Xuming Tan
China Institute of Water Resources and Hydropower Research

CONTENTS

3.1 TOPOGRAPHY, CLIMATE CHARACTERISTICS, AND IRRIGATION TYPES

China's topography is characterized by high terrain in the northwest and low terrain in the southeast. The mountains account for about 33% of the total area, the plateaus for about 26% of the total area, the hills for about 10% of the total area, and the plains and basins for about 31% of the total area. Most of the major rivers flow from the west to the east, along with the topography. The flood-affected plains, the downstream areas of the seven large rivers in China (i.e., Yangtze River, Yellow River, Huaihe River, Haihe River, Pearl River, Songhua River, and Liaohe River), and the areas along the southeast coast cover an 8% of the total land area. Historically, these regions were also the wealthy areas in China. Despite this, they are home to more than 40% of the population, cover 35% of the arable land, and generate 60% of the country's industrial and agricultural output (Tan, 2005).

China has a typical East Asian monsoon climate. Most regions have the advantage of the same periods of rainfall and heat, which is suitable for plant growth. The eastern region of China has a large volume of rainfall, and 60%–80% of its annual

DOI: 10.1201/9781003222736-4

precipitation occurs between June and September. The largest monthly rainfall may reach 30%–50% of the total annual precipitation. The annual precipitation decreases from 1,600 to less than 200 mm from the southeast coast to the northwest of the mainland. Therefore, rainstorm floods often occur in eastern China. In contrast, droughts occur throughout the country, including in the Guangdong and Guangxi regions with abundant rainfall and along the southeast coastal areas of China (Yao, 1987).

The differences in regional topography, climatic conditions, and water resources drive the invention and creativity of Chinese traditional irrigation engineering (Zhou, 1986). For example, in the arid region of the Turpan Basin in Xinjiang in western China, Karez was developed to take groundwater for irrigation. The Yellow River flows through Ningxia Plain and Inner Mongolia Plain. Therefore, a complete irrigation system was built in the region 2,000 years ago. Rain-fed agriculture has relied on field irrigation for at least 3,000 years in the North China plain, where surface water is insufficient. In the humid southeast coastal regions, rice agriculture has a history of 7,000 years with many types of traditional irrigation projects (Tan, 2017). Most water diversion projects distributed in plains and barrage ponds for water storage in hilly areas have more than 300 years of history and are still in use today.

3.2 THE PROCESS OF CIVILIZATION AND IRRIGATION DEVELOPMENT

Throughout China's 5,000-year history, agriculture has been the major support for politics and the economy (Chi, 1981). The rise and fall of a dynasty are closely related to food. Because of the specific natural environment of China, irrigation and technology often represent the degree of agricultural development and reflect Chinese political, cultural, and economic fundamentals in different historical periods.

3.2.1 The origin of civilization and water (about 4000 BC to 200 BC)

The first page of recorded Chinese history was about floods tamed by Yu the Great. Around 4,000 years ago, Chinese people living in the Yellow River and Huaihe River basins began to shift from mountain hunting and livestock husbandry to agriculture. Devastating floods lasting ten years put the indigenous people in a bind. Under the leadership of Yu, people fought against the floods, recovered from the disaster, and rebuilt their homes. After the success of flood control, the Xia Dynasty (about 2070 BC–1600 BC), also the first Chinese dynasty, was established. In 700 BC, China entered the Spring and Autumn Period (770 BC–476 BC) with the emergence of various schools and thoughts. In the Zhou Dynasty (1046 BC–256 BC), a form of feudalism was developed, and several relatively independent vassal states were built with numerous political and economic exchanges. Agriculture began to get wide attention in the vassal states.

3.2.1.1 The origin of irrigation and irrigation engineering

Irrigated agriculture is the symbol of the origin of Chinese civilization. Without any written record, it is not easy to know the exact time of the occurrence. China's irrigation can be traced from Hemudu Site in Zhejiang, dating back 7,000 years. Archaeological discoveries included ancient paddy, rice, and well remains, demonstrating the presence of irrigation in the rice civilization. Archaeologists also found an irrigation system dating back 3,600 years near the capital of the Shang Dynasty (about 1600 BC–1046 BC). From the remains of a 245 m long canal, significant differences among main, lateral, and field irrigation canals were observed. A stone dike for water diversion was found at the intersection of the main and lateral canals (Zhou, 2002). Changes in channel cross-sections indicated that field canals were separated from a lateral canal. A field was divided into several rectangles by the criss-cross channels, making significant elevation differences between canals and fields.

The West Zhou Dynasty (1046 BC to 476 BC) had well-developed agriculture. At that time, the economic center was located downstream of the Yellow River, i.e., the Shandong and Henan areas today. In the West Zhou Dynasty, an irrigation system was regarded as a "divine kingship" system, "Ida irrigation canals," where a piece of land was divided into a "井" shape, i.e., nine smaller pieces. A well was located in the central piece, while the remaining eight were arable land surrounded by canals. The first collection of Chinese poetry, *The Book of Songs*, depicted how farmers near the West Zhou Dynasty capital—Gao Jing (now southwest of Xi'an) took water from Biao Pond for irrigation. Biao Pond was a water storage pond connected with rice fields by canals. Around 400 BC, the ancient ritual text "The Rites of Zhou" (i.e., the Zhou ritual system) documented that the earliest bureaucracy appointed the official in charge of irrigation management as rice men, whose primary responsibility was to manage water storage ponds and canals.

Around 1000 BC, China was in the heyday of the Spring and Autumn Period, ruled by the Zhou Dynasty. At the time, the Chu, located in the Jianghuai plain, was the most prosperous state. In 605 BC, Sun Shuao of Chu built the Quebei project (as shown in Figure 3.1) on the Shi River, a tributary of the Huaihe River, and the Qisibei project on another tributary of the Guanhe River. These two water storage projects strongly supported Chu's agricultural development. Since then, for 2,000 years, these two projects have been continuously managed. The water storage capacity has now exceeded 100 million m^3, and the irrigation area has reached 12 million mu (1 mu = 1/15 ha).

3.2.1.2 *Wells and irrigation*

Utilizing wells was an important milestone event in civilization, which indicated that human settlements could be far away from rivers to seek more expansive spaces for survival and development. Wells became widespread water sources of irrigation and life in the Spring and Autumn Periods in today's Hebei, Shaanxi, Shanxi, Henan, Shandong, and other northern arid or semi-arid areas. It showed that the agricultural production in these regions had developed to a certain level.

It is said that Boyi—the successor of Yu the Great, was the inventor of the well. Well water was first used for domestic purposes. The invention of a water-lifting tool—shadoof—promoted the use of well water for irrigation. Well irrigation might originate in aristocratic gardens in the 4th century BC. During this period, fields with canals and wells were considered the ideal field system under royal rule. In the 1st century BC, well water had become a source of irrigation water with supporting irrigation canals. Around the 3rd century, windlass began to take water from deep wells for irrigation.

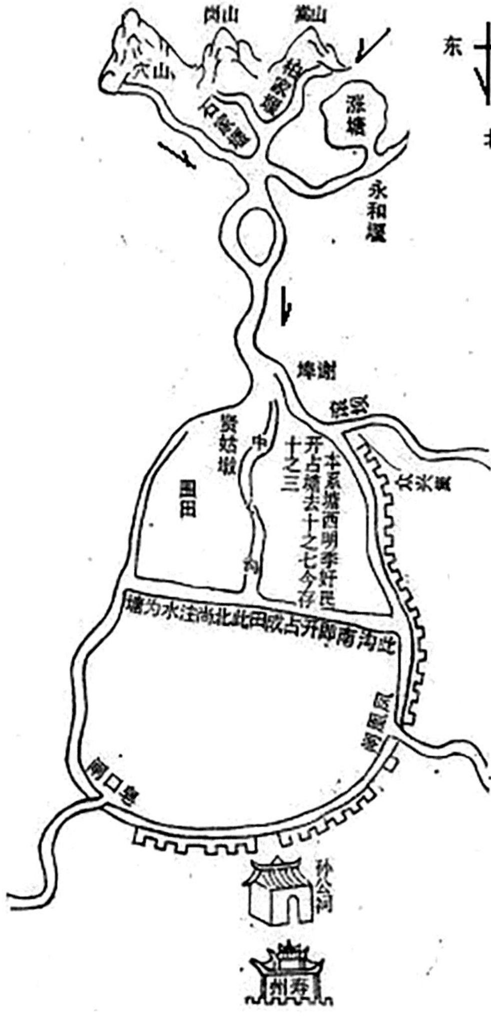

Figure 3.1 Schematics of Quebei project (drawn in the 19th century). This water storage project was built in 605 BC. With land reclamation, the water area nowadays is only 30% of that 2,000 years ago.

Deep well water was often stored in open ponds to increase the temperature and flowed into farmlands along canals. The well, the equipment used to take water from the well, and canals formed the well irrigation system. However, well irrigation was not widely used in the fields but was only used for small-scale manors, courtyard gardens, or landscapes. Until the 14th century (during the Ming and Qing Dynasties), the Chinese population increased to 100 million. When wheat began to be planted in large areas in China, the central and local governments allocated funds to support farmers in the northern semi-arid plains to develop upland well irrigation. Since then, well irrigation has become widely used in the fields (as shown in Figures 3.2 and 3.3).

Figure 3.2 Lifting water with shadoof (Palace Farming and Weaving figure in the 18th century).

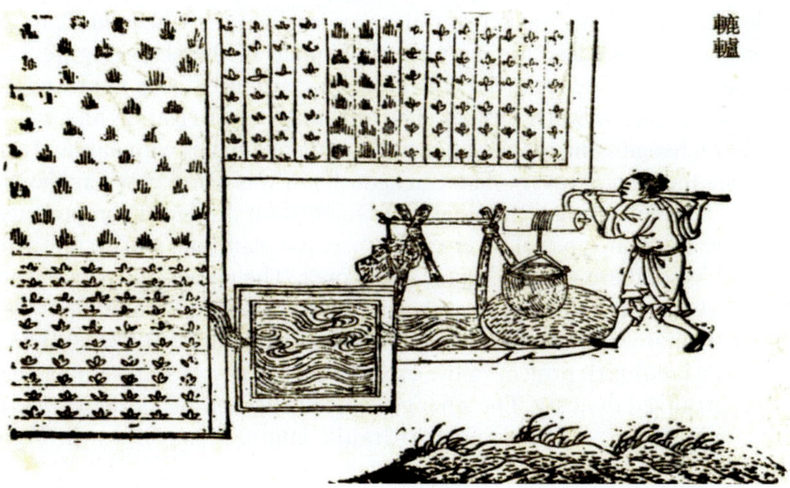

Figure 3.3 Windlass used for taking water from well: well irrigation (drawn in the 13th century) (During the 1st to 5th centuries, well irrigation was commonly used in gardens. The invention of the windlass made it easier to take water from deep wells. After the 13 century, wells became the primary water source for irrigation in farmlands of the North China plain)

3.2.2 Irrigation projects in the Qin and Han dynasties (3rd century BC to mid-3rd century)

From the 3rd century BC to the 1st century AD, China saw its first boom in the construction of large-scale water diversion projects during the Qin dynasty. In 422 BC, the Wei State built 12 canals of Zhangshui in the Haihe River basin. In 256 BC, the Qin State built Dujiangyan Weir on the Minjiang River in Sichuan (Tan, 2009). In 246 BC, the Qin State constructed the Zhengguo Canal on the Jing Rive—a third tributary of the Yellow River. These projects provided a stable source of water for the valley plains and alluvial plains, promoting agriculture and water transportation. The earliest urban and rural areas also developed in these regions.

The Dujiangyan Weir was a water conservancy project constructed by the state of Qin after its wars of unification. The system is located upstream of the Minjiang River and in the northwest of Chengdu, where the Minjiang River rushed down to the Chengdu Plain, thus providing favorable conditions for natural water diversion. The first step was to excavate a water inlet to bring water from the Minjiang River into the Chengdu Plain. The Baopingkou water inlet was the permanent inlet cut at the end of the Minshan Mountain and also the earliest critical project of Dujiangyan Weir. The successful construction of the Baopingkou water inlet opened up broad prospects for the development of Dujiangyan Weir. The Dujiangyan Weir has undergone continuous improvement and development processes. The critical water control works consisted of the water division project (i.e., fish head), water diversion project (i.e., Baizhang dike and 人-shaped dike), water regulation project (i.e., Feisha weir), and water inlet (i.e., Baopingkou). No later than the Tang Dynasty (618–907), the headworks already had their present scale (as shown in Figures 3.4 and 3.5).

In the past, the Dujiangyan Weir irrigation region almost covered the Chengdu Plain. The irrigation areas were 200,000 qing (a unit of area in China, 1 qing = 100 mu = 100/15 = 6.667 ha). Navigable waterways crossed the Chengdu Plain in different directions, which were also the main flood channels of the plain (Figure 3.5). After the main canal entered Chengdu, the canal system continued to be improved, forming the city's urban landscape water system. The advantages of irrigation and water transportation made Chengdu one of the important commercial cities with national economic prosperity since the Han Dynasty. The Chengdu Plain was also known as the "Land of Abundance." The engineering technologies of Dujiangyan produced comprehensive and long-term effects. Similar types of diversion dam, canal planning, and architectural style could be found in southern China, Japan, and Korea. The names of hydraulic and river components in these regions, such as bamboo cages and Macha, were similar.

Zhengguo Canal was built in the 1st year of the Qin Dynasty (246 BC). The Zhengguo Canal was a landmark project in the evolution from vassal states under the feudal system to a centralized dynasty. The Zhengguo Canal was named after its planner and constructer, "Zhengguo"—China's first hydraulic engineer (as shown in Figure 3.6) (Jiang, 2017).

The planning and design of the Zhengguo Canal were reasonable. The water taken from the Jing River flowed into the Luohe River along the main canal of the Zhengguo Canal. The main canal was 150 km long and was arranged from west to east along the highest line of the second terrace of the Weibel plain. The areas in the south of the main canal applied gravity irrigation. Water from the Jing River with high sediment

Figure 3.4 The statue of Dujiangyan creator, Li Bing (In 168, the officer of Dujiangyan weir management, "Dushui Yuan," constructed the statue, which was a symbol of Dujiangyan weir regulatory agencies in the Han Dynasty, 206 BC-220).

and organic matter content was used for irrigation, which turned saline fields and the barren Weibei Plain into fertile fields with "no bad year."

The successful project planning of the Zhengguo Canal laid the foundation for the later expansion of the irrigation area. During the Western Han Dynasty (140 BC–135 BC), a new canal (The Bai Canal) was constructed south of the Zhengguo Canal. The Bai Canal was named after the person who proposed the construction of this canal. Therefore, the Zhengguo Canal was later renamed the "Zheng Bai Canal." The Zheng Bai Canal benefited areas, including Jingyang, Sanyuan, and Gaoling counties in Shaanxi, with an irrigated area of over 4,500 qing (30,000 ha). In the Tang Dynasty (0–907), the Zheng Bai Canal continued to expand. In the 8th century, imperial relatives and influential officials in the capital Chang'an purchased land, ran manors in the irrigation region of the Zheng Bai Canal, and operated water mills on the main canal, resulting in a decrease of 6,200 qing (41,300 ha) irrigated areas. The Tang government had to promulgate the "Tang Shuibushi" to stop these actions and to protect the irrigation priority. The "Tang Shuibushi" then became the first irrigation law in China.

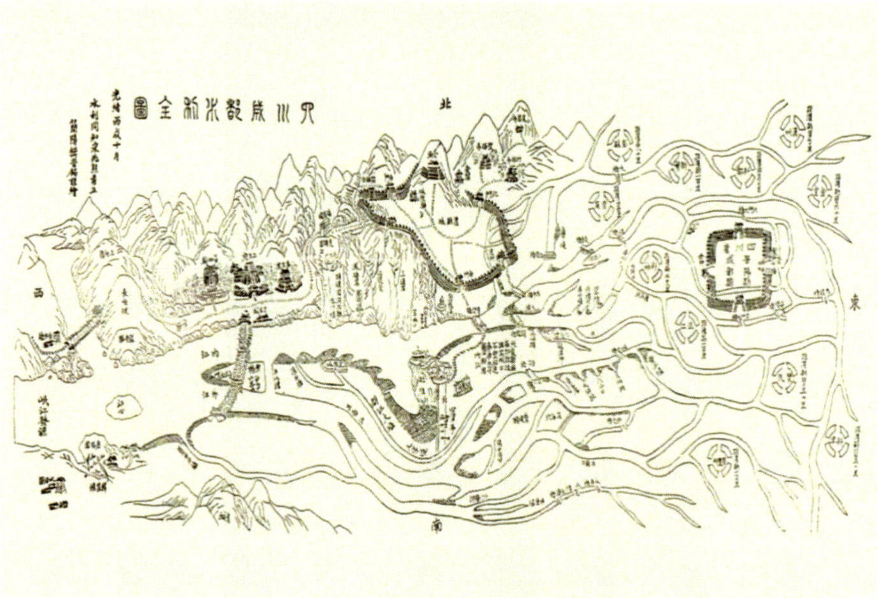

Figure 3.5 Dujiangyan and the canals on Chengdu Plain (drawn in 1886). (Since the 2nd century, the Dujiangyan irrigation system has been distributed in the plain).

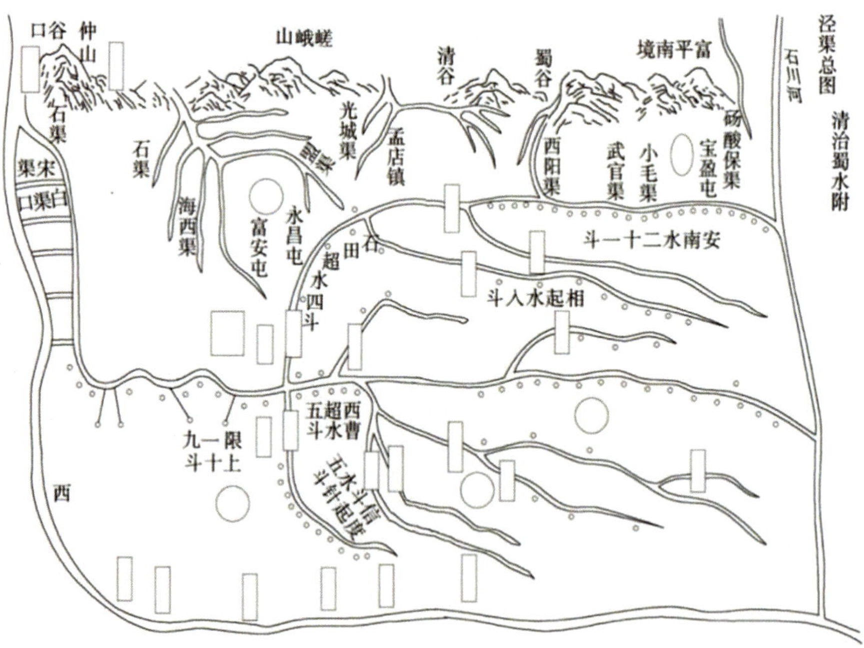

Figure 3.6 Zhenguo Canal and the related system (13th century). The figure shows the changes in inlets from the 3rd century BC to the 11th century. The Jing River irrigation project maintained the sustainable development of agriculture in the Guanzhong Plain for more than 2,000 years.

In the Song Dynasty (960–1127), it was difficult to extract water from the Zhengguo Canal's diversion headworks due to the Jing River's riverbed gullies. In Xining 5th year (1072), Emperor Shenzong allocated government relief funds to reconstruct the Zhengguo Canal headworks, but the reconstruction project was not completed because of the severe drought in Guanzhong. After 36 years, the reconstruction project was completed, and the rebuilt main canal extended for 2 km upstream. From the 12th to 15th century, the headworks were moved upstream several times. Finally, Zhengguo Canal could no longer obtain water from the river and used the spring water along the canal for irrigation. At the end of the Qing Dynasty (1911), only over 1,300 qing (8,700 ha) of irrigation area remained. In the 1930s, a project to take water from the Jing River (i.e., the Jinghui canal) was built. Upstream of the headworks of the Zhengguo Canal was built in the Song Dynasty, and an overflow dam was constructed with a length of 68 m and a height of 9.2 m. A three-hole intake was set up on the right bank, and the ancient canals were utilized below the headworks. Eventually, the system was partially restored for irrigation. Since the 1950s, the dam of the Jinghui Canal has been heightened several times with continuous improvements on the canal, and the modern irrigation area has reached 80,000 qing (533,000 ha).

3.2.3 The period for continuing irrigation progress and irrigation machinery invention (from the 3rd to 13th century)

China has entered an era during which local separatist regimes fought against each other. In the 3rd century, the powerful centralized rule in the Han Dynasty gradually declined. From the 4th to 5th century, frequent wars happened among the northern separatist regimes, while the Western Turks, Huns, and other ethnic groups invaded China. The Northern and Southern Dynasties (AD 402–589) were the most tumultuous era in the history of China, during which centralized control was temporarily lost. Therefore, it was a dynamic age rich in ideas and culture, and a period of the invention of irrigation and water conservancy facilities.

After the 3rd century, powerful families with large land areas had the opportunity to grow quickly without central control. As a result, the manorial agriculture in the north became prosperous, and the decline in population and the prosperity of the manorial economy created a demand for machinery. During this period, water mills and other hydraulic machinery for food processing, cylinder cars, and small equipment for irrigation and drainage began to be used in potential royal manors. Some nobles or imperial officials used these machines to show their status and wealth.

After the 5th century, the center of the agricultural economy shifted to the south. Weirs and dams block tides and store water upstream in the shorter rivers along the southeast coastal areas. With the rapid development of the southeast coastal areas, many weir dams were built, such as Tuoshan Weir in Yin county of Zhejiang and Mulanbei Weir in Putian of Fujian. These projects were used for irrigation and as a source of drinking water in urban and rural areas.

During the Tang and Song Dynasties (618–1270), the prosperity of commerce and the development of agriculture promoted the broad applications of hydraulic machinery and water powerlifting irrigation. Famous inventions of hydraulic machinery included cylinder cars used for irrigation or drainage (as shown in Figure 3.7), water pestles, water mills, and water spinning wheels used for industrial processing (as shown

Figure 3.7 Hydraulic lifting water: the cylinder cars in working (drawn in the 19th century). (The diameter of the cylinder car or water wheel is generally 2-3 m, determining the height to lift water. Cylinder car was commonly used in urban and rural water utilities before the 20th century)

in Figures 3.8 and 3.9). Hydraulic machinery was widely used by ordinary people, distant regions, and remote mountain areas. After the 13th century, the hydraulic spinning wheel and the hydraulic blower gradually disappeared.

In the 13th century, Wang Zhen wrote the first book to record agriculture and irrigation, *Agricultural Book*. In the book, the author systematically listed and described the irrigation machinery used at that time or already lost and the lost ancient water irrigation and hydraulic machinery for food processing.

Tuoshan Weir was built on the Yin river in Ningbo and was a water conservancy project with the integration of saline water blockage, freshwater storage, irrigation, water supply, and drainage (as shown in Figure 3.10). In the 7th year of the Tang Dynasty (833), the magistrate of Yin County, Wang Yuanwei, was in charge of the construction of Tuoshan Weir at the location 75 km away from the southwest of today's Ningbo city. The main body of Tuoshan Weir was a 130 m long, 8–9 m high spillway dam across the Yin river, which was used to resist saline water and retain freshwater. Downstream of the mainstream, three weirs were built: Wujin, Jiddu, and Xingchun. During the flood season, flood water was drawn to the Yongjiang River. These three weirs were opened at high tides to introduce the freshwater pushed by tides into irrigation canals, which helped block saline water and store fresh water. The Tuoshan Weir provided the main water source for the city of Ningbo in ancient times. Water brought in by the main canal was stored in "Ri" (sun) and "Yue" (moon) lakes, while water supply canals were distributed along the streets of the city.

Figure 3.8 Hydraulic machinery used for pottery (drawn in the 16th century). Pestle driven by a water wheel was used to mix clay, significantly reducing labor.

Figure 3.9 Multiple water mill (drawn in the 13th century). This was an ancient water mill set for grain processing. A water wheel could drive multiple water mills.

Figure 3.10 Tuoshan Weir and the related canal project.

Mulanbei Project was built on Mulan Stream, 4 km southwest of Putian city of Fujian province, during the Northern Song Yuanfeng 6th year (1083). The project continues to operate, irrigating over 10,000 ha of farmland in the Putian plain. Mulanbei Project was built at the tide-traced end to prevent tides, resist saline water intrusion, and retain freshwater from the mountain streams upstream of the dam. The key project of Mulanbei consisted of the overflow weir, watergate, sand flushing sluice, and diversion mouths at both banks of the river. The dam is 111.13 m long and 7.25 m high, along which there are 32 gates. So far, the project has served for about 1,000 years. There is no sediment in front of the dam, and the two inlets are still in normal use. The irrigation region of Mulan Stream was divided into Nanyang and Beiyang, which were controlled by the main canals on the left and right banks, respectively. Drainage water enters the sea at Sanjiangkou of Putian. Now the storage capacity of Mulanbei is 30 million m^3 with an irrigation area of 13,000 ha (as shown in Figure 3.11).

3.2.4 The culmination of irrigation development (from the 14th to 19th century)

In the 1270s, the Mongol army commanded by Kublai Khan matched a long way from the northern grasslands to the south and established a unified regime in China: the Yuan Dynasty. At the time, the 1,700 km long Beijing–Hangzhou Grand Canal was constructed to transport grain from the south to Beijing. The rice produced in the south was continuously shipped to Beijing through the canal, stimulating agricultural development in the south. Grain production in the Yangtze River's middle and lower reaches supported China's economic lifeblood.

Figure 3.11 Mulanbei in Putian of Fujian (Photo was taken in 2013).

During the Ming and Qing Dynasties (1368–1911), China's population exceeded 100 million, increasing the pressure on the food supply. With the increase in population, food production areas expanded rapidly. Even in the arid northwest and cold northeast, lands were reclaimed. In Hetao Plain (i.e., Ningxia and Inner Mongolia) upstream of the Yellow River, irrigation projects developed much faster than in any previous historical period. In Fujian, Hainan, and Taiwan in the southeast, and Yunnan, Guizhou, and Sichuan in the southwest, various patterns of irrigation engineering were developed under different local and natural environment conditions.

Before the 11th century, the main crops in China were rice in the south and millet grains in the north. With wheat, corn, and sweet potato from abroad into China in the 2nd century, rain-fed agriculture began to develop in alpine and sub-alpine regions, where initially only tribal peoples lived. In the 14th century, rice agriculture, originally in the south, was developed in hills and mountains. Terraces with irrigation systems resolved the problem of rain-fed irrigation. High-yielding rice fields also appeared in some areas with poor natural conditions. In mountain areas with slopes greater than 45°, the rice terrace system was a homogeneous system composed of vegetation, irrigation and drainage, and terraces. The system achieved ecological and agricultural benefits, and, more importantly, it better protected the region's natural environment.

In the Qing Dynasty, China's territory was extended to the northwest. During the Kangxi and Yongzheng periods (1662–1735), the northern region of Xinjiang began to cultivate farmland and build irrigation projects. In the 18th century, Xibe soldiers stationed in the Ili region constructed the Chabuchaer canal (Chabuchaer means granary in the Xibe language), which brought water from the Ili River to rice fields. This was the only rice agriculture area in Xinjiang at that time. In the 19th century, the Huangqu canal was built on the Kashgar river—a tributary of the Ili River. The irrigated area of the Ili plain reached 2,000,000 mu (133,300 ha), making it a granary in northern

Figure 3.12 Diversion project of Kashgar river of Ili River in Xinjiang (canal headwork built in the 17th century, water diversion without dam)

Xinjiang. The area is still the most prosperous agricultural region in Xinjiang, with continuous irrigation benefits (as shown in Figure 3.12).

The Turpan and Hami Basins in Xinjiang are connected to the Tianshan Mountains in the north, where smelting snow water infiltrates the ground through the gravel soil. Before the 19 century, the local Uighur landowners began to build Karez to irrigate vineyards. After the 19th century, the construction of Karez was supported by government funding. Since the 1980s, motor-pumped wells were commonly used, so the Karez was on the edge of disappearing. Using water resources with Karez contributes to the maintenance of an ecological oasis. The protection of Karez should be of great value from the ecological and cultural aspects (as shown in Figure 3.13).

China is one of the first countries in the world to build terraces. The rain-fed terraces in the north probably appeared in the 2nd century BC. After the 9th century, they were gradually developed in the areas south of the Yangtze River. With the population increase, lands in hills and mountain regions were gradually reclaimed, and terraces for growing rice appeared. Terraces were built from the bottom up, and rain-fed fields gradually evolved into rice terraces. Terraces are widely used in southern China, among which the Hani Terraces in Nanyang of Yunnan province, the Ziquejie terraces in Xinhua County of Hunan province, and the Longji Terraces in Longsheng County of Guangxi province are the most famous.

The Ziquejie terraces (as shown in Figure 3.14) are located in Xuefeng Mountain in the western part of Xinhua County. The terraces cover a total area of 60,000 mu (4,000 ha) and are distributed on slopes with a vertical height of 500 m and more than 1,000 terrace levels. With an annual rainfall of 1,200 mm, the area is a rain-fed agricultural area in southern China (Li et al., 2020).

The Ziquejie terraces are located at the northern end of China's sub-alpine rice terraces. The terraces were constructed in the 11th century with a history of 1,000 years.

Figure 3.13 Xinjiang Karez (culvert and open channel).

Figure 3.14 Ziquejie terraces in Xinhua of Hunan.

The earliest indigenous ethnic groups in the Xinhua region were the Miao, Yao, and Dong. In the 9th century, the entry of the Han brought the farming civilization from the Central Plains. Most likely, this was the birthplace of sub-alpine rice terraces in China. During the Ming Dynasty, agriculture practices were spread to the neighboring Guangxi, Guizhou, and Yunnan and were accepted by different ethnic groups.

The Ziquejie mountainous areas had a thick layer of red loam with high water content in the soil. The rich rainfall in the region ensured the basic need for rice growing. During spring and the early stages of rice seedlings, a small amount of supplemental water was needed for irrigation. In spring and dry years, spring water seepage was an important source of water for irrigation. The terraces were distributed along the contours, and water was diverted to each terrace through field ditches and streams. The small irrigation system at each water source consisted of soil pores, rock fissures, and mini ditches. Over the hillsides, terraces varied in slope, size, length, width, and height, facilitating slope stability, management, and reduction of soil erosion.

The Hani Terraces (as shown in Figure 3.15) are located at the southern foot of Ailao Mountain in Yuanyang County of Hani and Yi Autonomous Prefecture in Yunnan Province. The terraces of different shapes are connected, among which the largest rich field reached more than 1,000 mu (67 ha). The terraces are distributed on the cliffs with slopes ranging from 15° to 75°. At the most, more than 3,000 levels of terraces were constructed on one side of a hill. The terraces extend from the river valley to the mountains at an altitude of over 2,000 m above sea level, reaching the highest limit of rice growth.

Around the 6th century, the Hani people lived in the Ailao Mountains and began to grow rice in about the 16th century. Centered on the 198,000 mu (13,200 ha) of terraces in Yuanyang County, the terraces were extended to several surrounding

Figure 3.15 Hani terraces in Yuanyang of Yunnan.

Figure 3.16 Longji terraces in Longsheng of Guangxi.

counties, covering a total area of 1.05 million mu (70,000 ha). It concerned natural ecological concepts in the Hani people's thoughts. The mountains regulated by the villages were respected as holy mountains, and deforestation was strictly restricted. The Hani's rituals were related to the God of Water, and the ritual festivals were arranged for planting, flowering, and harvesting. These practices have become the DNA of the cultural heritage of the terraces.

The Longji terraces (as shown in Figure 3.16) are located in Longji Mountain (Ping'an village) in Guangxi. Longji mountain has an altitude of nearly 1,000 m, and the slope is mainly between 26° and 35°, with a maximum gradient of 50°. The terraces are distributed at altitudes ranging from 300 to 1,000 m, and the total area of Longji terraces is about 66 km^2. The Longji terraces were first built in the 13th century and completed in the 17th century, with over 650 years and a total terraces of 99,000 mu (6,600 ha). The steep mountains here make many terraces smaller, less than 1 mu. Locals said: "A frog jump can cross three terraces." In the deep mountain regions, forests, terraces, and villages are distributed along the mountains as follows:

- Forest is on the top of mountains for water conservation.
- Villages are in the middle.
- Terraces are around the villages and extend to the foot of mountains.

The Longji terraces distributed along the contours form a convenient gravity irrigation system, which meets rice irrigation needs and avoids soil erosion in mountainous areas. The construction of the Longji terraces provides valuable experience in achieving food self-sufficiency and maintaining a good regional ecological environment.

3.3 CONCLUSIONS

Irrigation plays an important role in China's civilization due to its unique natural and geographical conditions. The origin and development of irrigation engineering sometimes are of historical significance, representing the evolution and expansion of civilization in a region. There are many conservancy projects in China, all of which have a history of hundreds or even thousands of years. The continuation of projects is the continuation of management and culture, reflecting the wisdom of ancient people in dealing with the human–water relationship and their values for water resources. The management of irrigation projects and water administration promote a close connection between water and society and culture and the integration of water with regional folk, religion, and architectural culture.

Traditional irrigation projects are all precious cultural heritage, whether they are of different styles of water conservancy engineering, engineering and hydraulic structures with ecological value, or valuable management documents and archives. They are still operating and producing benefits per the traditional mode. Since the 20th century, with the rapid development of modern technology, we have gradually lost sight of ancient water conservancy projects. Many ancient irrigation works have been destroyed in the so-called modernization process. The motivation for quick success and short-sighted behaviors have driven people in some regions to make economic gains by converting or demolishing ancient water conservancy projects. To learn from historical wisdom and preserve irrigation civilization, it is necessary to protect the ancient irrigation heritage, which may be the best example of sustainable development.

ACKNOWLEDGMENTS

The article was modified based on a research report entitled "Sustainability of Chinese Civilization and Water Conservation". The report was submitted to ICID by CNCID in 2014. This work was supported by the Key Scientific Research Base Program of the State Administration for Cultural Heritage (2020ZCK207), IWHR Research & Development Support Program (JZ0199B212019, JZ1003A112020).

REFERENCES

Chi, Ch'ao-ting (1981). *Key economic areas in Chinese history.* China Social Sciences Press, Beijing (in Chinese).

Jiang, Chao. (2017). *Zhengguo Canal* (in Chinese).

Li, Yunpeng, Tan, Xuming, and Zhou, Bo. (2020). Philosophy and value in irrigation heritage in China. *Irrigation and Drainage*, 69(2): 153–160 (in Chinese).

Tan, Xuming. (2005). *History of irrigation and flood control in China.* China Water & Power Press, Beijing (in Chinese).

Tan, Xuming. (2009). *History of Dujiangyan Weir.* Sciences Press, Beijing (in Chinese).

Tan, Xuming (2017). *History of Chinese material culture (Volume Water).* Kaiming Press, Beijing (in Chinese).

Yao, Hanyuan. (1987). *An outline of China's water resources history.* China Social Sciences Press, Beijing (in Chinese).

Zhou, Kuiyi. (1986). *History of agricultural water resources in China.* China Water & Power Press, Beijing: 1–2 (in Chinese).

Zhou, Kuiyi. (2002). *History of science and technology in China (Volume Water).* Sciences Press, Beijing (in Chinese).

Chapter 4

Dujiangyan irrigation system and its over 2,200 years of sustainable development

Jiangang Liu, Bo Zhou, Jun Deng, and Xiaoming Jiang
China Institute of Water Resources and Hydropower Research
Research Centre on Flood and Drought Disaster Reduction of Ministry of Water Resource

Xuming Tan and Jianzhao Guan
China Institute of Water Resources and Hydropower Research

CONTENTS

4.1 INTRODUCTION

The Dujiangyan irrigation system is a world-renowned water conservancy project. Its headwork is located in the city of Dujiangyan. The irrigation system was first built in 256 BC. For over 2,200 years, it has lent major support to local economic and social development, and left a rich water culture legacy, making it a monument in China's history of water governance. The irrigation system provides an abundant water supply to the Chengdu Plain. Its canals bring irrigation, flood prevention and control, and water transport benefits. It contributes to transforming the Chengdu Plain into the Land of Abundance, where "water is under control, and man does not know what famine is." Thanks to the irrigation system, the Chengdu Plain has become the granary of

DOI: 10.1201/9781003222736-5

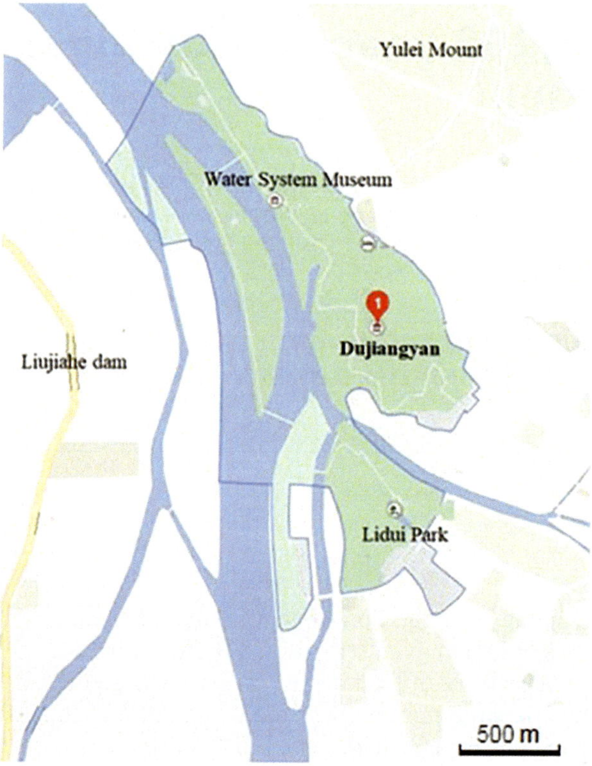

Figure 4.1 Location of the headwork.

Western China and the political, cultural, and economic center of Southwest China since the 2nd century (Wang, 2020; Zhang et al., 2013).

The Dujiangyan irrigation system is located where the Minjiang River enters the Chengdu Plain (Figures 4.1 and 4.2). The River, with a total length of 711 km, is a first-level tributary of the Yangtze River. The middle reach of the river in the Chengdu Plain runs 216 km, accounting for 30% of its total length. The river's gradient gradually drops from 10% to 1%, and its slope is around 8.2%. The left bank of the river is the ancient landslide belt of Er'wang Temple, and Mount Lidui is the product of geotectonic movement (Kuang, 2018).

The Dujiangyan irrigation system, nourishing 710,000 ha of farmland in 38 counties under seven cities of the Sichuan Province, as shown in Table 4.1 and Figure 4.3, occupies an exceptionally important status in the local and even national economic and social development (Tan, 1986, 2009).

4.2 HISTORY OF DUJIANGYAN IRRIGATION SYSTEM

All the rulers throughout the history of the Shu (Sichuan) region prioritize the construction of water conservancy projects. The Dujiangyan irrigation system has been constantly reinforced at different historical stages. (1) The period of creation and

Figure 4.2 A recent photo of the headwork of the Dujiangyan irrigation system.

Table 4.1 Irrigated area of Dujiangyan irrigation system

City	*County*
Chengdu City	Dujiangyan, Pi, Pengzhou, Wenjiang, Shuangliu, Tianfuxinqu, Xinjin, Longquanyi, Xindu, Qingbaijiang, Jintang, Chongzhou, Dayi, Qionglai, Jianyang, Wuhou, Qingyang, Jinniu, Jinjiang, Chenghua, Gaoxin
Meishan City	Pengshan, Renshou, Qingshen, Dongpo
Deyang City	Guanghan, Shifang, Mianzhu, Jingyang, Luojiang, Zhongjiang
Mianyang City	An, Fucheng, Santai
Suining City	Shehong, Daying
Ziyang City	Yanjiang
Leshan City	Kaiyan

improvement. In 256 BC, Li Bing, the governor of the Shu Prefecture, started to build the Dujiangyan irrigation system. The water of the Minjiang River was introduced into the heartland of the Chengdu Plain through the construction of the Fish Mouth Levee (Figure 4.4) and the Bottle Neck Canal (Figure 4.5).

The project was mainly used for flood prevention and control, water transport, and irrigation. In AD 662, the Flying Sand Weir (Figure 4.6) was completed, signifying the formation of the distribution of the three major projects of the headwork. (2) The period of stable development. Ancient China enjoyed prosperity during the Tang (AD 618–907) and Song (AD 960–1279) dynasties. As a result, the irrigated area of the Dujiangyan project quickly expanded to cover 12 counties. By the end of the 1940s,

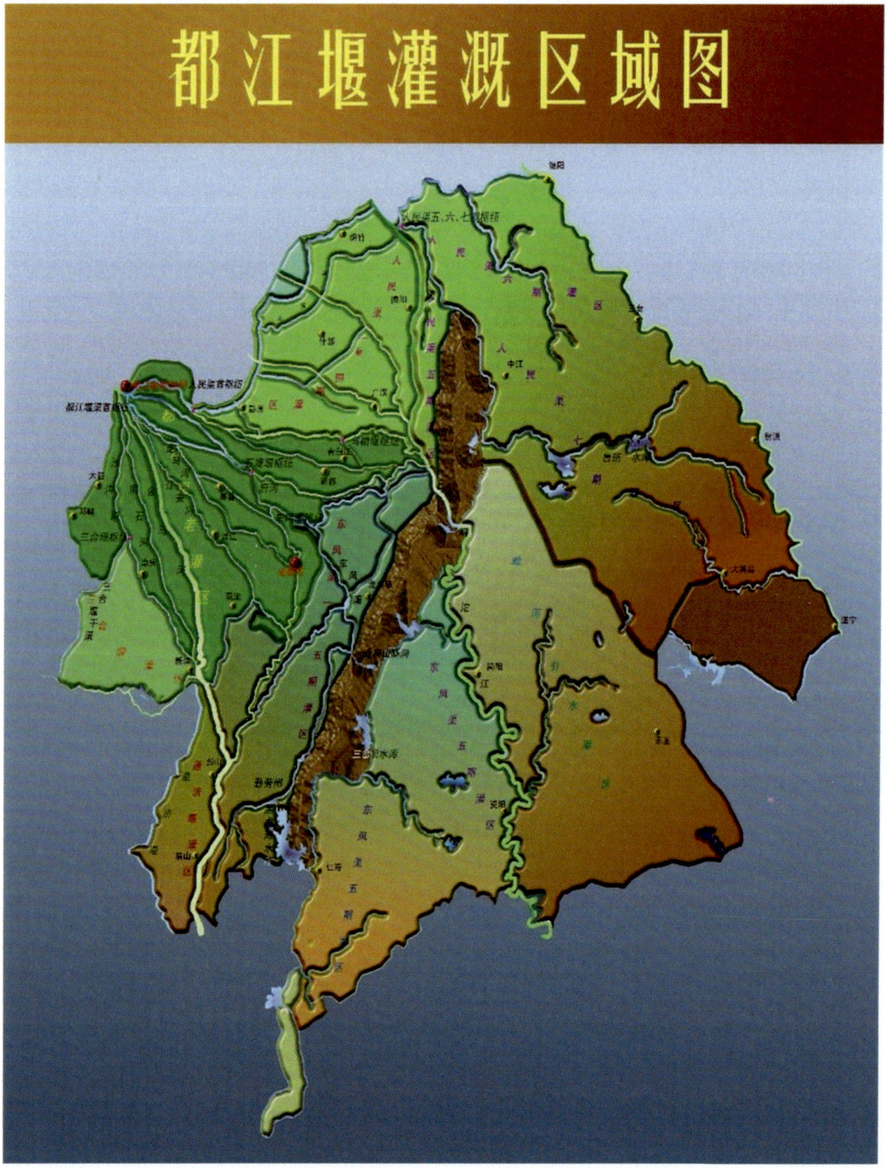

Figure 4.3 Irrigated area of the Dujiangyan irrigation system.

the project had provided irrigation water to 188,000 ha of farmland in 14 counties of the Chengdu Plain. (3) The period of fast development. From 1949 to the present, the Dujiangyan irrigation system has experienced large-scale transformations and rapid development. Its irrigated area has exceeded 710,000 ha, spanning 38 counties of seven cities. In the 2008 Wenchuan earthquake, the whole system remained largely intact except for only a few cracks in the Fish Mouth.

Figure 4.4 The Fish Mouth Levee.

Figure 4.5 The Bottle Neck Canal.

Figure 4.6 The Flying Sand Weir.

4.3 CHARACTERISTICS OF DUJIANGYAN IRRIGATION SYSTEM

The Dujiangyan irrigation system is an engineering paradigm for utilizing and improving natural conditions. Having existed for over 2,200 years, it represents the outstanding technological achievement of water conservancy projects in ancient China. As an engineering system consisting of headwork (Table 4.2 and Figure 4.7), water-diverting canals at various levels, ponds, weirs, and farmlands, it has created a water environment featuring dense networks of rivers, ponds, lakes, and swamps.

As the water-diverting hub of the irrigation system, its headwork mainly consists of the Fish-Mouth Diverting Levee, the Flying Sand Weir (spillway), the Bottle Neck Canal, the Baizhang Dike's auxiliary projects, and the Renzi Dike. By wisely adapting to the terrain of the Minjiang Riverbed, people achieved the goals of water diversion, sediment flushing, inlet flow control, and flood discharge with minimal engineering facilities. The headwork, having undergone various rehabilitations, is a witness to the sustainable development of the Dujiangyan irrigation system. Starting from the Fish Mouth Levee, the water flow is controlled by water diversion levees, and overflow weirs made of bamboo cages and timber piles. Though no gate has been installed, the water of the Minjiang River could reach the farmland and the residential communities smoothly. There are now 111 main and sub-main canals in the irrigation district with a total length of 3,664 km, 260 branch canals that run 3,234 km and each irrigating thousands of hectares of farmland, and field ditches below branch canals that run over 34,000 km.

4.4 MANAGEMENT EXPERIENCE OF DUJIANGYAN IRRIGATION SYSTEM

Multiple types of management have brought order and vigor to the Dujiangyan irrigation system at various historical stages. The management of the system involves a

Table 4.2 Inventory list of Dujiangyan irrigation system

Heritage item	*Name*	*Number*	*Location*
Headwork	Fish Mouth Levee	I	Dujiangyan city
I.	Flying Sand Weir	I	Dujiangyan city
I.	Bottle Neck Canal	I	Dujiangyan city
Main Canals and Sub-Main Canals	Total	III	–
I.	Puyanghe River	I	Dujiangyan city
I.	Baitiaohe River	I	Dujiangyan city
I.	Zoumahe River	I	Dujiangyan city
I.	Jiang'anhe River	I	Dujiangyan city
I.	Shagouhe River	I	Dujiangyan city
I.	Heishihe River	I	Dujiangyan city
Branch Canal		260	–
Field Ditches at Various Levels Below Branch Canal		Many	Irrigation system area
Important Projects of Branch Canals and Those of Lower level	Yangliuhe River	I	Dujiangyan city
I.	Baimayan Weir	I	Dujiangyan city
I.	Yangliuhe River	I	Wenjiang district
I.	Dalangyan Weir	I	Shuangliu district, Xinjin county
I.	Gufoyan Weir	I	Shuangliu, Pengshan, Renshou
I.	Jiulidi Dam	I	Jinniu district
I.	Zhulihuoyan Weir	I	Shifang County
I.	Qiangongyan Weir	I	Chongzhou City, Dayi County, Xinjin County, Qionglai County
I.	Huoshaoyan Weir	I	Gaoxin district
I.	Zhugejing Well	I	Qingbaijiang district
I.	Sanbodong Canal	I	Dujiangyan city
Buildings for Sacrifice and Worship	Er'wang Temple	I	Dujiangyan city
I.	Fulong Temple	I	Dujiangyan city
I.	Water Resources Administration Office	I	–

highly centralized administrative system. Relevant officials at provincial, prefectural, and county levels manage the headwork weirs, main canals, and branch canals. Lower water management bodies and water user associations are responsible for field ditches. Combining government and non-government resources, such a management system engages all beneficiaries, forcing them to take respective management responsibilities and fulfill corresponding obligations. This management system is still in use today. The project is currently operated under a management system that combines centralized management with hierarchical management and professional management with management by the public. The provincial water authority is in charge of the irrigation system. It sets up the Dujiangyan Administration to take charge of the system's centralized management and the headwork's specific management. Various administrative offices in the irrigation district are responsible for managing the water-diverting facilities on main, sub-main, and branch canals within their jurisdiction.

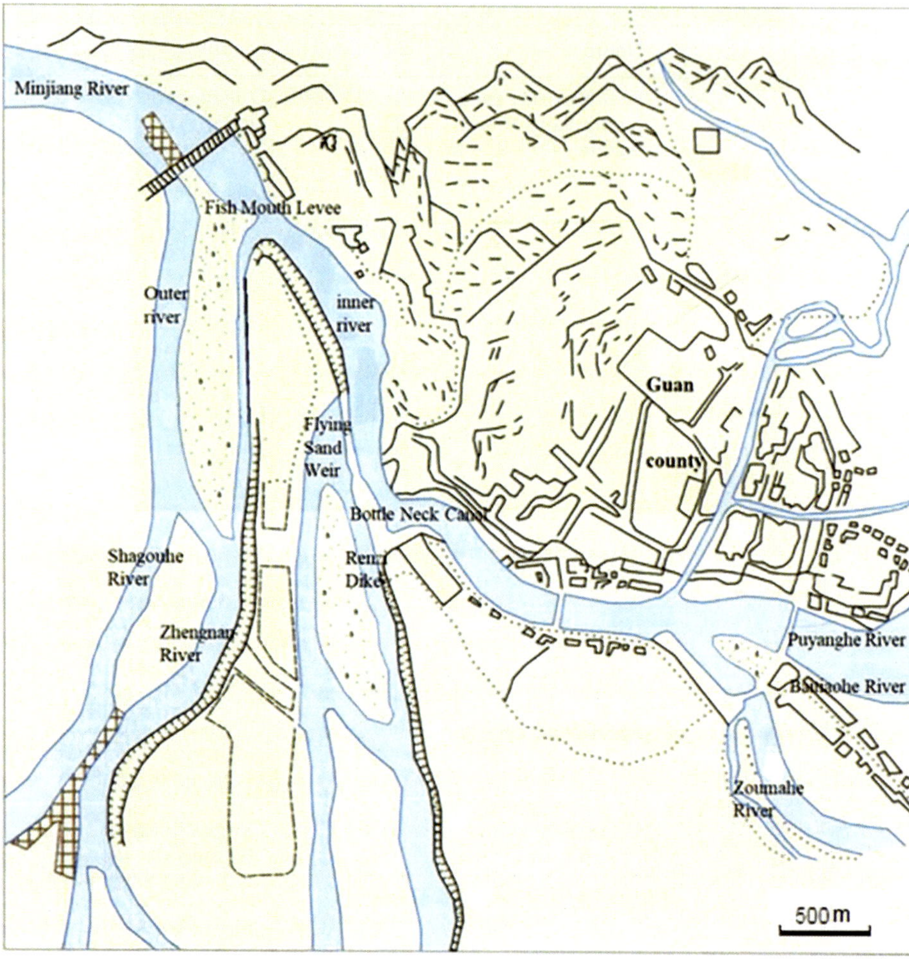

Figure 4.7 Headwork of the Dujiangyan irrigation system in 1910.

During its 2,200 years of sustainable utilization, Dujiangyan Weir has generated rich irrigation culture. One example is the magnificent Er'wang Temple. In the 3rd century, to encourage the development of agriculture, the central government of China enshrined Li Bing as the water god and worshipped him on the national level. A memorial temple was built for him on the left bank of the Minjiang River. Since then, Li Bing has been worshipped by generations of Shu people. After the 10th century, the memorial temple of Li Bing was turned into Er'wang Temple, which worships both Li Bing and Er'lang God—a deity of Taoism. The religion used the irrigation project to enlarge its impact and simultaneously build a bridge between the government and the local people to manage the project. The stele inscription at the Er'lang Temple records the key points of flooding control and annual repair, which serves as a reference for the weir managers. In the Qing Dynasty, government officials and weir managers met at Er'wang Temple to discuss management issues and solve water use conflicts. In ancient times, the government was responsible for managing headwork, main canal, and branch canals; the water users managed canals of the lower level. Local governments in the irrigation district contributed to the management by collecting water fees, distributing water, and organizing labor.

4.5 VALUES OF DUJIANGYAN IRRIGATION SYSTEM

The canal network of Dujiangyan Weir has created a lot of rivers and ponds in the Chengdu Plain. Starting from the Fish Mouth Levee, the water flow is controlled by water diversion levees and overflow dams made of bamboo cages and timber piles. Though no gate has been installed, the water of the Minjiang River could reach the farmland and the residential communities smoothly. This 2,000-year-old irrigation project is an outstanding example of engineering technology in ancient China.

4.5.1 Its enormous benefits have lasted till today

It is a brilliant example of water conservancy projects that positively affect the environment and achieve harmony between man and nature. The benefits of the irrigation system mainly fall into the following categories.

4.5.2 It was ahead of its time

It was ahead of its time in project formulation, engineering design, construction techniques, and structure dimensions. It was an example of an engineering marvel at the time of its construction. It was innovative in its ideas at the time of its construction. It contributed to the evolution of efficient and contemporary engineering theories and practices. The planning, design, and construction of the Dujiangyan irrigation system are both scientific and creative. The headwork system mainly consists of the Fish-Mouth Diverting Levee, the Flying Sand Weir (spillway), the Bottle Neck Canal, the Baizhang Dike's auxiliary projects, and the Renzi Dike. It is the water-diverting hub of the irrigation system. The Chinese people adopt proper measures to utilize local geographical and climatic conditions. They fully adapt to the terrain of the Minjiang River bed and achieve the goals of water diversion, sediment flushing, water transport, and flood discharge in a scientific way, and practice the development philosophy of the harmonious coexistence of man and nature.

4.5.3 It was unique in some positive and constructive way

The Dujiangyan Irrigation System was built in 256 BC. For over 2,200 years, it has lent major support to local economic and social development and left a rich water culture heritage, making it a monument in China's history of water conservancy projects development. The system provides abundant water supply to the Chengdu Plain through artificial canals, bringing the benefits of irrigation and water transport, thus transforming the Chengdu Plain into the Land of Abundance where "water is under control, and man does not know what famine is." Thanks to the irrigation system, the Chengdu

Plain has grown into the granary of Western China and the political, cultural, and economic center of Southwest China since the 2nd century. The construction of the Dujiangyan irrigation system at the end of the Warring States Period (475–221 BC) laid a solid foundation for the water development of the Minjiang River. Since the Western Han Dynasty (206 BC–AD 24), the irrigation district has rapidly developed. The headwork was gradually improved and perfected. Irrigation canals quickly extended on the Chengdu Plain and formed a dense river network. The canals in the irrigation district, similar to natural waterways, have provided irrigation water and the convenience of flood discharge and water transport. As a result, Chengdu Plain, with its economic prosperity, has gradually become the political, cultural, and economic center of Southwest China. Chengdu became the capital of local separatist regimes five times, from the Three Kingdoms (AD 220–280) to the Five Dynasties and Ten Kingdoms (AD 907–960). In contrast to the Central Plains of China, which were in turmoil during that period, the affluent Chengdu Plain was prosperous and full of vitality, and that is how it won the reputation of the Land of Abundance.

4.5.4 It is an outstanding example of operation and management over a long time

The management mechanism characterized by the combination of governmental and non-governmental resources and annual maintenance was adopted and is still in the Dujiangyan irrigation system. Such arrangements ensure the sustainable development of the irrigation district and provide a historical experience that people can draw from when managing present-day water conservancy projects. The system's water regulation technology and system have been constantly renewed, while its weir engineering technology and annual maintenance system have maintained their uniqueness. People summarized its principle and philosophy of water management as "digging the riverbed deep and building the weirs low," which evolved into the charming and rich water culture of Dujiangyan. The management of the irrigation system involves a highly centralized administrative system, in which the water officials at provincial, prefectural, and county levels manage the headwork, weirs, main canals, and branch canals, and a localized system in which rural water management bodies and water user associations take care of field ditches. Under such a mixed management system, all beneficiaries must take respective management responsibilities and fulfill corresponding obligations. At the same time, multiple management types have brought order and vigor to the irrigation system.

4.6 CONCLUSIONS

The Dujiangyan irrigation system is the world's most time-honored dam-less water diversion project. Having simultaneously achieved the goals of irrigation, flood prevention, control, and water transport, it is an engineering paradigm of the harmonious coexistence between man and nature and the sustained development of humanity. Having played a role in China's unification, it is a water conservancy project with great historical significance; meanwhile, it perfectly combines the liveliness of nature's creation and the scientific beauty of man's work. The project integrates the functions of flood prevention and control, irrigation, water transport, domestic water use, and ecological water use. With sustainability and enormous benefits, the Dujiangyan irrigation system—a brilliant environment-friendly example of water conservancy projects has helped achieve harmony between man and nature.

ACKNOWLEDGMENT

This paper is supported by the 2021 official business fee of the Research Centre on Flood and Drought Disaster Reduction of the Ministry of Water Resource (WH0166B012021), and it is based on a previous paper titled "Analysis on Dujiangyan irrigation system and its sustainable development experience" published in the 3rd World Irrigation Forum.

REFERENCES

Kuang, Liangbo. (2018). Study on heritage system, value, and protection of Dujiangyan Irrigation Project. *China Flood and Drought Management*, 9: 72–76.

Tan, Xuming. (1986). On the Evolution of the Dujiangyan Irrigation System before the Song Dynasty, *the Collection of Papers in Commemoration of the Fiftieth Anniversary of the Foundation of the Department of Water Resources History Study*. Water Resources and Electric Power Press, Beijing (in Chinese).

Tan, Xuming. (2009). *A History of Dujiangyan Irrigation System*. China Water and Power Press, Beijing (in Chinese).

Wang, Ruifang. (2020). A typical example of harmony between man and nature: present value of Dujiangyan and its conservation. *China Water*, 3: 33–36 (in Chinese).

Zhang, Shanghong, Yi, Yujun, Liu, Yan, and Wang, Xingkui. (2013). Hydraulic principles of the 2,268-year-old Dujiangyan project in China. *Journal of Hydraulic Engineering*, 139(5): 538–546.

Chapter 5

Zhuji Shadoof Irrigation System and heritage values

Yunpeng Li and Xuming Tan
China Institute of Water Resources and Hydropower Research

Shushu Guo
Development Research Center of the Ministry of Water Resources of P. R. China

Keping Fu
River and Lake Management Center of Pujiang County

Changhai Zhou and Changrong Zhou
Zhuji Water Conservancy and Hydropower Bureau

CONTENTS

5.1 INTRODUCTION

Jiegao, or Shadoof, is one of the most time-honored water-carrying devices, and it was extensively applied in irrigation in ancient Babylon and Egypt before the 15th BC (Needham, 1999, Singer et al., 2004). The earliest records about Jiegao as a tool for well irrigation could be found in Zhuang Zi's book written in the 4th century BC. The story told that Zi Gong, one of the students of Confucius, advised a farmer that "Jiegao is a wooden gadget that is heavy" in the back and light in the front and pumps water out of

DOI: 10.1201/9781003222736-6

the well efficiently. This gadget helps to irrigate 100 pieces of farmland in one day only with a limited workload (Zhuangzi, around 300 BC). The record proves that as early as 2,400 years ago, shadoof irrigation was already used in China. Because of its simple structure and low cost, well-shadoof irrigation has been widely used for more than 2,000 years of Chinese history, especially in areas abundant in groundwater (Zhou, 2002; Guo, 2013; Wang & Zhang, 1990). The *Agricultural Book* composed by Wang Zhen in Yuan Dynasty also described that Jiegao (Shadoof), set in the most farmland nearby water, is a universal tool in ancient times and "nowadays" (Wang, around 1300). Until the 19th century, this traditional irrigation mode is still in use in many rural areas of China (Figure 5.1).

Since modern times, with the rapid development of the economic level and the science and technology in China, traditional water-lifting machinery such as shadoof has been phased out. Nevertheless, in Zhaojia Town of Zhuji County in Zhejiang Province, shadoof–well irrigation with a particular scale is retained and embraced with unique natural and cultural backgrounds. This paper studies the history, heritage components, engineering characteristics, and heritage values of the Zhuji Shadoof Irrigation System through fieldwork and literature research. It explores objective reasons for the persistence of this ancient irrigation pattern at Zhaojia Town. The study provides a sound basis for science-based protection and rational utilization of Zhuji Shadoof Irrigation System Heritages.

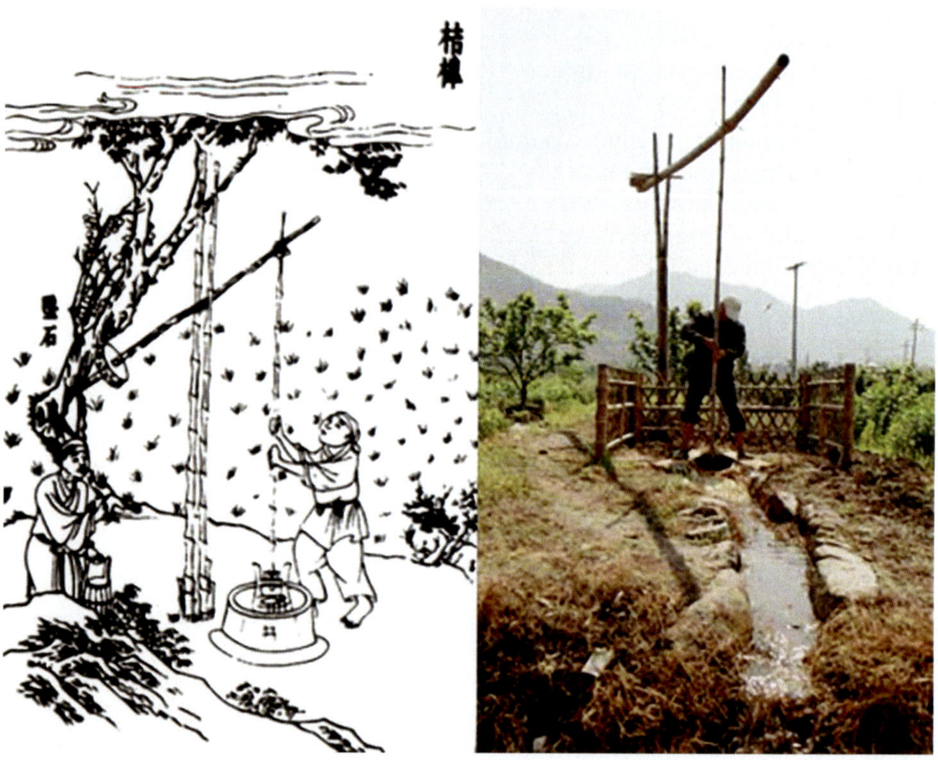

Figure 5.1 Drawing of Jiegaolifting water from well with Jiegao in Zhaojia Town of Zhuji. (From the book of Tiangongkaiwu written in 1637.)

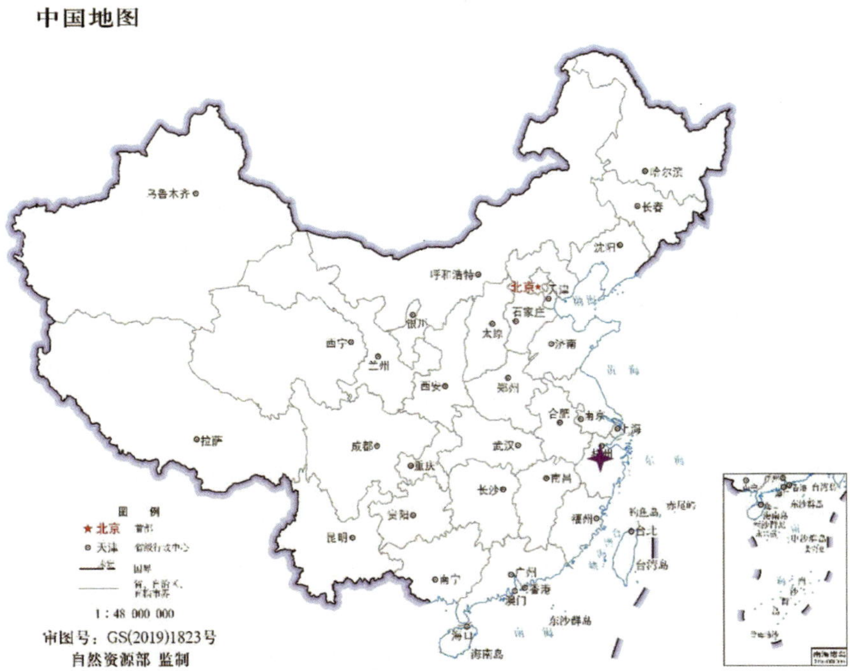

Figure 5.2 The location of the project in China.

Notice: This map has been provided by the authors, and the editors and publisher of this book disclaim any disputes regarding the political boundary shown in the South China Sea.

The Zhuji Shadoof Irrigation System is located at Quanfan and Zhaojia villages in Zhaojia Town, and on the alluvial basin of Huangtan Brook at the foot of the main peak of Zhoumagang of Kuaiji Mountain (Figure 5.2). Affected by the subtropical monsoon climate, the average annual precipitation in this area is 1,462 mm (Compilation Committee of Water Conservancy Chronicles in Zhuji County, 2007). Huangtan Brook has rapid flows and considerable fluctuation in water levels during the flood and drought seasons, resulting in a low guarantee rate of irrigation. However, the basin mainly contains sandy soils and is rich in shallow groundwater resources, enabling rapid replenishment; groundwater is buried at only 1–3 m in the drought and within 1 m in the rainy season. For the last hundreds of years, digging well and carrying water with Jiegao or shadoof have become the main irrigation mode for the two villages. The two villages now have more than 7,700 population and a cultivated land area of 5,331 mu, with an annual per capita income of ¥14,572 (2014).

5.2 ORIGIN AND HISTORICAL EVOLUTION

The history of Zhuji Shadoof Irrigation dates back to the Southern Song Dynasty (1127–1279). According to the genealogy, the inhabitants of Zhaojia Town of Zhuji are descendants of migrants from Central China in the 12th century. They primarily belong to two families—the He's and the Zhao's. Their ancestors dug well and carried water for irrigation and agricultural development; hence Quanfan and Zhaojia villages formed and continued to grow (Genealogy of Zhao's Family in Lantai, Jiyang, between 1875–1908; and Genealogy of He's Family in Tanxi, Jiyang, between 1911–1948). Due to the lack of records about the origin and early development of the shadoof irrigation system, the research was conducted through field investigation and literature analysis of the inscriptions and genealogies. The result shows that Shadoof and well irrigation has already become the main irrigation mode of the Zhao in the 17th and 18th centuries at the latest. At that time, the ancient people already had a scientific understanding of groundwater circulation.

There is a "Lantai Gushe Stele" engraved in 1809 inside the ancestral temple of Zhao's family. The stele states that the Zhaojia Village "prospered with well-plowed farmlands, which had bumper harvests even in years of drought thanks to the unique irrigation method here" (Lantai Gushe Stele, 1809). Zhao's genealogy states,

> All households dig wells for rice paddy irrigation in drought seasons. However, the wells are dried when the drought persists and will not be replenished until water flows around. Luckily, the lands are covered with sandy soils so that water can easily infiltrate underground.
>
> (Discussions of the ban on Yongkang Weir, 1840)

In the Kangxi years of the Qing Dynasty (1662–1722), Yongkang Weir was constructed on Huangtan Brook to store water and increase groundwater supply via infiltration (notes beyond the genealogy 1895). It shows that the ancient people already had a scientific understanding of groundwater circulation and systematic planning of the conservation and extraction of groundwater.

There are many shadoofs and wells in Zhuji historically, and they are called "Ao Well" by the local people. Ao Well also refers to the process of the well irrigation and water carriage. There used to be more than 8,000 wells in the region before the 1930s and 3,633 wells in 1985, with an irrigating area of 440 ha (Compilation Committee of Water Conservancy Chronicles in Zhuji County, 1994). However, many wells have been buried during urbanization for the past 30 years. Today, Ao Wells are mainly concentrated in Quanfan Village: in the core area of well irrigation, there are still 118 ancient wells, and the irrigation farmland is about 27 ha (Figure 5.3). The traditional Ao Well irrigation mode is well preserved until today. Yongkang Weir was flushed away in the middle of the 20th century, and after that, a new river dam was constructed on the upper reaches to take over its historical duties.

5.3 FEATURES ANALYSIS OF THE HERITAGE

Zhuji Shadoof Irrigation System has its natural and cultural basis, given its production and continuation. The composition of the Shadoof–well irrigation system, distribution of heritage, and irrigation management also shows the uniqueness of the heritage.

Figure 5.3 The shadoof wells in the core zone of Quanfan Village.

5.3.1 Components of Zhuji Shadoof Irrigation System

Zhuji Shadoof Irrigation System comprises two parts: the river weir located on Huantan Brook, which increases infiltration of surface water and the water-carrying capacity for irrigation, and the field Shadoof–well irrigation system consisting of more than a hundred irrigation units.

Each irrigation unit is an independent and well-established well irrigation unit constituted of one ancient well, one shadoof, one piece of farmland, and several field canals for irrigation or drainage. Such a unit is called "Jishui Field," with "Jishui" meaning "water-lifting." It is commonly known as "one field, one well" in Zhaojia Town. Each field area is about 1–3 mu (1 mu = 0.16 acre), but some fields are up to a dozen mu or even dozens of mu in recent years after integration. A well usually is 2–5 m deep and takes an inverted bell shape. The well mouth and bottom diameters are usually 1–2 and 1.5–2.5 m, respectively. The well walls are dry-laid with pebble stones, and some of the wells in the silt field have pinewood supports at the bottom. A filter layer is crushed sand and stone around the well wall (Figure 5.4).

A Jiegao, or Shadoof, comprises a pile named Ao Zhuang, a lever named Ao Cheng, a rod named Ao Gan, and a counterweight stone named Ao Stone. Jiegao is a very ancient Chinese word that explains the composition and characteristics of machinery. Likewise, the word Ao means the action of lifting water up from the well using Jiegao in Chinese. It was described as "bow under the pull, lift while loose" in the book *Zhuangzi* (Wang, proofread in 1961). Moreover, "Jie is the standing, and Gao is the pitching," according to the Agriculture Book (Wang, around 1300), which additionally

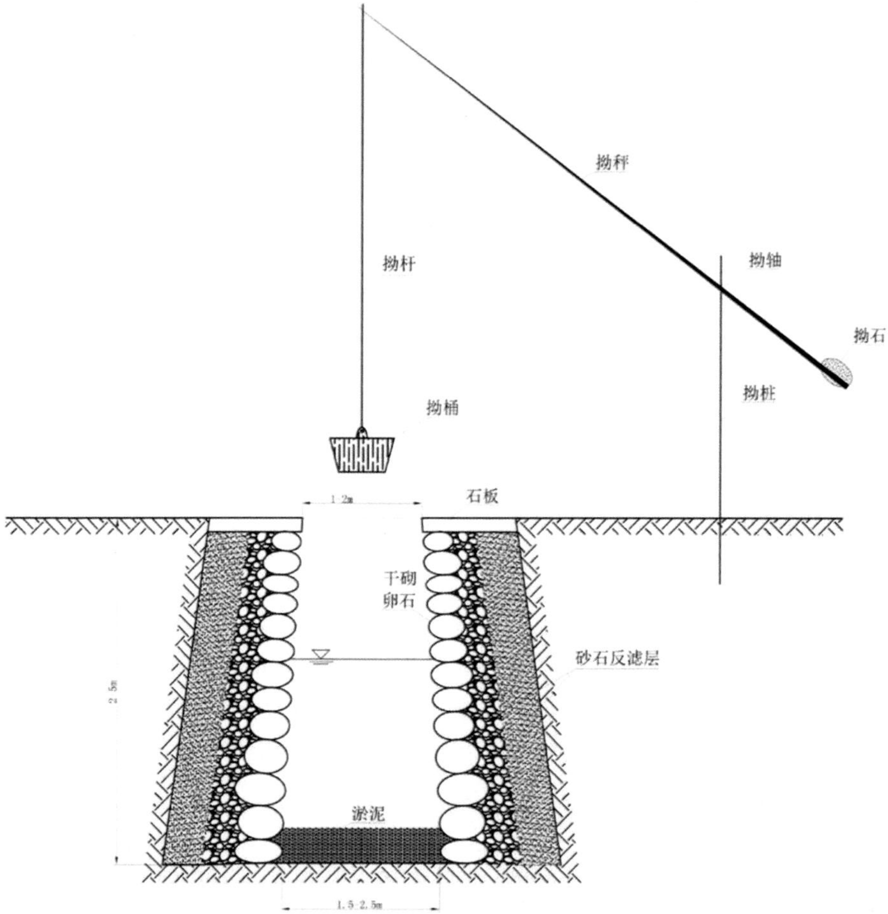

Figure 5.4 The project structure of the irrigation well and Jiegao. (Drawn by authors.)

quoted the book Shuowen as "Jie refers to ties that are fixed, and Gao like a lever by which things can be moved easily, the process should be slow down and keep stable" (Wang, around 1300). According to the description, Jie is the pile of the shadoof with the local name Ao Zhuang, and Gaois is the level of the shadoof with the local name Ao Cheng. A measurement in Zhaojia Town found that the Ao Zhuang is normally 4m tall and made of a pine trunk with more than 10cm diameter. Ao Cheng is normally 6.5m long and made of large bamboo. It binds a heavy stone to the thick end, 2m apart from the tie connecting Ao Zhuang and Ao Cheng, simultaneously binding Ao Gan-made with thin bamboo, normally 5m long to the thin end of Ao Cheng (Figure 5.4). The tailor-made water-carrying barrel comprises a wooden axle connected to the lower end of the pile. To lift water, the operator stands on two bamboo beams or a wooden plate set up at the well opening and pushes the rod so the barrel sinks into the well water. Using the lever reduces the effort needed to lift a barrel of water. A straw braid is located where the water flows out from the well to protect the barrel from damage.

A "rain factory" (a simple hut) is distributed among several wells for shelter from the rain, rest, and farming tool storage.

5.3.2 Management characteristic

Zhuji shadoof–well irrigation structures and facilities are constructed, maintained, and used by farmers. A household normally owns one well, and sometimes two or three households jointly own a well. Two wells near each other are called "interconnection wells," because the walls of the two wells are so close, and the groundwater fillers are almost the same. Suppose lifting water from one well; the water level and yield will be influenced immediately. The owners of interconnection wells always lift water with a shadoof by turns for half a day per family to keep the normal water-lifting efficiency. One well owned by several families is called an "alternate well," which supplies irrigation water for different fields by turns.

5.3.3 Irrigation benefits

Jishui Field could provide a stable harvest and became the economic pillar of a local peasant family. The He's and the Zhao's family ancestors migrated to Zhaojia Town in the 12th century. They settled down, dug wells, and carried water for irrigation to develop agriculture; hence, Quanfan and Zhaojia Villages were formed. According to the record in the epitaph of He Xingqi, who died between AD 1875 and 1908 (Figure 5.5), his family eventually went through a tough period when they suffered three funeral events and two wedding events over four years with the support of dozen mu fields irrigated by shadoof-well (Si, 1905). Most of the fields in Zhaojia Town depended on the shadoof–well irrigation for a long time, and there was a 6,600 mu irrigation area with shadoof

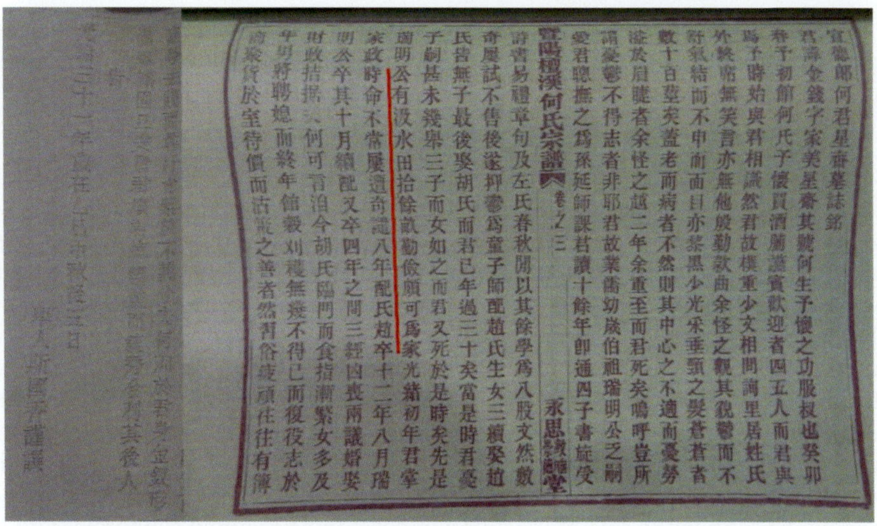

Figure 5.5 The record in the epitaph of He Xingqi (between AD 1875 and 1908).

until 1985. However, the irrigation range and benefits have decreased by a large margin compared with history. Today, there are 118 ancient Ao wells with more than 400 mu irrigating areas in the core area of the well irrigation in Quanfan Village. The rice paddies have been partly converted into orchards of the more valuable cherries, and agriculture still plays an important role in the income of local peasant families.

5.3.4 Cultural characteristic

Jiegao (or shadoof) in early China is of special meaning in Chinese culture. Zhuang Zi considered Jiegao a metaphor for the human behavioristic principle with the saying: "Didn't you see the shadoof? It bows under the pull and lifts while loose. It responds passively rather than actively, meaning pitching would not offend any people" (Zhuangzi, proofread in 1961). Moreover, in another story, when Zigong recommended the utilization of Jiegao for irrigation, it was rejected by the elderly of the State of Chu as an opportunist tool. However, the residents in the heritage area of the Zhuji Shadoof Irrigation System treat Ao Wells with a simple sentiment. Famers write the words "operating the shadoof and loving its sufficient irrigation water as a timely rain" to show their love for the heritage. A ballad tells the characteristics of the well irritation and the character of farmers in villages of Zhaojia Town, with the words "The resident of Quanfan with the surname of He and Zhao has hard head tipped on the neck. With one well in each field, three hundred barrels of water are lifted daily and returned to the quondam hole in the night". Ao well has also been written into local traditional drama, Miss Jiujin as a cultural symbol extracted libretto in Chinese: All of these reveal Zhuji Shadoof Irrigation System's deep historical and cultural foundation.

5.4 HERITAGE VALUE ANALYSIS

The Irrigation System of Zhuji is a living fossil of the ancient water-carrying device of shadoof. Its unique historical, scientific, technological, and cultural values should be acknowledged.

5.4.1 Historical value

Through thousands of years, water-carrying and irrigation devices have developed along with the human civilization from Jiegao (Shadoof) and Lulu at the early stage of human civilization to various types of watermills and then to the fossil-powered or electricity-powered water pumps after the industrial revolution. As the oldest water-carrying device, Jiegao has a special position in the human history of irrigation civilization. The Zhuji well irrigation project and Jiegao have been well preserved as unique water-carrying and irrigation modes. It can be seen as a living fossil that continuously provides irrigation benefits and witnesses the advancement of ancient irrigation civilizations and cultures. As a "witness" and "participant" of history, it highlights the supporting role of irrigation in developing regional economics and culture in an agricultural society.

5.4.2 Scientific and technological value

Zhuji Shadoof Irrigation System fully used regional natural conditions and played a full irrigation efficiency with the most simple type of ancient works. About 200 years ago, the villagers' ancestors developed a scientific understanding of groundwater circulation mechanisms. Based on the knowledge, an artificial weir was constructed on Huangtan Brook to store water and increase groundwater supply via infiltration. The rational arrangement of the well group makes every field at different levels get irrigation water. The ownership and right of use of every shadoof irrigation facility are identified with the field; with an equitable distribution of groundwater resources and suitable coordination mechanism of water extraction from the same well or same groundwater between different families through village regulation and agreement, the water dispute could be eliminated to a great extent.

5.4.3 Cultural value

Zhuji Shadoof Irrigation System blended with Yue culture during its development and evolution and derived the culture of Ao well featured with strong regional characteristics. Jiegao bears a special meaning to Chinese traditional culture and philosophy. It can be reflected in residents' lives and manifested by cultural elements such as local ballads or dramas. Additionally, residents of Zhaojia Town highly identify Ao well culture with a deep sense of pride.

5.5 CONCLUSIONS

This paper studied the history, heritage components, engineering characteristics, and heritage values of the Zhuji Shadoof Irrigation System through fieldwork and literature research from a perspective of irrigation heritage. The conclusions can be drawn as follows.

Firstly, the construction and continuous utilization of the Zhuji Shadoof Irrigation System in Zhaojia Town are based on its unique natural conditions. The landscape of the basin, the geological condition of overlaying sandy loam on the bedrock, abundant rainfall, and fluent groundwater circulation created the objective basis for shadoof–well irrigation. Huangtan Brook has rapid flows and considerable fluctuation in water levels during flood and drought seasons, resulting in a low guarantee rate of irrigation. Therefore, in the relatively closed small basin and low level of engineering technology and social economy in ancient times, traditional shadoof–well irrigation almost became the inevitable choice for agricultural development of Zhaojia Town due to its simple construction work and low cost.

Secondly, in modern times, despite the rapid development of the economy and science and technology in China, the ancient shadoof–well irrigation with a certain scale is still retained and utilized in Zhaojia Town. Four objective reasons are as follows: (1) The abundant and high-quality groundwater resources in the unique natural environment are still preferred for small-scale agricultural irrigation; (2) residents tend to continue to use shadoof irrigation facility with historical inertia due to its simple construction, low cost and easy to use; (3) shadoof irrigation with manpower could meet the demand of the small area of each field so that farmers are not willing to erect a cable for pumping water from the well with extra cost; (4) the local villagers have a profound cultural identity to Shadoof–well irrigation, which has been part of their agricultural lives. The local government has also paid attention to the heritage in recent years.

Thirdly, Zhuji Shadoof Irrigation System is a living fossil of the earlier irrigation civilization with unique historical, scientific, technological, and cultural value. It is a historical witness to the origin and spread of irrigation civilization. It also witnessed the development of the social history, including immigrants of He's and Zhao's families, and regional growth of agriculture, population, and culture. As the oldest irrigation mode, the persistence and utilization of shadoof–well irrigation at Zhaojia Town of Zhuji County are valuable and rare. There is research value in the environmental suitability of irrigation type, scientific planning and design of the well distribution and structure, simple and efficient management, and even the cognition and utilization of the mechanism of groundwater circulation by the ancient people. Jiegao bears a special meaning to Chinese traditional culture and philosophy in history. Ao well culture with a strong regional characteristic has derived from the evolution process of the Zhuji Shadoof Irrigation System.

Fourthly, the existence and continuation of the Zhuji Shadoof Irrigation System are facing challenges of modern development, including the shrinkage of farmlands and the low agricultural income for farmers caused by rapid urbanization, which has weakened their positivity of agricultural work. Considering the unique and outstanding heritage values, the research team suggests the local government protect the living fossil through specialized planning, which will consider heritage protection with new rural construction, transformation, and upgrading of the agriculture economy and development of the cultural industry as a whole. This planning should be scientific and include multiparty participation to promote the protection and sustainable development of the irrigation heritage. Zhuji Shadoof Irrigation System was approved to be included in the World List of ICID Heritage Irrigation Structures.

ACKNOWLEDGMENTS

This work was supported by the Key Scientific Research Base Program of the State Administration for Cultural Heritage (2020ZCK207), IWHR Research & Development Support Program (JZ0199B212019, JZ1003A112020).

REFERENCES

Compilation Committee of Water Conservancy Chronicles in Zhuji County. (1994). *Water Conservancy Chronicles of Zhuji County*. Xi'an Map Publishing House, Xi'an, China (in Chinese).

Compilation Committee of Water Conservancy Chronicles in Zhuji County. (2007). *Water Conservancy Chronicles of Zhuji County*. Local Press, Beijing, China (in Chinese).

Discussions of ban on Yongkang Weir (永康堰禁议) (1840). held by Zhao's Ancestral Hall in Zhaojia Village, Zhaojia Town, Zhuji County, Zhejiang Province, China (in Chinese).

Genealogy of He's Family in Tanxi, Jiyang, pressed in the Republican period of China (1911–1948), held by He's Ancestral Hall in Quanfan Village, Zhaojia Town, Zhuji County, Zhejiang Province, China (in Chinese).

Genealogy of Zhao's Family in Lantai, Jiyang, pressed in Guangxu year of Qing Dynasty (1875–1908), holding by Zhao's Ancestral Hall in Zhaojia Village, Zhaojia Town, Zhuji County, Zhejiang Province, China (in Chinese).

Guo, Tao. (2013). *A history of water science and technology in ancient China.* China Building Industry Press, Beijing, China (in Chinese).

Lantai Gushe Stele (兰台古社碑) (1809). held by Zhao's Ancestral Hall in Zhaojia Village, Zhaojia Town, Zhuji County, Zhejiang Province, China (in Chinese).

Needham, Joseph. (1999). *Science and civilization in China*, translated by BAO Guobao, et al. Science Press, Beijing, and Shanghai Ancient Books Publishing House, Shanghai, China.

Notes beyond the genealogy (谱外杂记) (1895). held by Zhao's Ancestral Hall in Zhaojia Village, Zhaojia Town, Zhuji County, Zhejiang Province, China (in Chinese).

Si. Guoxiang. (1905). *Epitaph of Xuandelang HE Xingqi*, held by He's Ancestral Hall in Quanfan Village, Zhaojia Town, Zhuji County, Zhejiang Province, China (in Chinese).

Singer, C., Holmyard, E. J. and Hall, A. R. (2004). *A history of technology*, translated by Qian Wang & Xizhong Sun, Shanghai Science and Technology Education Publishing House, Shanghai, China.

Wang, Jialun & Zhang, Fang. (1990). *History of farmland water conservancy in China.* Agriculture Press, Beijing, China. (in Chinese)

Wang, Zhen. (around 1300) *Agricultural books*, Version of Guangxu years of Qing Dynasty (in Chinese).

Zhou, Kuiyi. (2002). *A history of science and technology in China (Section of Water).* Science Press, Beijing, China (in Chinese).

Zhuangzi. (1961). *Zhuangzi*, proofread by Xiaoyu Wang. Zhonghua Book Company, Beijing, China (in Chinese).

Chapter 6

Introduction of the Beijing–Hangzhou Grand Canal and analysis of its heritage values

Jingdong Cai and Jing Peng
China Institute of Water Resources and Hydropower Research

CONTENTS

6.1 INTRODUCTION

The (Beijing–Hangzhou) Grand Canal is a waterway system stretching on the eastern plain of China for 1,794 km and connecting five major rivers. Its history can be traced back to the 5th century BC, and the unified form on a national scale was created twice in two different shapes. Given the ups and downs of the immense territory, the uneven spatial and temporal distribution of rainfall and water resources, the connectivity with rivers along the way, and the harassing Yellow River, the canal represents the world's highest level of hydraulic advancement and water project management before the Industrial Revolution. While passing on enormous heritage values to future generations, the canal also offers rich experiences and lessons for today's civil engineering development.

DOI: 10.1201/9781003222736-7

6.2 HISTORY

Also known as the Beijing–Hangzhou Grand Canal, its oldest section (the Hong Gou that linked the Yellow River near Kaifeng to the Sishui and Bianshui rivers) is believed to have been constructed in the 5th century BC. In the southern section, the work of the earliest section (Han Gou from south of Yangzhou to north of Huai'an in Jiangsu) began in 486 BC when King Fuchai, ruler of the State of Wu, ordered a canal to be constructed for both of trading and military purposes. The Han Gou connected the Yangtze River to the Huaihe River through existing waterways, lakes, and marshes within three years.

Various sections were connected in AD 609 during the Sui Dynasty (AD 581–618) into a Y-shaped canal with Luoyang (in West China) as the center. It consists of two major sections: the north portion connecting Luoyang and Beijing is called Yongji Channel, while the southern Luoyang–Yangzhou section is called Tongji Channel.

During the Yuan Dynasty (AD 1271–1368), China's capital was moved to Beijing, eliminating the need for the canal to go westward to Luoyang or Kaifeng. For the first time, a section was dug to connect Beijing with Hangzhou to form a north-south waterway that resembles the course in modern times.

6.2.1 Infancy and evolvement of the canal

The earliest canals were dug with a military purpose for political ambitions. The earliest canals in China were constructed about 2,500 years ago when the kings of the warring states started to put their political ambitions into action. From north to south, the Yellow River, Huaihe River, and Yangtze River are the three major west-east rivers in the central part of China. During the warring period, states with the ambition of unifying fragmented China needed to consider the transportation issues for soldiers and military supplies. As a result, King Fuchai (of the state Wu with its capital in today's Suzhou) dug the Han Gou canal (connecting Yangtze River northward to Huaihe River) in 486 BC to gain easier access northward. The history was recorded in Zuo Zhuan—a chronicle from 722 BC to 468 BC: "In the autumn of 486 BC, Han Gou was dug to connect Yangtze River and Huaihe River" (Yang Bojun, 1981: 1652). For a similar reason, King Hui (of the state Wei) started digging the Hong Gou canal (connecting the Yellow River southward to the Huaihe River) in 361 BC, the second year after he moved the capital to Daliang (today's Kaifeng, Henan).

In warring times, the military purpose might be special for separated states to dig canals. Still, the connected canal was created and maintained in ancient China's unified dynasties for political and economic reasons. Although the political center has been in the north for the larger part of Chinese history, the economic center started shifting to the south during the Sui (AD 581–618) and Tang (AD 618–907) dynasties. By the Song (AD 960–1279) dynasty, South China had completely outweighed the north economically and become the nation's breadbasket. Under such circumstances, a connected canal from south to north would well serve the purposes of transporting grains to the north, maintaining the nation's unification and manifesting the ruling of the emperors.

6.2.2 Route of the canal

Sections were linked into a national canal to connect key cities. Still, the ending point was always the capital city, serving the purpose of transporting grains from the breadbaskets to the prosperous core of the nation.

In history, the Beijing–Hangzhou Grand Canal stretches in two shapes: Various sections were connected in AD 609 during the Sui Dynasty (AD 581–618) into a Y-shaped canal that has Luoyang (in West China with easy river access to Xi'an) as the center, and consists of two major sections: The north portion connecting Luoyang and Beijing is called Yongji Channel, while the southern Luoyang–Yangzhou section is called Tongji Channel (Figure 6.1). During the Yuan Dynasty (AD 1271–1368), China's capital was moved to Beijing, eliminating the need for the canal to go westward to Luoyang or Kaifeng. For the first time, a section was dug to connect Beijing with Hangzhou to form a north-south I-shaped waterway that resembles the course in modern times (Figure 6.2).

Figure 6.1 The Y-shaped Grand Canal in Sui, Tang, and Song dynasties.

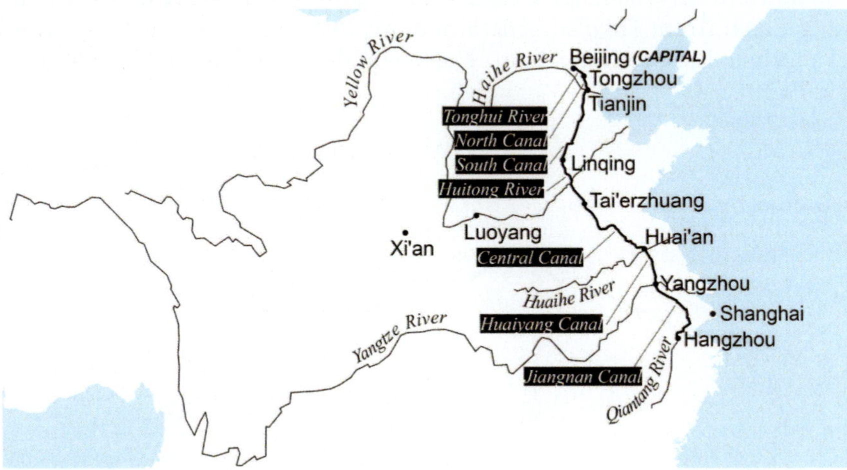

Figure 6.2 The I-shaped Grand Canal in Yuan, Ming, and Qing dynasties.

6.2.3 Interaction between canals and rivers

The canal is crossed by five major rivers in China, namely Haihe, Yellow, Huaihe, Yangtze, and Qiantang, from north to south. As a result, the interaction between the canal and rivers at the joints has been a critical issue for over five centuries of canal operation. Among them is the long-time arduous efforts to control the adverse influence, such as the high sediment concentration and the Yellow River's violent flow upon the canal.

Almost since its inception, the canal has been severely troubled, disturbed, and even disrupted by the west-east Yellow River—the world's muddiest river and China's most troubling river in history. It was particularly so during the over 700 years when the Yellow River broke its southern bank, rushed southward into the Huaihe River in the South Song dynasty (AD 1128), and occupied the river course to sea until the Qing dynasty (AD 1855). To maintain the navigation function of the canal, during the 400-year engineering activities, the world's most complex controlling structure (the Qingkou–Hongze Lake Project) at that time was constructed with the 60 km long dam—the Gaojia Weir to impound clean water in Hongze Lake (with a total storage capacity of 3.5 billion m^3) for scouring the sediment brought in from the muddy Yellow River.

6.3 GEOGRAPHICAL FEATURES

The canal starts at Beijing and passes through Tianjin and the Provinces of Hebei, Shandong, Jiangsu, and Zhejiang to Hangzhou, linking the Haihe River, Yellow River, Huaihe, Yangtze River, and Qiantang River. Its greatest height is reached in the mountains of Shandong in Nanwang at a summit of 42 m and the lowest point at about –7 m, forming a total elevation difference of about 50 m, as is illustrated in Figure 6.3. Ships did not have trouble reaching higher elevations after the pound lock was invented in the 10th century during the Song dynasty (AD 960–1279).

From north downward, the Grand Canal is divided into seven major sections: Tonghui River (from Beijing down to Tongzhou), North Canal (from Tongzhou down to Tianjin), South Canal (from Tianjin up to Linqing), Huitong River (from Linqing up and then down to Tai'erzhuang), Central Canal (from Tai'erzhuang down to Huai'an), Huaiyang Canal (from Huai'an generally down to Yangzhou), and Jiangnan Canal (from Yangzhou down to Hangzhou). The longitudinal sections and lengths are illustrated in Figure 6.3.

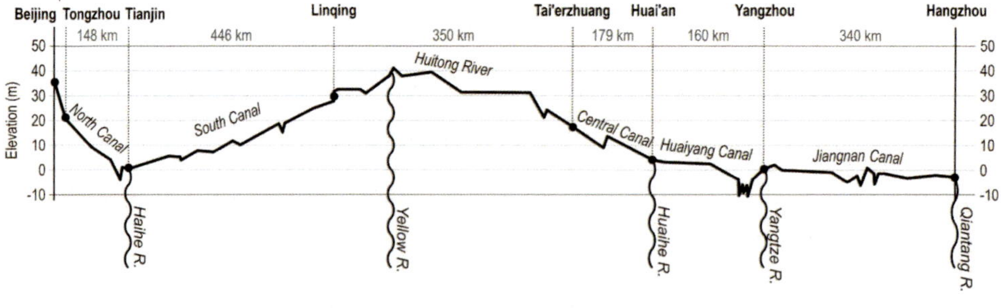

Figure 6.3 Longitudinal sections of the canal.

From this figure, it can be found that, though considered a unified waterway, the canal has higher elevations of water-dividing points prominently in Beijing and Shandong Province (with Nanwang in Shandong Province as the summit). It caused the water to flow southward from Beijing and Nanwang and northward from Nanwang.

6.4 ENGINEERING ACHIEVEMENTS AND HERITAGE VALUES

The Grand Canal flows across China's five major river basins with different geographical features and water distribution patterns. It overcomes an elevation difference of about 50 m. It has been constantly threatened and damaged by the sandy Yellow River that has kept changing its course and inundating immense regions on both the sides for more than 700 years. Its engineering achievements and heritage values (Tan et al., 2009) are mainly exhibited in four aspects: (1) engineering structures on watercourses, (2) projects for maintaining water sources, (3) subsidiary facilities, and (4) cultural facilities.

6.4.1 Engineering structures on the watercourse

Engineering structures are the major part of the canal consisting of (1) the watercourses of main and branch channels, (2) dykes and embankments, locks, dams, culverts, and the combined deployment of these structures, (3) structures for flow reduction, (4) bridges, etc. For example:

Regulation sluice: The Tonghui River (connecting Beijing and Tongzhou) has a gradient of 1.0%–1.2% and an elevation difference of 10–30 m. To overcome the geographical obstacles, 24 sluices were built during the Yuan Dynasty (AD 1206–1368). Through the joint operation of sluices and dams, unpowered boats were enabled to overcome the waterway gradient and reach the capital city's downtown area. At the same time, the volume and depth of canal water could be controlled section by section. The relics (Figure 6.4) can still be found today.

Consecutive regulation sluices were also deployed and operated on the over 400-km watercourse of Huitong River (from Linqing to Zaozhuang in the hilly region of

Figure 6.4 Relics of sluices on the canal section of Tonghui River in Beijing (Tan et al., 2012: 38).

Figure 6.5 Daicun Dam. (http://blog.sina.com.cn/s/blog_62cad0410100kavx.html. Retrieved on June 15, 2016.)

today's Shandong Province) to control water distribution and a channel depth of multiple watercourses, overcoming a gradient of 2% and an elevation difference of about 20 m (the highest point of the canal in Nanwang).

The joint operation of sluices has eased the sediment deposit from the west-east Yellow River into the north-south canal.

Weir: For more than 500 years, Gangchen Weir, Jinkou Weir, and Daicun Weir, still in use today (Figure 6.5), were constructed along the Huitong River. Together with dykes and diversion channels, they formed a series of lakes that covered an area of 1,496.6 km^2 with a storage capacity of 2.353 billion m^3—the world's largest "water tank" (similar to a modern reservoir) before the Industrial Revolution.

Pound lock: A pound lock, or chamber lock, with two gates, was invented during the Song dynasty (AD 960–1279) at the end of the 10th century by a civil servant named Chiao Wei-Yo, who was in charge of transport in the Huaihe region (Pierre-Louis Viollet, 2005) and commonly built on Huaiyang Canal (connecting Huaihe River and Yangtze River) and Jiangnan Canal (connecting Yangtze River and Qiantang River) for boats to reach higher elevations. It functions similarly to modern ship locks.

One of the representative structures is the pound lock in Zhenzhou (in today's Yangzhou, Jiangsu Province), completed in AD 1026 during the Song dynasty. With three gates and two chambers, the lock is 610 m long. At its peak, about 480,000 tons of grains are transported annually through the lock (Shen Kuo, 1088).

Curved channels: Curved channels were designed to reduce longitudinal gradient by 50%, decreasing the elevation difference while improving the safety of navigation.

6.4.2 Projects for maintaining water sources

The canal is located in the plain area of East China, from the monsoon temperate zone (annual average precipitation: 500–600 mm) in the north to the monsoon subtropical zone (annual average precipitation: 700–1,200 mm) in the south. Because of the uneven precipitation distribution among seasons (60%–80% of the rain falls from June to September) and regions caused by climatic characteristics and elevation differences, it is important to guarantee water supply to the canal channels. As a result, projects for maintaining water sources, including water diversion channels, dams, locks, embankments, and the deployment of their combinations, are significant components of the canal.

The West Mountain Spring Diversion Project (Figure 6.6) is one of the representative projects for maintaining water sources. Beijing, located in North China, has an extremely uneven temporal precipitation distribution, rivers with high sediment concentration, and a large river channel gradient. Without water supply from external sources, canal operation would be impossible. Springs oozing out in the West Mountain and mountain torrents were collected into the Kunming Lake in Beijing's Summer Palace to supply the canal section of Tonghui River (connecting the ancient Tongzhou City), and the complete set of supporting structures (for drawing, storing, and diverting spring water) has fed the urban water system in Beijing and contributed to the forming of the urban landscape.

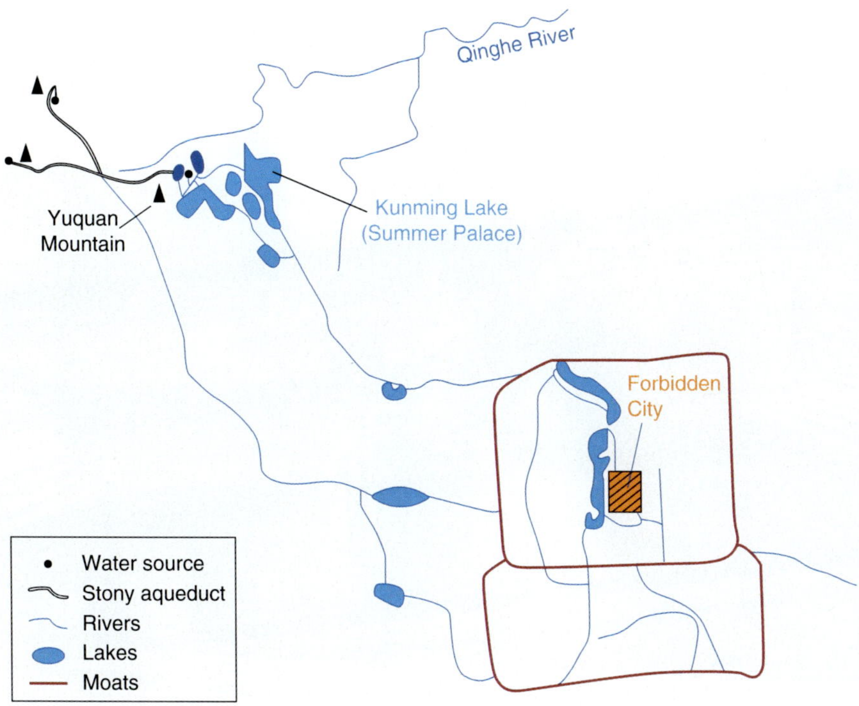

Figure 6.6 Water source project for the canal section of Tonghui River in the Qing dynasty (Tan et al., 2012: 31).

6.4.3 Subsidiary facilities

Navigation is the most important function of the Grand Canal. It is realized through the joint operation of the canal, water source projects, transportation facilities (harbors, piers, shipyards, towpaths, and bridges), and management facilities (piles for gauging water level, water lifting devices, and administrative agencies). Markets near bridges and prosperous socio-economic regions were formed along the canal. The ancient towpaths in Shangyu and Shaoxing in Zhejiang Province can still be seen today (Figure 6.7). The Water Criterion Stele, erected in the canal channels, is China's earliest existing facility for gauging water levels (Figure 6.8).

Figure 6.7 Towpath in the canal. (http://bbs.zgkqw.com/forum.php?mod=viewthread&tid=735977; http://www.huitu.com/photo/show/20130709/094351409200.html. Retrieved on June 15, 2016.)

Figure 6.8 Water Criterion Stele in Ningbo. (http://you.ctrip.com/travels/ningbo83/1949882.html. Retrieved on June 15, 2016.)

6.4.4 Cultural facilities

During the long-time operation of canals, worship of the water god has gradually formed along the canal channels, creating unique water culture. According to a survey conducted in AD 2006, a water god and sacrifice venue exist in every canal section. Statues of water gods (goddesses) were erected, temples built, and sacrifice ceremonies performed. Typical examples are explained as follows.

- **Temple of the Dragon King**: Dragon King is a deity in Chinese mythology commonly regarded as the divine ruler of an ocean and can manipulate the weather and bring rainfall. Along the canal, many temples, such as the one built in Kunming Lake in Beijing (Figure 6.9), were constructed to pray for peace, favorable weather, and peaceful rivers and lakes.
- **Temple of meritorious persons**: The related history is thus recorded by deifying historical persons.
- **Stelae for documenting water projects**: Stelae were erected at the site of large water projects to document the projects' objective, construction process, and management.
- **Iron bulls**: Officially and uniformly fabricated, these iron bulls (Figure 6.10) were enshrined to pray for the peace of the rivers and canals.

From north to south, the canal is a collection of almost all water structures, representing the highest level of planning, construction, and management of the water sector in ancient China while sustaining socio-economic development and improving the regional water environment. The canal's engineering achievements and heritage values are

Figure 6.9 Temple of the Dragon King in the Kunming Lake, Beijing. (http://blog.sina.com.cn/s/blog_656a9edf0101exza.html. Retrieved on June 15, 2016.)

summarized in Table 6.1, with examples of projects and sites. It is evident that the Grand Canal is rich in all types of river projects and heritage values and has showcased the great wisdom of ancient Chinese people in river harnessing and river course engineering.

Figure 6.10 Iron bull in Shaobo, Jiangsu Province. (http://blog.sina.com.cn/s/blog_62cad0410100kavx.html. Retrieved on June 15, 2016.)

Table 6.1 Representative projects and heritage values of the Grand Canal

	Categories	*Examples of project*	*Examples of heritage values*
1	Projects for controlling watercourses	• Main and branch channels • Dykes and embankments, locks, dams, culverts, and the combined deployment of these structures • Structures for flow reduction	• Joint operation of locks and dams enables unpowered boats to overcome the waterway gradient (1%); • Joint operation of locks enables the control of volume and depth of canal water section by section; • Regulation sluices (on Huitong River) enable the control of water distribution and a channel depth of multiple rivers; • The world's largest "water tank" (similar to modern reservoir) before the Industrial Revolution: construction of dams and diversion channels for over 500 years has formed a lake that has a total surface area of 1,500 km^2 and storage of 2.353 billion m^3 today; • Joint operation of locks has eased the sediment deposit from the west-east Yellow River into the north-south canal; • Pound lock was invented during the Song Dynasty (AD 960–1279) for boats to reach higher elevations; • Curved channels were designed to reduce longitudinal gradient by 50%, decreasing the elevation difference while improving the safety of navigation

(Continued)

Table 6.1 (Continued) Representative projects and heritage values of the Grand Canal

	Categories	*Examples of project*	*Examples of heritage values*
2	Projects for maintaining water sources	• Water diversion channel • Dams, locks, embankments, and the deployment of their combination	• The Kunming Lake in Beijing's Summer Palace, which was for storing water, has become an important landscape of the ancient capital city; • Groundwater was utilized as the water source for the Tonghui Channel (in Beijing). The complete set of supporting structures (for drawing, storing, and diverting spring water) has fed the urban water system in Beijing and contributed to forming the urban landscape.
3	Subsidiary facilities	• Transport facilities (bridges, towpaths) • Management facilities (water piles, water lifting devices, administrative agencies) • Transportation facilities (shipyards, piers, and harbors)	• Shangyu Ancient Towpath • Shaoxing Ancient Towpath • Markets near bridges and prosperous socio-economic regions were formed • Harbor facilities for storing and transferring cargo • Water Criterion Stele: China's earliest facility for gauging water level
4	Cultural facilities	• Temples and sacred statues for water gods • Inscriptions in stone slabs • Sacrifice venues	• Temple of the Dragon King • Temple of the meritorious persons (history recording) • Tianfei Temple (for worshipping the goddess of navigation) • Stelae for documenting water projects • Iron bulls (for stabilizing canal water)

6.5 FUNCTIONING IN PRESENT DAYS

Transportation of the Grand Canal was officially stopped on July 2, 1901, during China's Qing Dynasty because of the central government's inability on the one hand and some natural factors on the other hand. Since then, only limited transporting capacity remained in parts of the canal. It was not until the founding of the new China in 1949 was the Grand Canal reborn.

Today, the Grand Canal plays an important role in water diversion (via the eastern route of the South-to-North Water Transfer Project), transportation (from the southern end northward to the southern bank of the Yellow River), and irrigation water supply and tourism. With its inscription on the UNESCO World Heritage List in 2014 (UNESCO, 2016), this ancient canal has been given a new life. It continues to perform an essential function in the socioeconomic development of the riparian region.

6.6 CONCLUSIONS

The Grand Canal witnessed the evolution of China's long history and natural landscape and showcased the most advanced water planning and civil engineering before the Industrial Revolution. Driven by military, economic, and political purposes, the canal was built and constantly upgraded according to the local geographical, topographical, and social characteristics. With rich natural, engineering, and cultural heritage values, it has been operated, improved, and maintained for more than 2,000 years and has been revamped into the eastern route of China's South-to-North Water Transfer Project that has been diverting water from the Yangtze River to North China since AD 2013. In observing natural and social laws and maintaining a harmonious relationship between humans and water, the canal offers great lessons for long-distance navigation, interregional water transfer, water and sediment regulation, and flood control. That is why it can endure the long history of dynasty alteration and natural disasters, including the troublesome Yellow River, and live till now.

ACKNOWLEDGMENTS

This article is based largely on the authors' previous paper, first published in the Proceedings of the 20th IAHR APD Congress (Colombo, 2016) and later in the *Journal of Hydro-Environment Research* in Volume 26 October 2019. The original articles have been expanded and rearranged to fit the IAHR monograph.

REFERENCES

Shen, Kuo. (1088). *Dream Pool Essays* (in Chinese).

Tan, Xuming, Yu, Bing, Wang, Yinghua, and Zhang, Nianqiang. (2009). Characteristics and core components of the heritage of the Grand Canal in China. *Journal of Hydraulic Engineering*, 40(10): 1219–1226.

Tan, Xuming, Wang, Yinghua, Li, Yunpeng and Deng, Jun. (2012). *Heritage components and value assessment of China's Grand Canal*. China Water & Power Press.

UNESCO. (2016). World Heritage Convention. The Grand Canal. http://whc.unesco.org/en/list/1443 (accessed January 12, 2016).

Viollet, Pierre-Louis. (2005). *Water engineering in ancient civilizations: 5,000 years of history*. Translated by Forrest M. Holly Jr., 2007. Taylor & Francis Group.

Yang, Bojun. (1981). *Chunqiu Zuo Zhuan*. Chung Hwa Book Co (in Chinese).

Chapter 7

Tuoshan Weir

An ancient estuarial river regulation project

Tibing Xu, Hanbin Gu, Chengcheng Wang, and Zhichao Yin
Ningbo University

CONTENTS

7.1 INTRODUCTION

Tuoshan Weir is an ancient project constructed to regulate and control water resources in ancient Ningbo, now a modern coastal city in Zhejiang, China (Figure 7.1) (Chen, 2014; Zhang, 2014; Gu et al., 2018). The weir played an essential role in water resource distribution for the western part of Yin-Feng plain and ancient Ningbo.

In the Yin-Feng Plain of the Yong River basin, there are more than 800,000 mu paddy fields (mu is an area unit in China, and 1 mu is equal to 666.67 m^2) (see Figure 7.2). The Yong River is rich as an important water source, which has two tributaries: the Yao river as the west branch and the Fenghua River as the south. The two rivers form the Yong River, flowing northeast to the sea across the Zhenhai district. The three rivers divide the Yin-Feng Plain into the east and west. The western part is the Yinxi Plain, in the Yin River basin. The Yin River is the largest branch of the Fenghua River, originating from Siming Mountain, with a catchment area of 382 km^2.

The Yong River had a very small slope. Seawater in the dry season would surge back to the Yin River, resulting in a mixture of fresh and saline water in the plain area, which was unsuitable for drinking and irrigation. Hence, seawater intrusion has

DOI: 10.1201/9781003222736-8

Figure 7.1 Geographical location of Ningbo.

become a serious problem that affects agriculture and people's lives. However, this situation has been remedied by constructing a hydraulic project—the Tuoshan Weir (Figure 7.3) and later improved through its maintenance works in the ancient Ningbo area (Shang et al., 2010).

The construction of Tuoshan Weir was proposed by a magistrate in the Tang Dynasty (618–907)—Yuanwei Wang. He was born in Langya in Shandong Province and was appointed as the Yin Magistrate during the Tang Dynasty in the middle Taihe period. Chinese historians boasted of him as an incorruptible and upright officer who "enabled the Yin County to be a safe and prosperous place for many years by working diligently and tirelessly, sticking to honesty and integrity, against laziness and inaction, and punishing the grafters and rioters." He was concerned about the agricultural production of the local community and people's livelihood.

To better use freshwater for irrigation, after fully considering the local topography, a weir was constructed at the exit of the mountain valley to prevent saline water intrusion and store fresh water. This weir then became the head project for the canal system of the city center, and it was also used to divert water from the upper reaches of the Yin River to the Nantang River. Banks and sluices were also constructed to separate the water from the sea and the inner-city river network. According to Wang's planning, 70% of water in the weir could be diverted to the inner-city river, while the

Figure 7.2 Yin-Feng Plain.

remaining 30% could flow to the sea in normal conditions. In the case of flooding, 70% of water could be discharged into the sea, and the remaining 30% could flow to the inner rivers. In addition, the Wujin, Jidu, and Xingchun floodgates were constructed downstream of the Nantang River to discharge the flooding water and prevent tidal saline water intrusion. The canal system diverted the freshwater in the Nantang River to the entire Yinxi plain. Besides irrigating the 240,000 mu paddy fields, the water was diverted to the city center from the south mouth and stored in Sun Lake and Moon Lake for the residents' drinking and washing purposes.

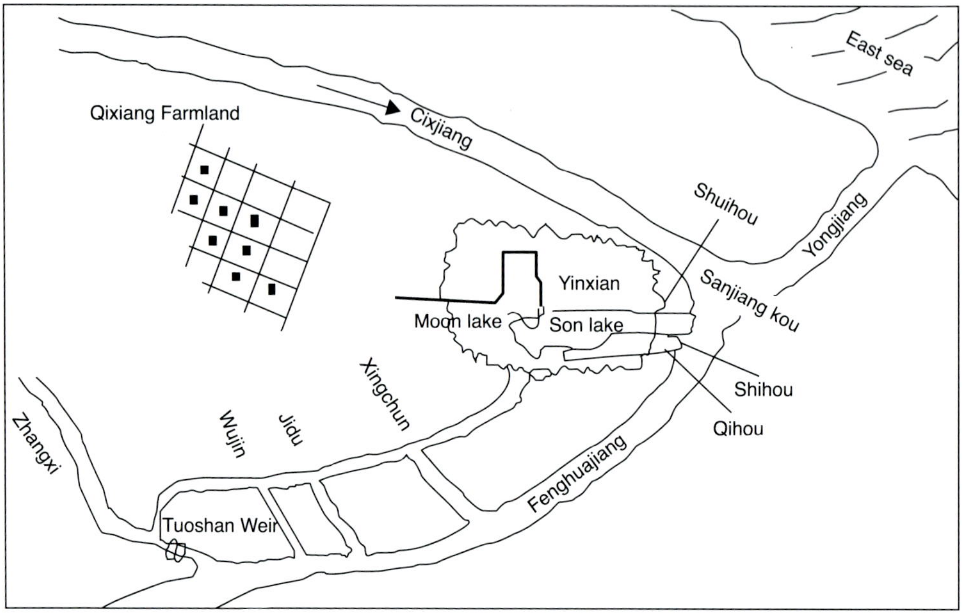

Figure 7.3 Tuoshan Weir in the ancient water system.

7.2 CONSTRUCTION BACKGROUND AND HISTORY

In AD 833—the 7th Daihe year of the Chinese Tang Dynasty—the Tuoshan Weir was proposed and built by the local magistrate Yuanwei Wang between the Siming Mountain and the Tuo Mountain in the mountainous upstream region of the Yin River. The weir is composed of dressed stones that form 36 steps both upstream and downstream. The length of the weir top is 42 zhang using 80 dressed stones (Zhang is a dimension unit in ancient China, approximately equal to 3.33 m.). It is recorded that the weir is hallowed and consolidated by large timbers. Three floodgates, Wujin, Jidu, and Xingchun, were later erected in the Nantang River to protect the city from flooding. Later during the Song Dynasty (960–1279), three discharge gates were built northeast of Songpo City for water drainage. Thus, a complete irrigation system was formed, including weirs, canals, and gates. In the initial operation of the system, sedimentation was insignificant; thus, dredging work once a year was sufficient. In the Southern Song Dynasty (1127–1279), sedimentation became a severe problem for the irrigation system. A three-orifice sluice was constructed more than 40 zhang away from the weir in the 2nd Chunyou year (AD 1242) at the Weixian to manage the sediments. To monitor the regulation performance of the sluices and gates in the irrigation system, a water gauge was equipped under the Ping Bridge in Ningbo city by Qian Wu in the 1st Kaiqing year (AD 1259) to measure the water depth. The irrigation system was maintained and repaired frequently throughout the later dynasties, including Yuan, Ming, and Qing. In the 15th Jiaqing year (AD 1536) during the Ming Dynasty (1368–1644), the height of the weir was increased by one more zhang to the present height. In the 7th Xianfeng year (AD 1857) in the Qing Dynasty (1636–1912), most of the projects were repaired, and the sediments were dredged to make the flow in the canal smooth in 1914. Currently, the weir is 134.4 zhang long and 4.8 m high, but most of the weir body is buried in the soil, no longer serving irrigation.

7.3 STRUCTURE AND DESIGN OF THE TUOSHAN WEIR

The Tuoshan weir is a stone-laying structure with features in terms of four aspects: the method to maintain stability, impermeable technique, arc-shaped layout, and cross-section configuration (Ye, 1982; Shu, 2011; Li, 2015). Shang et al. (2006) conducted field measurements of the project to reveal its scientific and engineering values.

7.3.1 Site selection

It was wise to choose the exit of the Siming Mountain valley as the site since this project has benefited Ningbo city, which experienced rapid growth in later years. Due to the extension of Tuoshan mountain, two islands were morphed by rivers, playing a key role in the stability of the weir.

7.3.2 Weir layout

The weir axis was slightly curved upstream, which is similar to the arch dam in hydraulic engineering. The two sides' foundations can support the forces exerted by the water. This layout was analyzed by Shang et al. (2010), as shown in Figure 7.4.

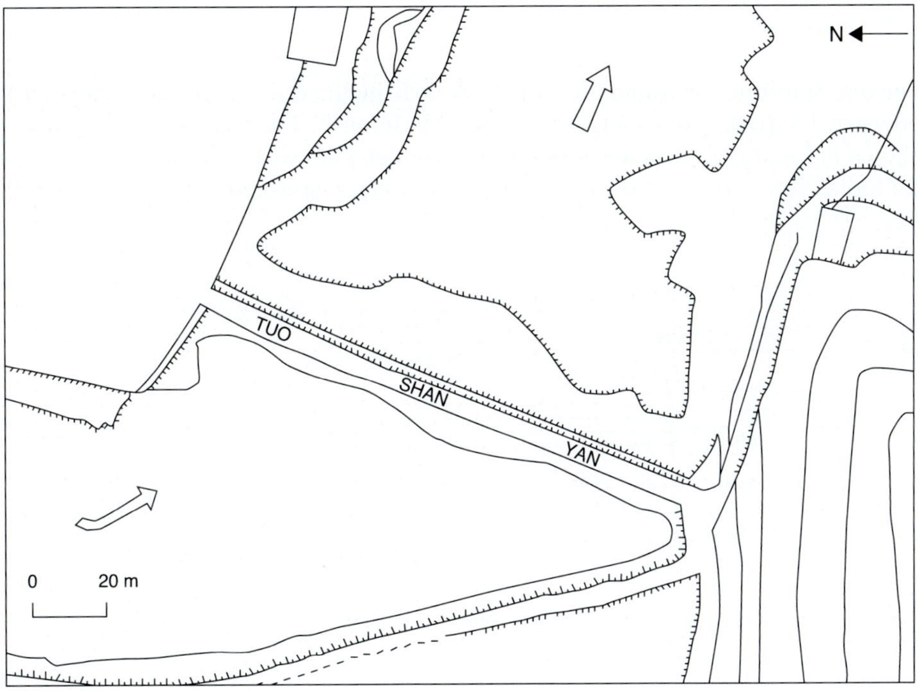

Figure 7.4 Structure layout of the Tuoshan Weir.

7.3.3 Weir structure

Figure 7.5 shows the structure of the weir. The top of the weir was constructed of 36 dressed stones with a length of 2–3.98 m and a width of 0.45–0.61 m. The top stone was equipped in 1857 during the Qing Dynasty (Wang and Chen, 1996).

The weir was designed with upstream and downstream aprons using strip stones and layered stair stepping. The downstream apron can reduce the local souring during the flooding season.

7.3.4 Impermeable inner structure

The weir was constructed using the internal clay wall (Figure 7.5) to prevent water infiltration, which was found and confirmed during the maintenance of the project in a test (Shen and Chen, 1995). The clay wall is 1.5 m deep in the center of the weir. The flagstone was used outside the internal clay wall to support the weir (Wang and Chen, 1996). This technique is similar to the modern methods in hydraulic engineering when building a dam.

7.3.5 Weir height

The height of the weir is 4.8 m. It mysteriously stratifies the requirements to regulate the flooding, divert fresh water for irrigation, and prevent saline water intrusion into the upstream region.

7.3.6 Weir stability

To ensure stability, the foundation of the weir is inclined upstream, as measured by modern techniques, with an angle of 5°–10°, averaging 8°. The cross-section of the weir is shown in Figure 7.5. Moreover, ten rotted wooden piles with a diameter of 1.2–1.3 m were found downstream of the weir. The wood was taken from the Chinese ash in the

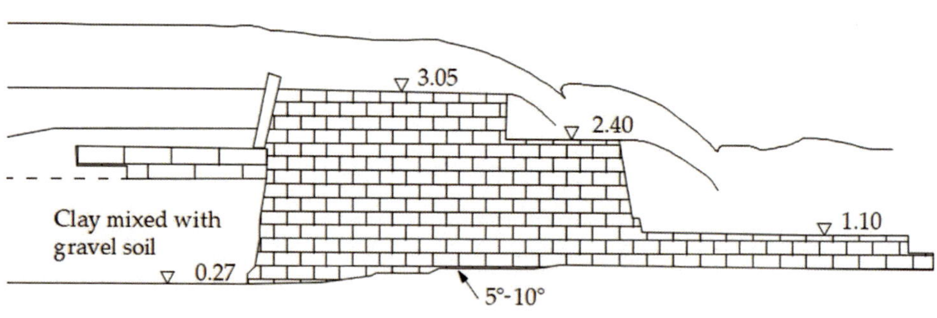

Figure 7.5 Cross-section of the Tuoshan Weir structure.

nearby mountain. The length and amount of these piles have still been debatable, yet the wood was inserted for the weir stability against sliding.

7.4 TUOSHAN WEIR TODAY

The main body of the weir remains in good condition (Figure 7.6), still used to prevent saline water intrusion, store freshwater, and divert fresh water for irrigation. Some modern facilities were also built to upgrade the weir's function.

This project was declared a Major Historical and Cultural Site Protected at the National Level in 1988. It was recognized as one of the ten new tour places by Ningbo City and the national science education base in 1994. It was also selected for the World Irrigation Engineering Heritage sites by the 66th Executive Council of the International Irrigation and Drainage Commission on October 14, 2015 (Shu and Fang, 2009).

The Yinjiang Town, where the Tuoshan weir locates, holds the Tuoshan Temple Fair on March 3, June 6, and October 10 of the lunar calendar each year. These are special ceremonies in memory of Yuanwei Wang and those who have contributed to the project (Chen and Wang, 2015).

This ancient hydraulic project for estuarial river regulation could be recognized as a typical coastal reservoir applied to restore freshwater, prevent saline water intrusion and flooding, and irrigate local farmland (Huang, 2016; Gu et al., 2019). By studying the project, we could learn from the past management experience of coastal water resources to serve the present.

Figure 7.6 Today's Tuoshan Weir captured in a different season.

7.5 CONCLUSIONS

This study introduced the Tuoshan Weir—an ancient estuarial river regulation project by presenting its background, history, and development from the Tang Dynasty to nowadays, with a review of its structure and design. The weir has significantly prevented seawater intrusion, diverting fresh water for irrigation and supplying water for household use. It has played a substantial role in the development of Ningbo City, thus having significant engineering and scientific implications for today's hydraulic projects.

ACKNOWLEDGMENTS

The article is based on the previous paper, *Tuoshan Weir System: an ancient hydraulic project for estuarial river regulation*, published in the Proc., 21st IAHR-APD Congress, Yogyakarta, Indonesia. The original article has been improved and modified to fit the book.

FUNDING

This research was partially funded by the Open Research Foundation of Key Laboratory of the Pearl Estuarine Dynamics and Associated Process Regulation, the Ministry of Water Resources, and the Zhejiang Provincial Natural Science Foundation of China. LGJ19E090001 and LQ19F060004, and the Ocean Wave Energy Application Project approved by the Ningbo Municipal Water Conservancy Bureau.

REFERENCES

Chen, S. (2014). *An introduction of Tuoshan Weir.* Impression of Yinjiang. DOI:10.16710/j.cnki.cn33-1272/d.2014.18.003 (in Chinese).

Chen, W. and Wang, Y. (2015). County head constructing Tuoshan Weir. *China Three Gorges.* 6: 106–109 (in Chinese).

Gu, H., Guo, Q., Lin, P., Bai, L., Yang, S., Sitharam, T. G., and Liu, J. (2019). Feasibility study of coastal reservoirs in the Zhoushan islands. *China Journal of Coastal Research.* 35(4): 835–841 (in English).

Gu, H., Guo, Q., Lin, P., and Zhang, Y. (2018). *Tuoshan Weir System: An ancient hydraulic project for estuarial river regulation.* Proceedings of the 21st IAHR-APD Congress, Yogyakarta, Indonesia (in English).

Huang, W. (2016). A history of seven thousand years water conservancy construction in Ningbo. DOI: 10.16710/j.cnki.cn33-1272/d.04.014 (in Chinese).

Li, Y. (2015). Three ancient Chinese projects selected for irrigation work heritage. *Half Month Water Story.* http://www.chinanews.com/cul/2015/10-14/7568032.shtml (in Chinese).

Shang, Y., Yang, Z., Li, L., and Chen, S. (2010). Highlights of one weir in eastern China constructed 1177 years ago. *Journal of Engineering Geology.* 18: 455–462 (in Chinese).

Shang, Y. J., Yang, Z. F., Li, L. H., Wang, S. J., and Chen, S. G. (2006). Scientific and engineering highlights of an 1171-year-old weir in Eastern China. *Agricultural Water Management.* 81(3): 371–380 (in English).

Shen, Z. and Chen, W. (1995). More about the mystery of Tuoshan Weir in the Tang Dynasty. *Science and Technology Review*. 11: 19–20 (in Chinese).

Shu, Y. (2011). Yinzhou Tuoshan Weir. https://wenku.baidu.com/view/1fd26e135f0e7cd1842536c2.html (in Chinese).

Shu, X. and Fang, L. (2009). Discussion on tourism exploitation of ancient hydraulic engineering Ningbo Tuoshan Weir. *Market Modernization* (in Chinese).

Wang, Y. and Chen, Y. (1996). Structure analysis of ancient hydraulic engineering Tuoshan Weir. *Zhejiang Hydrotechnics*. 4, 58–60 (in Chinese).

Ye, Z. (1982). Tuoshan Weir strategy. *China Water Resources*. 1, 58–61 (in Chinese).

Zhang, L. (2014). Tang dynasty Tuoshan Weir. *Heilongjiang Water Resources*. 2: 22–23 (in Chinese).

Part II

Historical water projects and traditional water technologies in Japan

The Teizan Canal in Miyagi Prefecture (above, courtesy of Ministry of Land, Infrastructure, Transport and Tourism) and a wall painting illustrating navigation through the canal during Edo era (below, Chapter 14).

DOI: 10.1201/9781003222736-9

It is estimated that rice cultivation in Japan was introduced from the lower reach of the Yangtze River in China around 300 BC. Thereafter, rice cultivation spread from western Japan to the northernmost part of the main island during the Yayoi period (until AD 300). However, due to poor irrigation technologies, once a major flood hit the paddy field area, it was heavily damaged with earth and sand, resulting in extremely unstable life. Although there was a highly limited number of large-scale water projects until the Kamakura (1185–1333) and the Muromachi (1336–1573) periods, the Sayama-ike reservoir is selected in this book to introduce a historical project in the ancient time. Subsequently, from the latter half of the Warring States Period (16th century) to the Edo Period (1603–1868), the foundation of the government became more stable, enabling large-scale projects, including grandiose castle constructions and water projects both for irrigation and flood management. All of the articles invited in this part, except one introducing the Sayama-ike reservoir and another one explaining the JSCE Heritage Award, are taken from this period. They include a flood control project by structural countermeasure in Kyushu, Okayama, Kyoto, Toyama, and Kanto areas, and an aqueduct project with an invert siphon in Kanazawa area. In addition, a shore-parallel canal for inland navigation in Tohoku area is introduced to discuss its effectiveness for reducing tsunami energy during the 2011 tsunami.

Chapter 8

Civil Engineering Heritage Award in Japan

Hitoshi Tanaka
Tohoku University

Nobuyuki Tamai
The University of Tokyo
NPO for Wise Learning on Tatsumi Aqueduct

CONTENTS

8.1 INTRODUCTION

It is extremely important to look back and recognize historic technology, which has achieved large contributions to social life till now, so that the next generation will inherit it. For this reason, numerous Japanese associations in the field of engineering have provided an award system for technologies in each field, e.g., the Japan Society of Mechanical Engineers, the Institute of Electrical Engineers of Japan, the Chemical Society of Japan, and the Japan Society of Civil Engineers (JSCE). Among these, JSCE Civil Engineering Heritage Award was founded in 2000. The purposes of awarding the JSCE Civil Engineering Heritage Award are summarized as follows:

1. To appeal to society (evaluating the cultural value of a Civil Engineering inheritance and recognition in society);
2. to appeal to civil engineers (recognition of past engineers' works, and promotion of responsibility and consciousness for creating new projects);
3. to utilize in community planning (promoting recognition that Civil Engineering inheritance can be a core as a local property by coexisting with its nature, history, and culture);
4. to promote relief of Civil Engineering inheritance from the risk of loss.

The award has been selected from projects contributing to national and regional development with high technological advancement. Candidates for the award are

DOI: 10.1201/9781003222736-10

recommended in the following categories in the Civil Engineering field: (1) transportation (road, railroad, port, river navigation, aviation, light beacon, etc.); (2) natural disaster prevention (river improvement, protection against tide, protection against wind, etc.); (3) agriculture, forestry, and fishery (irrigation, land reclamation, drainage, forestry, fishing port, etc.); (4) energy (power generation, coalfield, mining, etc.); (5) sanitation (water and wastewater, etc.); (6) industry (industrial water, shipbuilding, etc.); and (7) military. In principle, the project still needs to exist and function. In addition, it is necessary to have passed more than 50 years since the completion of the project. Candidates are initially selected in each regional branch, subsequently examined by the Civil Engineering Heritage Committee, and finally decided by the board of directors in JSCE. Although the awardees are limited to projects implemented in Japan, projects in Taiwan during the Japanese occupation are also eligible since 2009.

8.2 AWARDEES OF JSCE CIVIL ENGINEERING HERITAGE AWARD

As of January 2020, 448 civil engineering projects have been awarded (JSCE, 2021). Table 8.1(a) shows the number of awards each year, whereas, in Table 8.1(b), the number

Table 8.1 Number of commendations for civil engineering heritage selected by Japan Society of Civil Engineers (JSCE)

	(a)	*(b)*	*(c)*							
	Number of awarded	*Number of water-related*	*Distribution by region*							
Year	*projects*	*projects*	*Hokkaido*	*Tohoku*	*Kanto*	*Chubu*	*Kansai*	*Chugoku*	*Shikoku*	*Seibu*
2000	10	2		1					1	
2001	10	3	1		1					1
2002	15	2		1			1			
2003	17	1								1
2004	17	4			1		2	1		
2005	20	3			1	1	1			
2006	25	2			1	1				
2007	23	1								1
2008	23	3		1						2
2009	25	2					1	1		
2010	27	5		2	1			1		1
2011	22	3			1					2
2012	25	6			2	1		1		2
2013	21	6		1	2			2		1
2014	22	4	1			1		1		1
2015	21	3			1			2		
2016	24	4		1	2	1				
2017	23	2				1				1
2018	24	3			1	1				1
2019	28	4					1		1	2
2020	26	4			1		1			2
Total	448	67	2	7	15	7	7	9	2	18

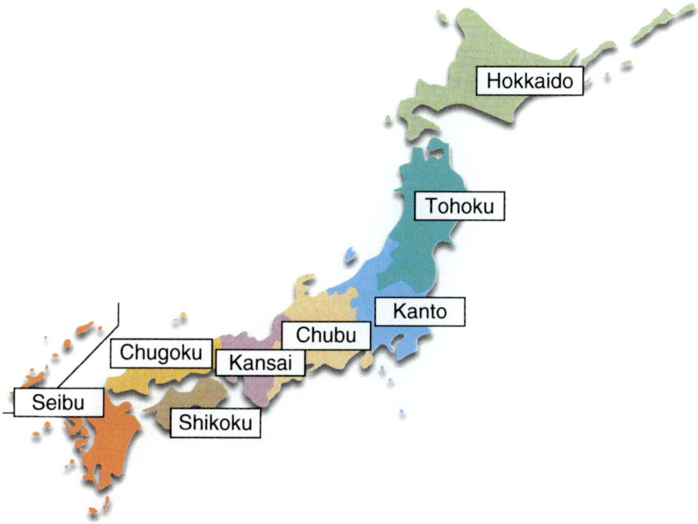

Figure 8.1 Regional branches of Japan Society of Civil Engineers (JSCE).

of water-related projects is extracted from Table 8.1(a). Here, by this book's criterion, hydraulic projects implemented before 1900 were selected in place of the criterion of JSCE, which is more than 50 years since its completion. Table 8.1 shows that about 15% of the projects are water-related projects that have spent more than 100 years.

As shown in Figure 8.1, JSCE has eight regional branches in Japan. To examine the regional distribution of the selected civil engineering heritages, Table 8.1(c) shows the number of civil engineering heritage nominated from each branch.

Figure 8.2 shows an award monument of the Nobiru Port in Miyagi Prefecture, awarded in 2000. The contributions of Dutch engineers were highly remarkable in this project, as will be described in more detail in the next section.

8.3 STATISTICAL CHARACTERISTICS OF AWARDED PROJECTS

8.3.1 Overall statistical characteristics

In Table 8.2, the water-related projects shown in Table 8.1 are classified into four categories based on each function. Since some projects have multiple purposes and cannot be narrowed down exactly to a sole function, one of the most major functions is selected in Table 8.2. According to this table, the number of projects for water resources is the largest, mainly located in the metropolitan area and the western part of Japan. The second-largest number is natural disaster prevention purposes (flood control and sabo), which are more evenly distributed throughout the country than water resources projects. In addition, it is noted that most of the projects related to port construction and reclamation are found mainly in western Japan. Especially those related to reclamation are observed in the western part of Japan. In this area, the development of new rice fields has been promoted since the early modern period by implementing reclamations. Since these reclaimed lands are flat and wide, they became a large-scale and highly productive rice-growing area, as fundamental facilities such as irrigation and drainage canals were developed. On the other hand, the number of projects related to river transportation is extremely limited, as seen in Table 8.2.

Figure 8.2 Monument of Japan Society of Civil Engineers (JSCE) Civil Engineering Heritage Award (Nobiru Port, Miyagi Prefecture).

Table 8.2 Classification and regional distribution of water-related projects that received Japan Society of Civil Engineers (JSCE) Civil Engineering Heritage Award

No.	*Function*	*Number*	*Distribution by region*							
			Hokkaido	*Tohoku*	*Kanto*	*Chubu*	*Kansai*	*Chugoku*	*Shikoku*	*Seibu*
1	Navigation	3			1		1	1		
2	Natural disaster prevention (flood control and Sabo)	23		1	5	6	3	4	1	3
3	Water use	26	1	5	8	1	2	1	1	7
4	Port and reclamation	15	1	1	1		1	3		8

8.3.2 Dutch engineers' contributions

To modernize Japan, the Meiji government invited about 2,300 specialists in various fields such as education, medicine, law, and civil engineering from Europe and the United States (JSCE, 1942). There were about 120 civil engineers, most of whom were from the United Kingdom. However, Dutch engineers had been invited for flood control and port construction projects (National Land Policy Agency, 2000; Iwamoto and Hein, 2018). The civil engineering heritage selected by JSCE shown in Table 8.3 includes several projects highly contributed by these Dutch engineers. The spatial distribution of these projects is shown in a map in Figure 8.3. It is observed that the projects cover all over Japan.

Of these, the contribution of Cornelis Johannes van Dorn (1837–1906), who was responsible for the Asaka Canal, has been reported in detail by Fujita and Nemoto (1991). Van Dorn was born in Hull, the Netherlands, in 1837 and was qualified as an engineer in 1860. After that, he came to Japan in February 1872 in response to the invitation of the Japanese government and was appointed as the Director of Civil Engineering. He resigned from his post in February 1880, and he was awarded the Order of the Rising Sun, Gold Rays, and returned to the Netherlands in May of the same year. He passed away in Amsterdam in 1906 when he was 69 years old.

Asaka Area (currently western Koriyama City), located in the center of Fukushima Prefecture, was a vast barren wilderness with poor water access. A large amount of irrigation water was indispensable for the development of this region. For this reason, the Asaka Canal excavation project was implemented. In January 1879, Van Dorn submitted a design document for Asaka Canal excavation work to the Director of Civil Engineering, and the work began on October 28, 1879. After a three-year construction period, this project completed 366 km long waterways. The completion of the project led to the development of Koriyama City, which used to be a small village with only 5,000 in the early Meiji Era, to become a core city in this region with a population of 340,000 after 125 years. At present, Van Dorn's statue stands at the foot of the bridge he designed on Lake Inawashiro (Figure 8.4). The statue, constructed in 1931, is still watching the development of the Koriyama area.

Table 8.3 Projects by Dutch engineer awarded Japan Society of Civil Engineers (JSCE) Civil Engineering Heritage Award

Dutch engineer involved	*Award year*	*Project*	*Location*
Van Doorn (1837–1906)	2000	Nobiru Port Construction	Tohoku
de Rijke (1842–1913)	2000	Ohtani River Sabo Dam	Shikoku
Mulder (1848–1901)	2001	Misumi West Port	Seibu
Van Doorn	2002	Asaka Canal	Tohoku
de Rijke	2004	Mt. Haruna Sabo Dams	Kanto
Escher (1843–1939) and de Rijke	2004	Mikuni Port Breakwater	Seibu
de Rijke	2004	Holland Sabo Dam	Kansai
de Rijke	2005	Kiso River Mouth Jetties	Chubu
Mulder	2006	Tone Canal	Kanto
de Rijke	2007	Nakashimagawa Revetment	Seibu

Figure 8.3 Distribution of projects by Dutch engineer awarded Japan Society of Civil Engineers (JSCE) Civil Engineering Heritage Award.

Van Dorn's achievements are not limited to the Asaka Canal project. In April 1872, he set up the first water gauge in Japan on the Tone River. He was also involved in the renovation of the Tone, Yodo, and Shinano Rivers, as well as the construction of numerous ports such as Nobiru Port (Figure 8.2) and Hakodate Port (Matsuura, 1995). He paved the way for modern civil engineering by incorporating Western scientific methods into Japanese technology that relied solely on their experiences at that time.

8.4 CONCLUSIONS

Among the hydraulic projects awarded civil engineering heritage by the Japan Society of Civil Engineers, the projects completed before 1900 are summarized in this paper. One of the features is that the contribution of engineers from the Netherlands was remarkable in the early Meiji Era. In addition, by classifying these projects in terms of function, water utilization-related and disaster prevention-related projects were the majority. In this book, therefore, the invited papers from Japan are mainly selected from water utilization and disaster prevention fields.

Figure 8.4 Van Doorn's statue in Inawashiro Town, Fukushima Prefecture.

REFERENCES

Fujita, T. and Nemoto, H. (1991). A study on the achievement of Van Doorn on the plan for Inawashiro Canal (Asaka Canal). *Proceedings of Civil Engineering History*, 11: 219–228 (in Japanese).

Iwamoto, K. and Hein, C. (2018). Cross-cultural engineering: the role of Dutch civil engineering in modern port planning in Japan (the 1870s-1890s). *Proceedings of the 18th International Planning History Society Conference*, Yokohama, Japan.

Japan Society of Civil Engineers. (1942). *After the Meiji Era, Japanese civil engineering and foreign engineers* (in Japanese).

Japan Society of Civil Engineers. (2021). JSCE Civil Engineering Heritage Award. https://www.jsce.or.jp/contents/isan/ (accessed on August 26, 2021) (in Japanese).

Matsuura, S. (1995). Services of Dutch engineers who arrived in Japan early in the Meiji era: through views of development in Holland. *Water Science*, 222: 45–61 (in Japanese).

National Land Policy Agency. (2000). *Engineers who created the land.* Kajima Institute Publishing Co., Ltd., 334 p. (in Japanese).

Chapter 9

Sustainable development of Sayama-ike reservoir

The historical value in East Asia

Tetsuya Sumi
Kyoto University

Koichi Koyamada
Osaka Prefectural Sayama-ike Museum

CONTENTS

9.1 INTRODUCTION

Sayama-ike Reservoir (latitude 34°30′24″ and longitude 135°33′02″) is located in Osaka-Sayama City, Osaka Pref., Japan (Figure 9.1). It has an earthfill dam-type bank that dams up two rivers running from south to north between Senboku Hills and Habikino Hills (Figures 9.2 and 9.3) (The Osaka Prefectural Sayama-ike Museum). In August 1982, the lower reaches of the Nishiyoke River were attacked by a large flood and got damaged. To prevent further damages, upgrading work on the reservoir, called the Great Renovation of Heisei (1988–2001) to increase the reservoir capacity to store floodwater was carried out (Kanamori et al., 1994; JCOLD, 2009, 2012). The reservoir bottom was excavated by 3 m, and the dam body was heightened by 1.1 m. This work has increased the storage capacity volume from 1.8 to 2.8 million m^3 for flood mitigation (Figure 9.4). As a result, it has a significant effect on flood control in the downstream region, and, at the same time, it is still distributing water to the downstream region as an irrigation pond. The archaeological surveys during the upgrading work carried out from 1987 to 2000 brought us much information about the historical civil engineering works for irrigation (Figure 9.5a and b).

DOI: 10.1201/9781003222736-11

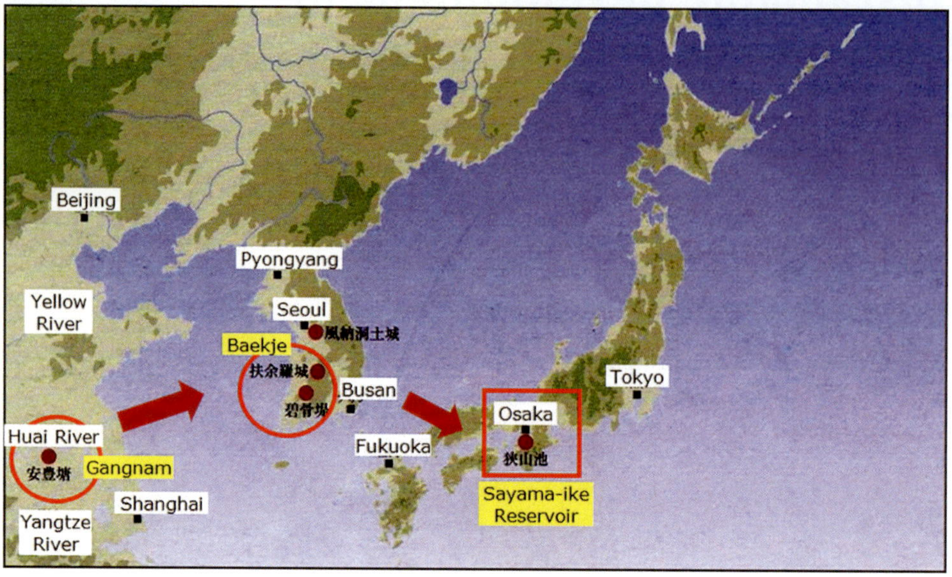

Figure 9.1 East Asian dam-type pond road.

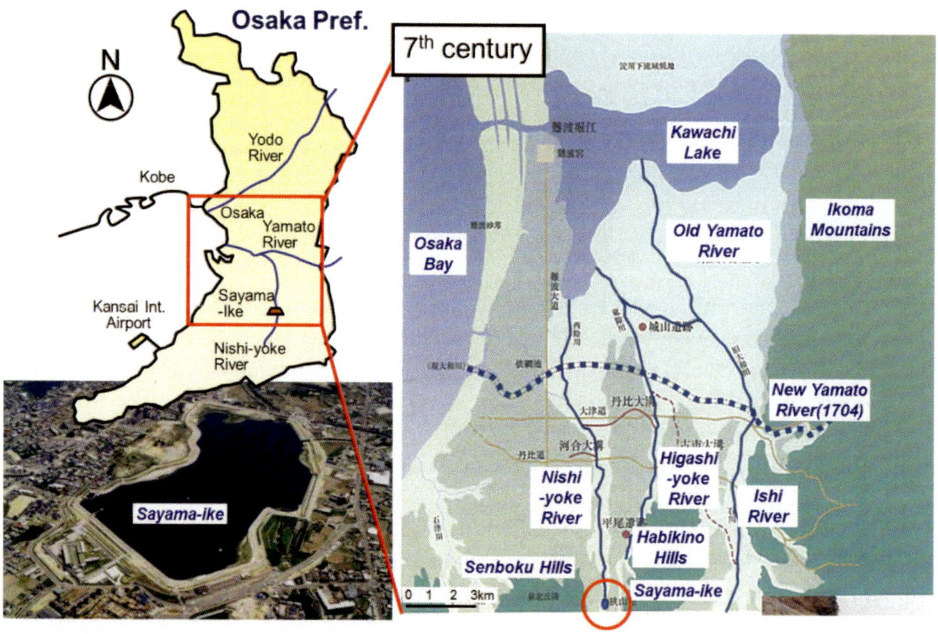

Figure 9.2 Location of Sayama-ike Reservoir.

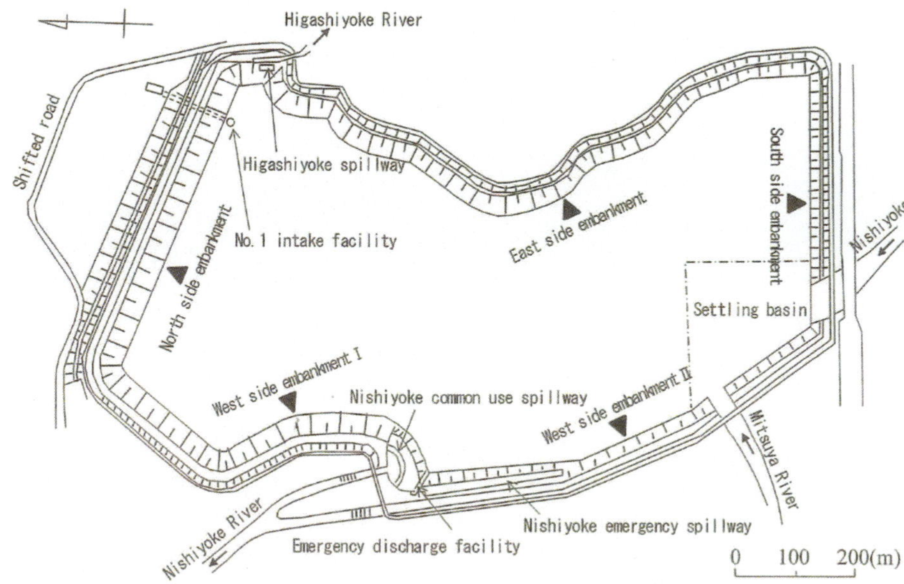

Figure 9.3 Plan view of Sayama-ike Reservoir.

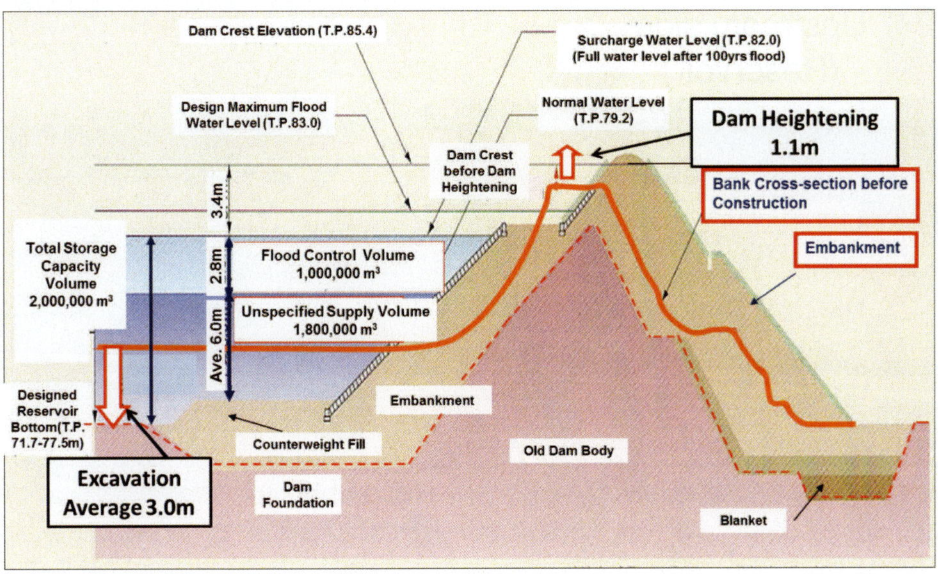

Figure 9.4 Schematic view of upgrading of Sayama-ike Reservoir.

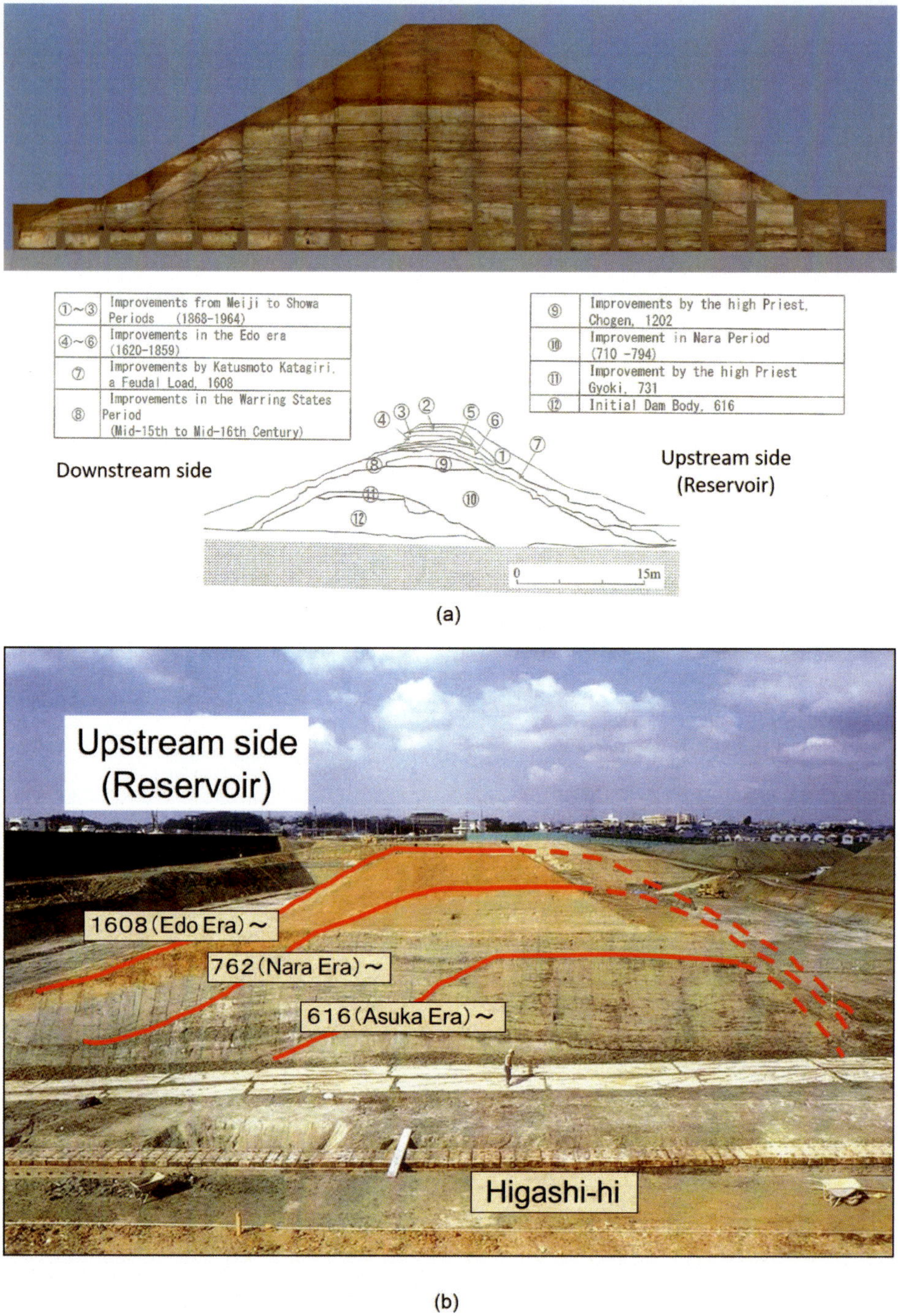

Figure 9.5 (a) Improvement history of the old dam near the center sluice. (b) Picture of improvement history of the old dam near Higashi-hi.

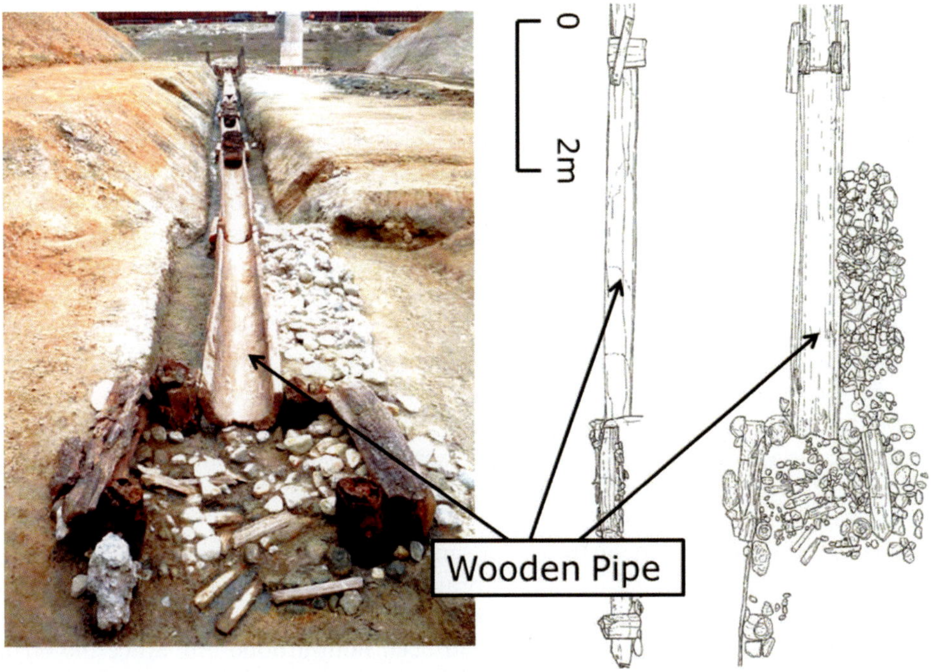

Figure 9.6 Ancient wooden pipes.

The Sayama-ike was built for reclaiming wastelands to rice fields. It has been supplying water to the rice fields of the lower reaches through wooden pipes connected to an intake at the north bank of the reservoir. Under the eastern part of the bank, ancient wooden and other wooden pipes placed just above the older ones in 1608 (Figure 9.6) were found. In the middle part of the bank, the intake tower was built in the early 20th century, and another old intake installed earlier at the beginning of the Edo period (1603–1868) was excavated. Huge stone blocks were placed at both sides of these intakes, which were revealed to be sarcophaguses (stone chambers) in the Kofun Period (3rd–7th century), to prevent local scour. They were recycled as stone pipes for water supply in 1202 during the Kamakura Period. In the western part of the bank, an intake installed in the early Edo period was also found. From the birth of Sayama-ike, the intakes installed in the eastern and middle parts have been supplying water to the areas between Higashiyoke and Nishiyoke Rivers. Water from the western intake was supplied through the Nishiyoke River in the Edo period.

After upgrading the Sayama-ike to the multipurpose one by adding a flood control function, it was changed to "the Sayama-ike Dam" in 2002 (Kuraku et al., 2019). This paper summarizes its history in different centuries based on the data discovered and current challenges for sustainable irrigation system management, including the dam.

9.2 HISTORY OF THE SAYAMA-IKE IN EAST ASIA AND ITS HISTORICAL VALUE

9.2.1 The Ancient Times (AD 600–800)

In the Asian monsoon region, where seasonal fluctuations in rainfall are remarkable, there is a history of building and operating dam-type reservoirs that secure water and reduce flood damages. Sayama-ike reservoir is one of them. Information on the dam-type pond was transmitted from China (Gangnam) to the Korean Peninsula (Baekje Dynasty) and Japan (Figure 9.1). In China, it appeared in the Liangzhu culture period, known as the rice cultivation civilization of the Yangtze River Delta, following the Hemudu culture and Kuahu Bridge culture (about 3100–2900 BC). The technology was introduced from Gangnam, China, to Baekje, Korea, during the Three Kingdoms period (around the 4th and 5th centuries). Sayama-ike is historically the end-point of the "East Asian dam style pond road." Among the dam-type ponds in China and Korea, the Sayama-ike reservoir, where excavation surveys found various types of civil engineering heritage, provides the most abundant history and civil engineering information.

The reservoir area at its birth was presumed to be approximately 260,000 m^2—two-thirds of the present, and its water storage volume was about 800,000 m^3, which is one-third of the present one. The dendrochronological analysis of the ancient wooden pipes proved that this reservoir had been built around 616. Since the birth of Sayama-ike, the full-scale reclamation of the terrace between Nishiyoke River and Higashiyoke River started by utilizing water from this reservoir through the irrigation ditches. Accompanied by the reclamation, human settlements and ancient temples were placed one by one. At present, parts of the ancient fields and main ditches for water supply have been unearthed. Based on the expansion of the settlements, the irrigation area by water supply from the Sayama-ike is estimated to be approximately 6 km from north to south and 1 km from east to west. This irrigation area is quite similar to the one irrigated by the main ditches to which water was supplied from the main intake of Sayama-ike in the Edo period.

According to the ancient documents named *Shoku-Nihon-gi*, the ancient state had built Sayama-shimoike Reservoir in 732, now named Taima-ike, at the lower part of nearby the Sayama-ike. It suggests that the Sayama-ike was under the control of the State because of its significance for reclamation tactics. In 762, the ancient state restored Sayama-ike by employing 83,000 persons. Its bank was heightened from 6 to approximately 9.5 m, and the bottom width widened from 27 to 54 m. This redevelopment work established the prototype of the Sayama-ike nowadays. The reservoir area is presumed to have been approximately 350,000 m^2, and the water storage volume is approximately 1,700,000 m^3. Archaeological surveys showed that expanded settlement sites in those days were similar to those in the 7th century.

9.2.2 The Middle Ages (AD 800–1200)

Even though the exact time when the first and middle intake and pipes were built is still unknown, they were probably built in the Heian Period (794–1185). In 1202, a monk

Figure 9.7 Stone monument with an inscribed document showing the history of the reservoir redevelopment work by the monk Chogen.

named Chogen, who had devoted himself to restoring the Todaiji Temple, which had been burnt down in the civil war of the Middle Ages, led the redevelopment work of the Sayama-ike as well. He replaced the ancient wooden pipes with stone ones. Through the archaeological survey in the course of the renovation work, a stone monument was found. A document referring to the background of the redevelopment work, its process, and the names of the people engaged in that work while obeying the Buddhism doctrine (Figure 9.7) was inscribed.

It shows that the upgrading work was carried out on the farmers' demand benefited from the water supply from Sayama-ike. Plotting the villages of those farmers showed that the village area is almost identical to the one benefiting from the irrigation water through the middle and the western intakes in the Edo period. Archeological analysis of the transition of the settlement pattern of the area also shows that the reclamations of the left terrace of Nishiyoke River, where few settlements existed in the ancient times, were carried out actively.

Considering this evidence, the upgrading work by Chogen is presumed to have made it possible to form a vast irrigation network. Careful examination is still needed whether the total irrigation system, which was available in the Edo period, had started or not, in those days when private manors and official territories had been mixed up in a complicated manner. Upgrading the irrigation system and water supply network by the people who followed Chogen would be the future subject.

9.2.3 The Edo Period (AD 1600–1800)

In 1608, at the beginning of the Edo period, the successor of Hideyoshi Toyotomi, who had unified the country to be Japan, ordered Katsumoto Katagiri to restore the Sayama-ike. Katagiri installed the three wooden intakes and the pipes unearthed by the archaeological surveys (Figure 9.8a and b). This seems rational for vertical weirs, and three intakes were built along the ridges and slopes embedded under the dike. It enabled intake of water from several levels with high water temperature warmed

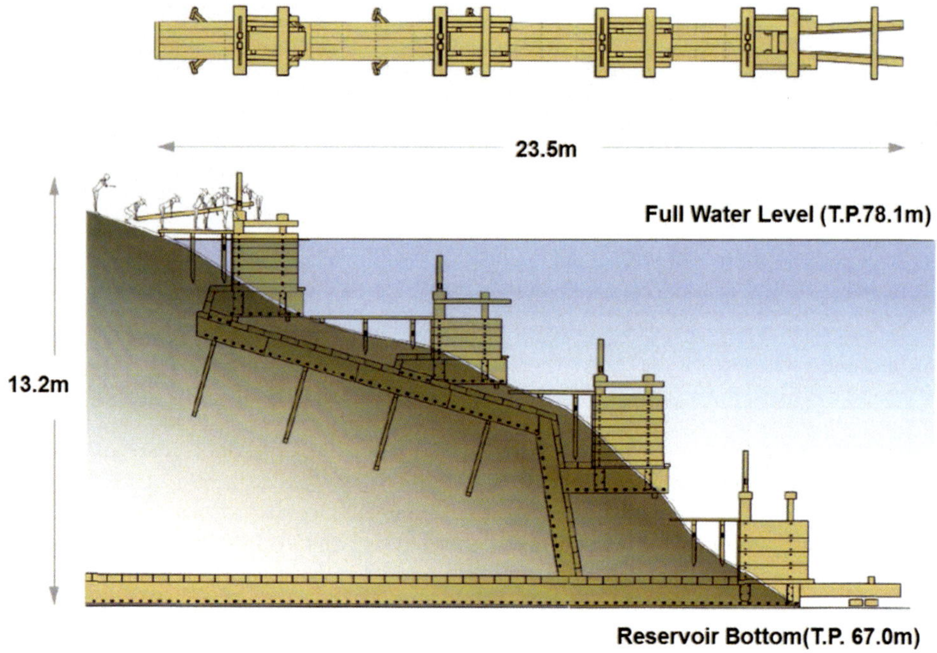

(a)

(b)

Figure 9.8 (a) Ancient intake structures 'Shakuhachi-hi'. (b) Picture of ancient intake structures 'Shakuhachi-hi'.

by sunlight. Moreover, it is easy to open and close. This intake structure was called 'Shakuhachi-hi,' derived from the fact that the shape of the pot resembles a traditional woodwind instrument in Japan. It is believed that the appearance that many intakes are open looks similar to that of shakuhachi's finger holes.

He also expanded the northern bank to be approximately 600m long. The western drainage was rebuilt, the eastern one was newly built, and the bank was heightened. As a result of this upgrading work, the reservoir volume became approximately 510,000m^2, and its water storage volume became three times larger than the original one. This work changed the environment surrounding the reservoir. It formed the irrigation network where the Sayama-ike supplied water to the smaller reservoirs in a cascade called Koike (Secondary) and Magoike (Tertiary) ponds. This network can be detected by the illustrated documents of the Edo period.

The main purpose of the irrigation system in the Edo period was not to irrigate the lower fields directly as in the ancient times but to supply water from the Sayama-ike to the lower reservoirs built in each village to keep their water storage full (Figure 9.9).

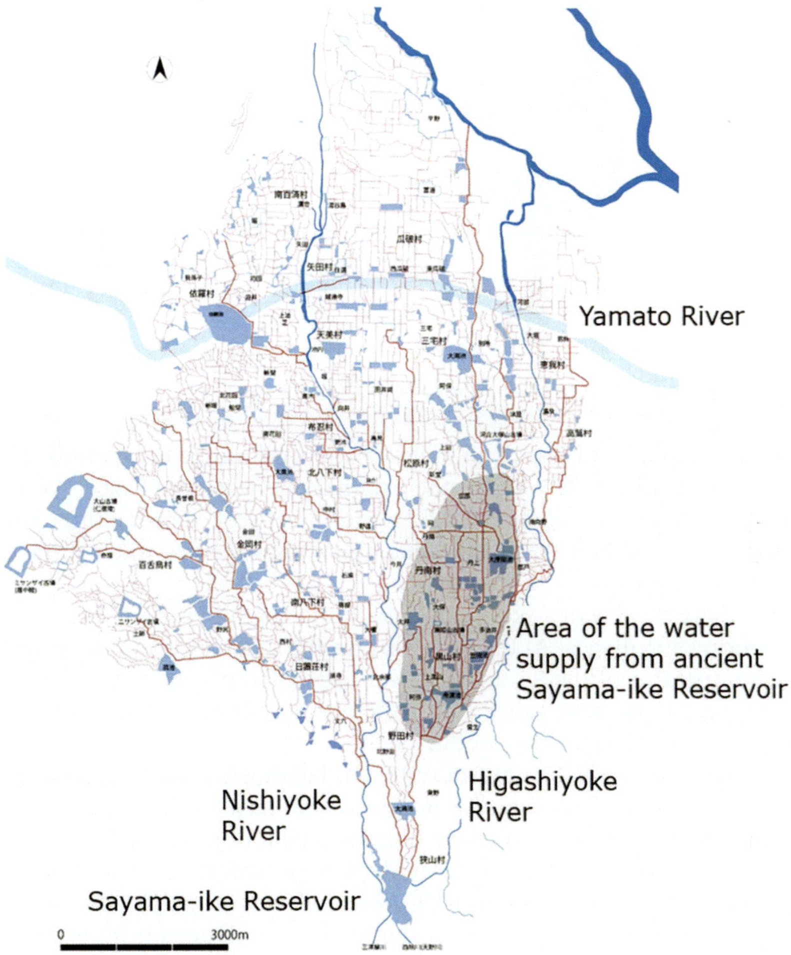

Figure 9.9 Schematic view of the irrigation network of the Sayama-ike and downstream reservoirs.

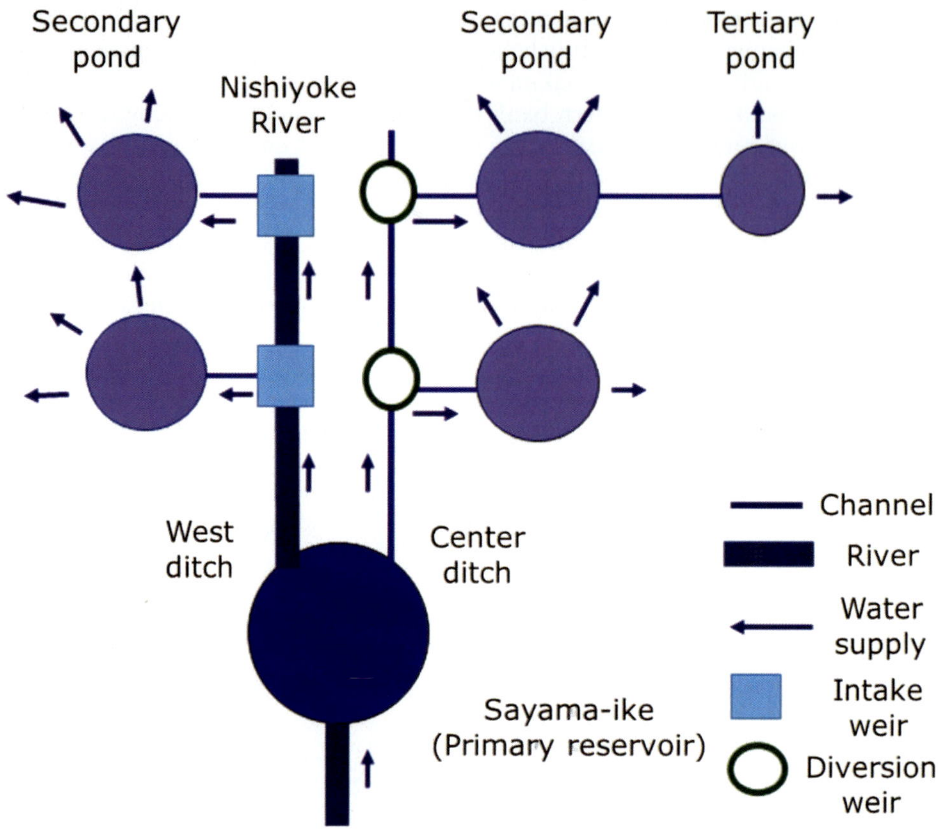

Figure 9.10 Old picture of irrigation network in the Edo period.

Figure 9.10 shows an old picture of an irrigation network in the Edo period. The modern network still works widely where the original one was established (Figure 9.11). This figure is a schematic diagram of the irrigation network showing three categories of irrigation ponds and how the Sayama-ike functions as primal water storage to stabilize the water supply function.

The area where the Sayama-ike supplied water through two main ditches was over 4,000 ha covering 80 villages. At the same time as the redevelopment, *Ikemori* (the hereditary administrator of the Sayana-ike) was appointed officially. Private managers were also engaged in the task called *Bansui* (a local rule for deciding the time span of providing water according to rice product yield).

Bansui is still a living water control system in Japan, enabling farmers to get water fairly. Many documents describing *the Bansui* system have been kept in the Tanakas, who was appointed as *Ikemori*. After the redevelopment in 1608, the Sayama-ike bank was restored more than ten times longer in the Edo period. Construction work on the Yamato River to change its course in 1703 prevented water supply to the river's northern part. In addition, increasing costs for the serial redevelopments of the Sayama-ike resulted in unbearable burdens to the farmers, so they gave up obtaining water from the reservoir. Thus, the irrigation area of the Sayama-ike has decreased to 3,000 ha, covering only 38 villages.

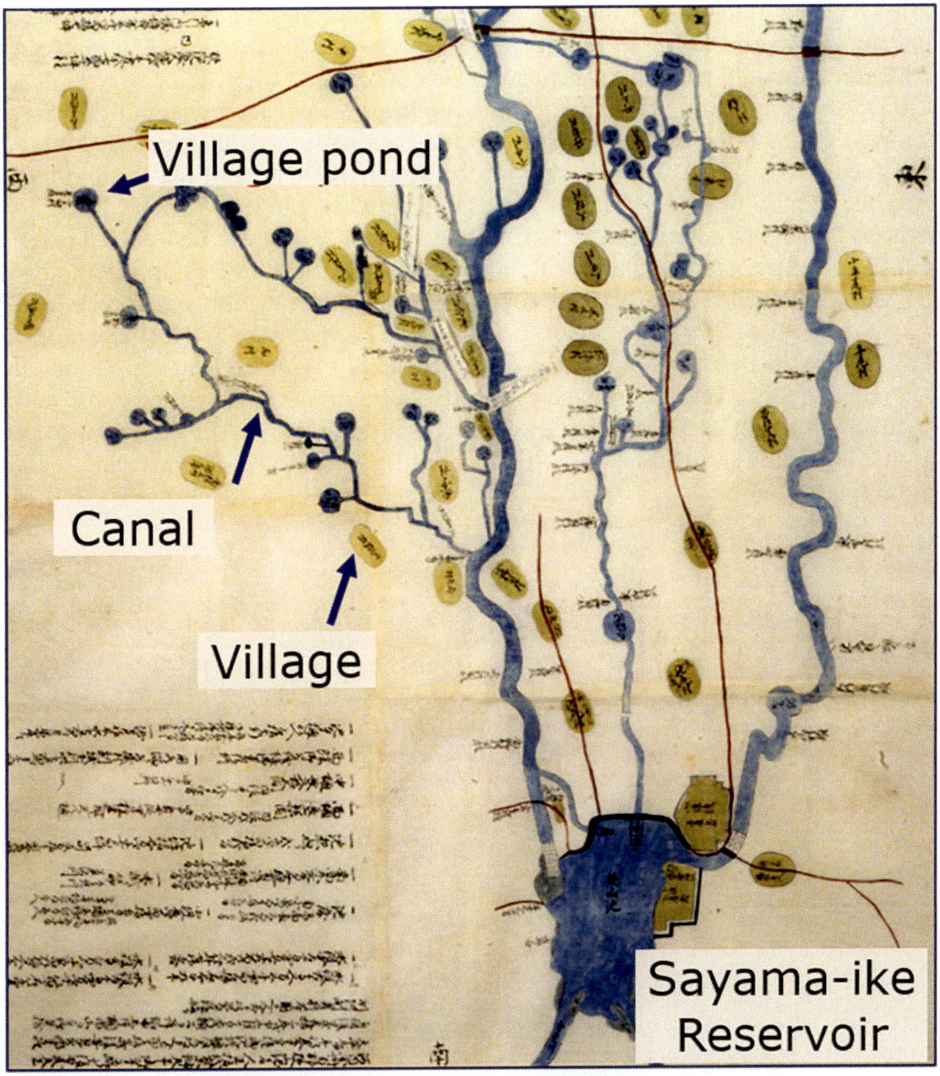

Figure 9.11 Contours of irrigation area of the Sayama-ike Reservoir in different centuries.

9.2.4 The modern times (1870–present)

Since the Meiji Era (1868–1912), various redevelopment works have been carried out at the Sayama-ike to increase agricultural products and prevent water disasters. However, the reservoir area has been decreased to approximately 390,000 m^2, and the water storage volume has also been decreased to approximately 1,800,000 m^3 because of the decreased reservoir depth by sedimentation from the upper streams. The total irrigation area of the lower region has shrunk to approximately 2,500 ha due to the land-use changes.

Through the Edo period, the Tanakas, the *Ikemori*, managed the Sayama-ike under the control of the Edo government, and the members of *Mizushima-Sodai* (representative of the lower villages benefiting from the reservoir) took part in *Bansui*. In the Meiji Era, the union of the villages concerning the Sayama-ike was established. Land Improvement District of the Sayama-ike reservoir was established in 1949, succeeding the inherited water management of over 400 years.

In recent years, the Sayama-ike has been largely upgraded to a multipurpose dam by adding a flood control function to the existing irrigation pond due to flood damage caused by the heavy rain that hit the downstream area of the pond in 1982. Based on the flood control master plan (Figure 9.12), the inflow flood peak discharge of 240 m^3/s under the target rainfall event of 100 years return-period can be reduced to 130 m^3/s by the gateless flood control operation. In this upgrading project, which was called the Great Renovation of Heisei (1988–2001), a flood control capacity of 1 million m^3 was added to the irrigation water capacity of 180 million m^3 by raising the bank by 1.1 m and excavating the reservoir bottom approximately 3.0 m (Figure 9.13). Figure 9.14 shows Nishiyoke's common use spillway for flood control. As a result, it has a significant effect on flood control in the downstream region, and, at the same time, it is still distributing water to the downstream region as an irrigation pond.

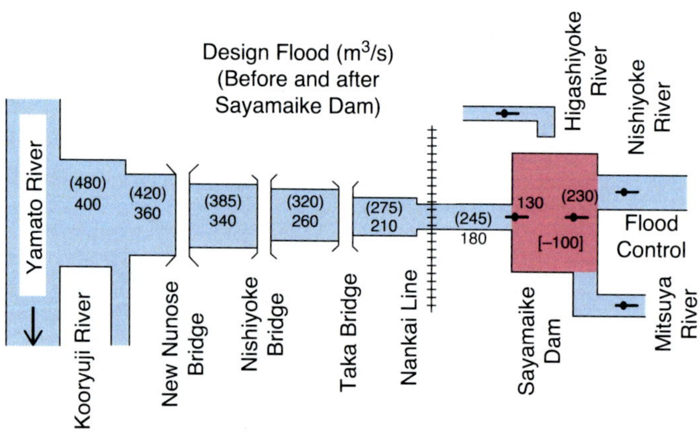

Figure 9.12 Flood control master plan.

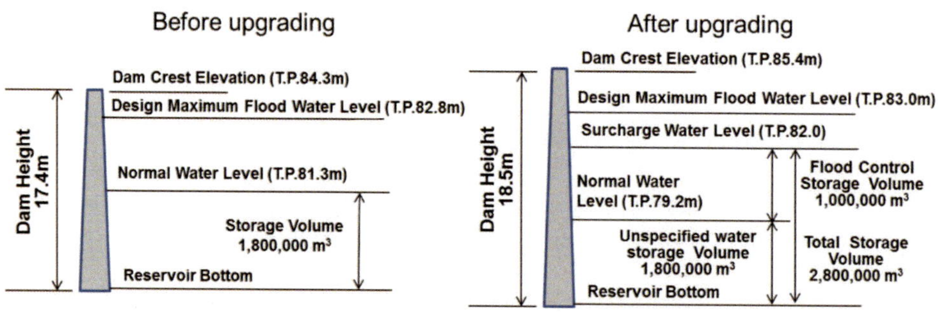

Figure 9.13 Storage and water level change of the Sayama-ike Dam.

Figure 9.14 Nishiyoke common use spillway.

Many civil engineering remains have been excavated during the Sayama-ike Reservoir rehabilitation, and these civil engineering techniques have been passed down through these historical heritages. During and after this upgrading project, Osaka Prefectural Government promoted a public awareness campaign to inform the history of the irrigation system and the necessity of its upgrading to the local community. To promote the campaign, Osaka Prefectural Sayama-ike Museum (Figure 9.15) was opened in 2001 to display the history and the civil engineering technologies concerning the water control and irrigation system, including the Sayama-ike. It creates symbolic value for the city by using the charm of the national historic site Sayama-ike. In the Museum, a preserved real piece of embankment cut from the old embankment—a total length of 64 and 15m high (Figure 9.16), and unearthed remains such as an East wooden pipe, a Central wooden pipe, a sarcophagus, and a wooden framing, in the various times are on display. The Museum was designed by the world-famous architect Tadao Ando. The Museum preserves the documents concerning reservoir management, which the Tanakas and the Ikemori have kept, and exhibits some yearly for public education.

Figure 9.15 The Osaka Prefectural Sayama-ike Museum.

Figure 9.16 A preserved piece of the dam embankment exhibited in the Museum.

9.2.5 Future issues

The Sayama-ike reservoir and others still supply water to their lower reaches for irrigation. Farmers benefiting from the system share the cost of maintaining the reservoirs and their network. As the farmers get older and give up keeping their farmlands, however, managing the system's functions is becoming more challenging. It is an important question how we shall sustain the reservoirs under the increased cost due to the decreased number of farmers. To solve this problem, we must examine the practical ways of using the reservoirs, such as clean-up campaigns by neighboring people and volunteers with help from local governments, securing necessary incomes by the solar power generation utilizing open spaces, increasing the number of farmers by engaging with food industries to secure constant consumption of agricultural products, and so on. The local governments, societies, and people must cooperate to sustain the reservoirs for not making the farmers alone and unaided.

9.3 CONCLUSIONS

The Sayama-ike Reservoir has been working to supply water to the lower fields for 1,400 years, which could be possible by the continuous redevelopment works through the long years. Irrigation has promoted the reclamation of the wasteland in the lower region. As a result, many settlements have been formed in the local society. The completed irrigation network that the Sayama-ike supplied water to the smaller reservoirs in cascade has formed a valuable landscape of rice production. This landscape is a common cultural property among the nations relying on rice in the East Asian region. We have to hand it to the next generations. Additionally, under the changing climate, the wise use of water for multiple functions, including flood control, is one of the important challenges to directly contribute to downstream communities, which enhances public involvement for long-term reservoir sustainability.

ACKNOWLEDGMENTS

This article is based mainly on the Sustainable development of irrigation system with Sayama-ike reservoir, authored by us and published in the *Journal of Hydro-environment Research*, Volume 26, October 2019, pages 8–13 and has been rewritten fitting this monograph.

REFERENCES

Japan Commission on Large Dams (JCOLD). (2009). *Current Activities on Dams in Japan*, 114–118.

Japan Commission on Large Dams (JCOLD). (2012). *Dams in Japan*.

Kanamori, W., Ando, M., Kimura, M., Nishizono, K. and Shimizu, H. (1994). *Raising of the Sayama-ike Dam*, ICOLD, Durban, Q70-R32, 503–516.

Kuraku, Y., Koyamada, K., Sumi, T. and Takei, Y. (2019). Sustainable development of irrigation system with Sayamaike reservoir, *Journal of Hydro-Environment Research*, 26: 8–13, https://doi.org/10.1016/j.jher.2019.08.001

Osaka Prefectural Sayama-ike Museum, Japan: http://www.sayamaikehaku.osakasayama.osaka.jp/

Chapter 10

Why all the tributaries of the Chikugo River flow into the old main streambed even after the cut-off channels were constructed

Koichiro Ohgushi
Saga University

Wataru Kawahara
Saga Prefecture

CONTENTS

10.1 INTRODUCTION

The Chikugo River is a primary river on Kyushu Island, Japan, with a trunk length of 143 km and a catchment area of 2,860 km^2, as shown in Figure 10.1. The basin lies within four prefectures: Kumamoto, Oita, Fukuoka, and Saga. Because of the severe meandering of its watercourse, many cut-off channels have been constructed from the Edo era through the Meiji, Taisho, and Syowa periods to increase its conveyance (Kase River Irrigation and Drainage Office, 1973; Kishihara, 2016). Table 10.1 shows a list of eight existing cut-off channels along the Chikugo River. There are four major cut-off channels of particular interest. These are the Sakaguchi, Komorino, Tenkenji, and Kanashima channels (Chikugo River Construction Office, 1976). Construction of the Sakaguchi cut-off

DOI: 10.1201/9781003222736-12

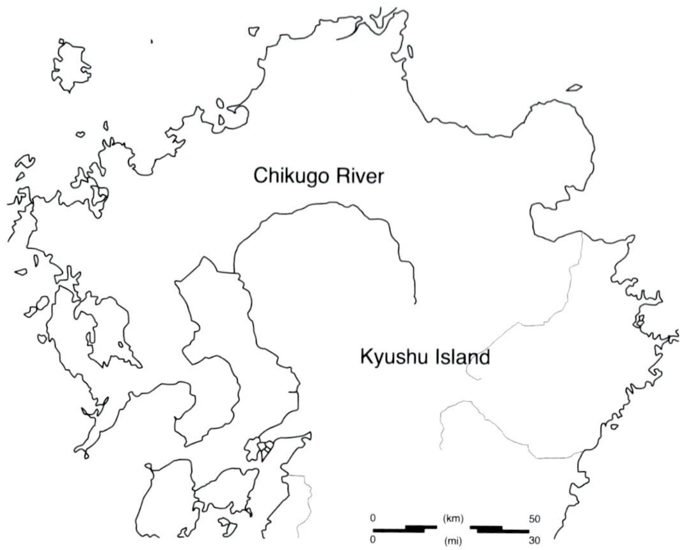

Figure 10.1 Kyushu Island, Japan, showing the location of the Chikugo River.

Table 10.1 Chikugo River cut-off channels

Cut-off channel name	*Distance from river mouth (km)*	*Shorted length (m)*	*Construction period*	*Tributaries flowing into old meandering reach*
1 Kanashima	36.6–39.2	2,870	1932–1951	Kose River
2 Komorino	27.0–28.9	2,650	1929–1950	Homan R, Akimistu R, Daigi R, Todoroki R, Yasuro R.
3 Senoshita	25.0–26.1	2,750	1606–1619	Numa River
4 Tenkenji	19.0–20.0	1,490	1927–1958	
5 Sakaguchi	16.6–18.0	1,780	1927–1956	Hiro River
6 Shimoda	15.8–16.6	4,750	Warring State Period	Shozu R, Kiritohshi R, Iryu R.
7 Ukishima	11.3–12.2	2,550	1610–	Tade R, Sanbonmatsu R, Baba R.
8 Dokaijima	9.2–10.5	1,750	1592–1610	Jobaru R, Sagae R
Total shortened length		20,590 m		

channel took 29 years, starting in 1927. About 920 thousand m^3 of soil was excavated, while about 782 thousand m^3 of soil was used for embankments. It costs about 10.97 million yen. The Komorino cut-off channel was constructed in 1929, taking 21 years. It costs about 6.50 million yen. Construction of the Tenkenji cut-off channel began in 1927. It took 31 years, costing about 4.15 million yen. The Kanashima cut-off channel was constructed in 1932 for 19 years, costing about 20.25 million yen. All costs above are given at the time of construction. The monetary value of 1 yen at that time equals about 636 yen. For these projects, humans and machinery were used to excavate, and 20-ton steam locomotives were used extensively to carry the excavated soil. However, with these cut-off channels having been built, many old meandering waterways still exist, and none of the tributaries flowing into the Chikugo River connect directly to its present mainstream.

A map of the Hizen region in the Keicho period (1596–1615) shows that Chikugo tributaries such as the Iryu, Kiritohshi, and Shozu rivers flowed separately into the Chikugo River during that time, as shown in Figure 10.2, left cell. By the Shouhou period (1645–1648), however, a part of the Shozu River that flowed south had disappeared and moved further to the southwest, reaching the old meandering waterway of the Chikugo River at the same point as two other tributaries, as shown in Figure 10.2, right side (Culture Heritage Online, 2018). In the intervening years, from 1632 to 1643, a riparian technical group, including Chief Retainer Hyogo Naridomi, constructed the 12 km long Chiriku Levee from Chiriku to Sakaguchi on the right side (facing downstream) of the Chikugo River. The rerouting of the Shozu River is thought to have been implemented in this period (Eguchi, 1977). Tributaries flowing into the lower reach of the Chikugo River do not join it directly. Instead, they mainly flow into its old meandering path. The reason this system was developed has not yet been sufficiently clarified.

The downstream part of the Chikugo River is lower than the surrounding area, so water flows through the lowest area of the plain, and it is difficult to use that water. For this reason, "Ao" intakes have been carried out for a long time, utilizing the large tidal difference of the Ariake Sea. "Ao" means freshwater flowing from upstream, which is lifted to the surface because the seawater in the Ariake Sea is denser, and the seawater runs up under the fresh water. Therefore, historically, during high tide in the Ariake Sea, fresh water was taken for agricultural use from rivers and waterways in the downstream region of the Chikugo River using sluices and pumps. Usage of Ao intake water ended when the region's river management was unified by completing the Chikugo estuary weir in 1984 (Igarashi, 1996; Teramoto et al., 2016).

In this area, boats were used for mass transport of goods until railroads and roads were developed after the Meiji era, as depicted in Figure 10.3. The Sagae River connected east and west from Saga Castle town to the Ariake Sea through the Chikugo River (Miyachi, 2009; Ohgushi and Nagahata, 2016). Many goods were transported

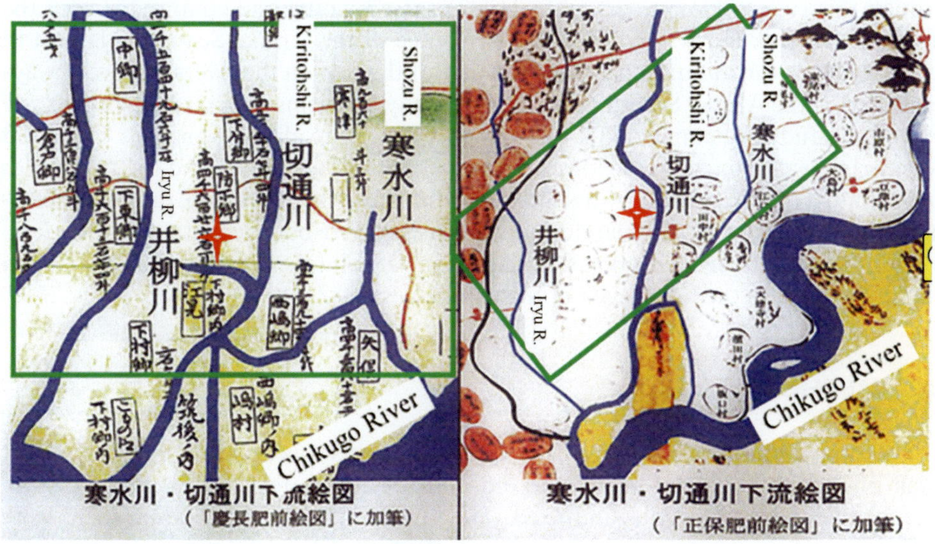

Figure 10.2 Downstream map of the Shozu and Kiritohshi rivers.

Figure 10.3 Timber rafting.

through this river network, and many boats crossed, going and coming between the banks of the Chikugo River. The tidal difference of the Ariake Sea is the largest in Japan, about 5–6 m at Suminoe. The influence of the tide on the rivers and waterways connected to the Ariake Sea seemed to cause many problems with river transport.

In this study, we conducted a one-dimensional unsteady flow analysis of the old meandering watercourse in the lower region of the Chikugo River and the tributaries flowing into it to clarify these hydraulic interactions and their function, taking the presence or absence of the old meandering waterway and tributaries into consideration. In addition, how river transportation and freshwater intake is related to this system was investigated in the tidal reach of the tributaries.

10.2 METHODOLOGY

In this study, a one-dimensional unsteady flow analysis was undertaken on the lower reach of the Chikugo River and the old meandering waterway. The period from June 20 to July 31, 2007 was analyzed. The study was conducted when there were no floods, and the spring tide had occurred. We focused on characteristic moments such as high, low, ebb, and flood tide at the confluence point of the Iryu and Kiri-tohshi rivers. At these moments, the hydraulic interaction of the old meandering waterway and the tributaries were examined. Figure 10.1 shows the old meandering waterway, analyzed rivers, and associated boundary points. Riverbed topographical data used for the analysis were prepared based on survey data from the Chikugo River Office of the Ministry of Land, Infrastructure and Transport, and the River and Sabo Section of Saga Prefecture. The calculated cross-sections were set at 200 m intervals. The time step was set to 1 second, and Manning's roughness coefficient was set to 0.020 for the Chikugo, Hayatsue, and Morodomi rivers and 0.032 for other small rivers.

To set the boundary conditions, measured flow rates just downstream of the Chikugo estuary weir were used at the upstream boundary of the Chikugo River, and measured water levels at the mouths of the Chikugo and Hayatsue rivers were used for the downstream boundaries. For the Shozu, Kiritohshi, and Iryu rivers, measured water levels at observation stations were converted to flow rates using the cross-sectional shape and Manning equation, as shown in Figure 10.4. Figure 10.5 shows measured flow rates just downstream of the Chikugo estuary weir and measured water levels of three tributaries' upstream boundaries and the mouth of the Chikugo River.

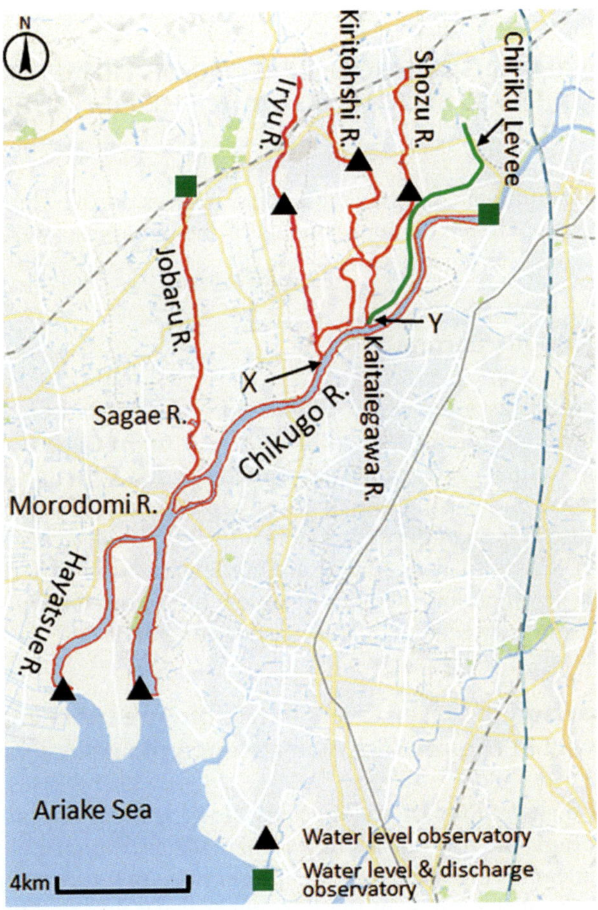

Figure 10.4 Analyzed rivers and observatories.

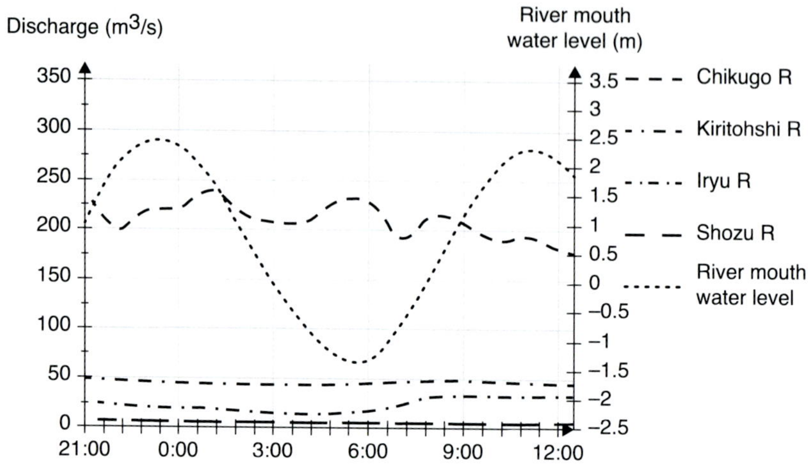

Figure 10.5 River discharges and water levels at the river mouth during the analyzed period (July 17–18, 2007).

10.3 RESULTS AND DISCUSSION: EFFECTS OF TRIBUTARIES ON WATER LEVEL FLUCTUATION IN THE MEANDERING WATERWAY

In order to verify the influence of the tributaries' inflows on the water level in the old meandering waterway, virtual channel models of the tributaries were set up, and the case models were compared using water levels.

10.3.1 Case used for comparison

Figure 10.6 shows the cases for which calculated water levels were compared. Case A models the current situation (left cell of Figure 10.6). In Case B, the flow path of the Iryu River is modified and connected directly to the Chikugo River. Case C is a model in which the path of the Shozu River is changed to connect directly to the Chikugo River. Case D is a confluent model in which both the rivers flow directly into the Chikugo River. Cases E, F, G, and H are models that do not include inflow from the Kiritohshi River, which has the highest inflow of the three tributaries. Table 10.2 shows the relationship between the inflowing rivers and each case.

10.3.2 Comparison of water levels in the old meandering waterway in the presence or absence of tributaries

The longitudinal distributions of water levels along the old meandering waterway were compared at high and low tide for cases where water from one of three rivers that flow into the old meandering waterway was present or absent. Among the three tributaries, the influence of the inflow of the Kiritohshi River was the largest. Figure 10.7 compares the water level distributions along the old meandering waterway calculated

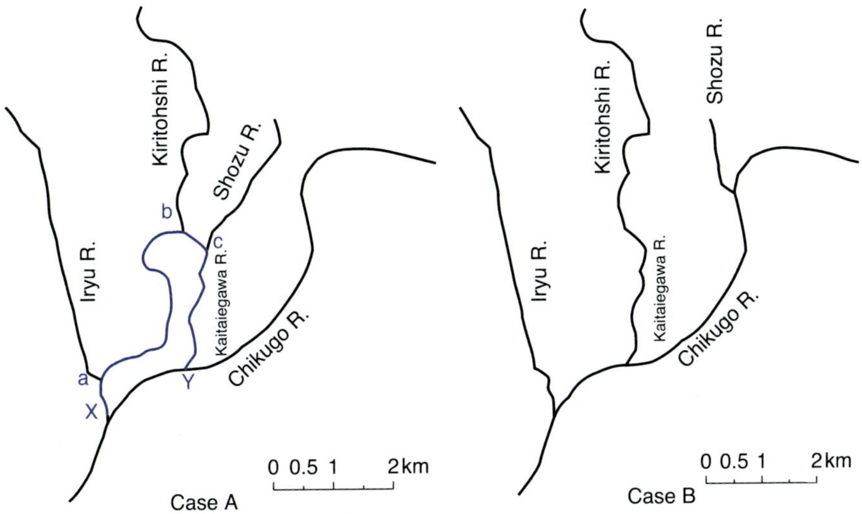

Figure 10.6 Model cases A and B used for comparison.

Table 10.2 Simulated cases. Presence and absence of tributaries' inflow into the old meandering waterway

Name of tributary	*A*	*B*	*C*	*D*	*E*	*F*	*G*	*H*
Shozu R.	○	○	×	×	○	○	×	×
Kiritohshi R.	○	○	○	○	×	×	×	×
Iryu R.	○	×	○	×	○	×	○	×

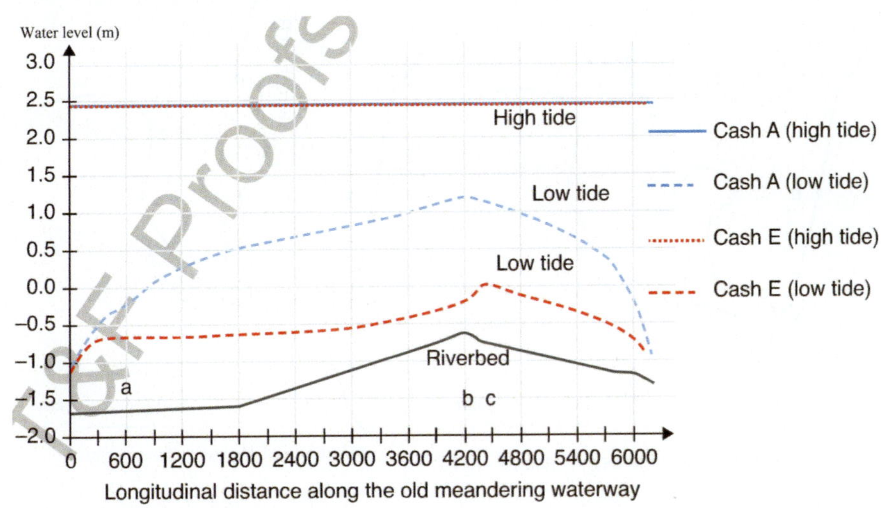

Figure 10.7 Water surface profile differences due to inflow from the Kiritohshi River to the meandering watercourse (Case A: inflow assumed, Case E: no inflow).

under this condition. The abscissa of the figure shows the part of the flow path of the old meandering waterway between X and Y in Figure 10.6 (The points where the Iryu, Kiritohshi, and Shozu Rivers flow into the Chikugo River along the way are taken as a, b, and c, respectively.). The maximum water level difference is about 1.4m, approximately 400m downstream from the confluence of the Kiritohshi and Chikugo rivers. Since water level decreases are also seen at the confluence, the water level of the Chikugo River can also rise due to the river's flow into the old meandering waterway. This seems advantageous for transfer points and other river transport issues on the Chikugo River.

10.3.3 Effects of presence or absence of the old meandering waterways on water levels in the tributaries

In order to estimate how much the water levels of the tributaries themselves change because some amount of water is stored as the tributaries flow into the old meandering waterway, two cases were compared. One is the current state where the tributaries flow into the old meandering waterway. The other is a case in which the Shozu, Kiritohshi, and Iryu rivers flow directly into the main channel of the Chikugo River without temporary storage in the old meandering waterway.

Figure 10.8 shows the comparison. In case I, the route of the Shozu River is moved to the south, so that it connects directly to the Chikugo River at point Z without flowing into the old meandering waterway. Also, in case I, the Kiritohshi River flows from point Y to the Kaitaiegawa River, avoiding the western half of the meander. In contrast, in case A, it flows from Y along the west-half of the route, bypassing the eastern half of the meander. In case A, we also considered the confluence with the Iryu River. The cross-sections between point Z on the Shozu River and its confluence with the

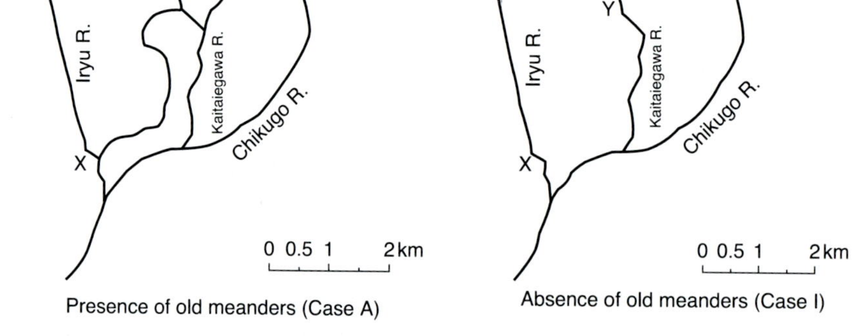

Figure 10.8 Cases taking the presence or absence of tributaries' inflows to the old meandering waterway into account.

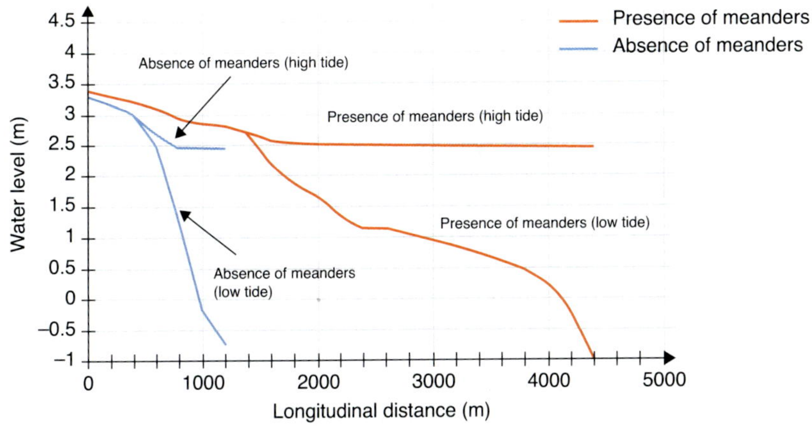

Figure 10.9 Effects of the presence or absence of old meanders on the Shozu River's water surface profile.

Chikugo River and the cross-sections from point Y on the Kiritohshi River to its confluence with the Kaitaiegawa River were linearly interpolated.

Figure 10.9 compares the calculated results for the water levels of the Shozu River when the old meandering waterway is present and when it is not. The abscissa of the figure shows the distance downstream from the water level observation point shown in Figure 10.4. The route of the Shozu River was switched to the old meandering waterway at location Z, shown in Figure 10.8. The calculations show that the water surface profile for the Shozu River is moderated because of the existence of the old meandering waterway. It is thought that switching the river's route increased the flow path length, which resulted in a milder riverbed slope. In addition, comparing the water level differences at high and low tides showed that the difference is smaller when the old meandering waterway is present (Figure 10.9). It is thought that using the old waterway buffers the influence of the tide.

For both the Shozu and Iryu rivers, water levels are lower when there is no temporary storage in the old meandering waterway, while the Kiritohshi River becomes higher when there is no temporary storage. This is explained as follows: Storing water in the old meandering waterway increases the water level, but the water comes from the Kiritohshi River. Moreover, letting the three rivers flow into the old meandering waterway benefits the Kiritohshi River as a means of flood control, and benefits the Iryu and Shozu rivers by improving fresh water intake and allowing river transportation.

A comparison of the longitudinal distribution of the water level for the Kiritohshi River with and without the old meandering waterway is shown in Figure 10.10. When the meanders are available, the water level drops because the water diverges to the west and east. The old meandering waterway functions as a buffering device by providing temporary storage against the tide of the Ariake Sea.

10.4 CONCLUSIONS

We conducted a one-dimensional unsteady flow analysis in the downstream area of the Chikugo River. We examined the significance and function of the old meandering

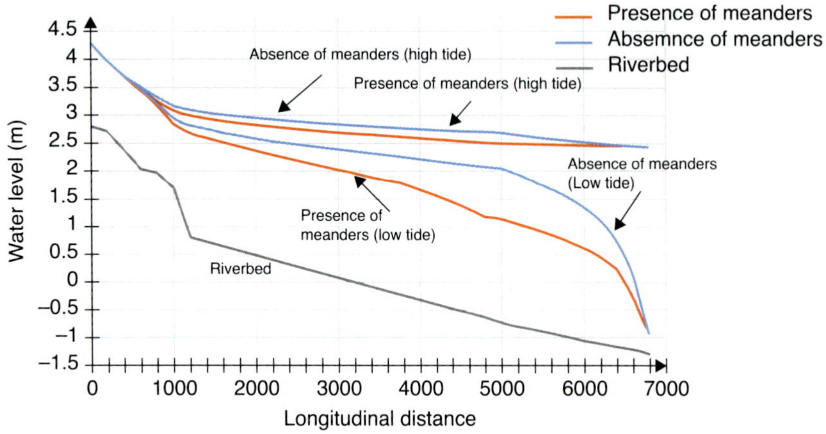

Figure 10.10 Comparison of water surface distribution along the Kiritohshi River depending on the presence or absence of old meanders.

riverbed, and the interactions between the tributaries and the old path. The findings are as follows:

1. Inflows to the old meandering waterway from the Kiritohshi, Shozu, and Iryu rivers, in that order of influence, raise the water level in the old meandering waterway significantly.
2. Plural tributaries into the old meanders have the effect of raising each other's water levels.
3. Re-routing the tributaries into the old meandering riverbed was done for river transportation and a freshwater intake.
4. As the Kiritohshi River flows into the old meanders, it raises the water level of the Shozu and Iryu rivers through them.
5. The old meandering waterway has the effect of buffering the influence of the tide on the tributaries.

From these findings, the old meandering waterway was deemed to have been used, because the direct inflow of the tributaries to the Chikugo River increases the water level fluctuations and affects river transport and freshwater intake, even though the flow rate for the Chikugo River was increased by constructing many cut-off channels. By utilizing the old meandering waterway, it was possible to secure a lengthened period with high water levels, thereby extending the time that freshwater intake could be done and reducing the flow rate to better accommodate river transport. This beneficial use of water flow in this area is based on understanding the characteristics of the large tidal difference for the Ariake Sea and indigenous knowledge about utilizing the natural terrain, such as using the meandering waterways to their maximum advantage.

ACKNOWLEDGMENTS

This research was conducted in part by the River Foundation under a research grant in FY2016 titled "Research on regional integrated flood control in the clan government era of the Chikushi Plain (research representative: Nobuyoshi Kishihara)." In carrying out the research, the Chikugo River Office of the Ministry of Land, Infrastructure, Transport and Tourism, and the Saga Prefectural Public Work Office provided valuable data and documents. We deeply appreciate the people and organizations concerned.

REFERENCES

Chikugo River Construction Office, Kyushu Regional Construction Bureau. (1976). *50-Year History of the Chikugo River* (in Japanese).

Culture Heritage Online. (2018). http://bunka.nii.ac.jp/heritages/heritagebig/149314/1/1

Eguchi, Tatsugoro. (1977). *Water and Soil of Saga Plain.* Water Management by Naritomi Hyogo, Supervised by Yonezo Miyachi, Shinpyosha (in Japanese).

Igarashi, Tsutomu. (1996). Reconstruction of the irrigation system in the Ariake northern lowland - Regarding the reorganization of "Ao irrigation" as tidal irrigation. *Research on Lowland Technology*, 5: 59–74 (in Japanese).

Kase River Irrigation and Drainage Office, Kyushu Agricultural Administration Bureau. (1973). *A History of Irrigation and Drainage of Kase River* (in Japanese).

Kishihara, Nobuyoshi. (2016). Research on regional integrated flood control in the clan government era of Chikushi Plain (1) Regional integrated flood control in Saga Plan. *Research on Lowland Technology*, 25: 19–26 (in Japanese).

Miyachi, Yonezo. (2009). Creeks connecting the sea, the river, and the land, water use cultivating Saga Plain, Integration of River Engineers. *Culture of Water*, 32: 34–39, Mitsukan Water Culture Center (in Japanese).

Nabeshima Houkoukai. Hizen map of Keicho period (1596–1614).

Nabeshima Houkoukai. Hizen map of Shouhou period (1644–1647).

Ohgushi, Koichiro and Yuki Nagahata. (2016). *A hydraulic study on the use and application of Jikken-Bori-Kawa canal, Saga City, Japan*, Proceedings of the International Symposium on the History of Indigenous Knowledge, Nov. 11–18, Saga, Japan.

Teramoto, Noriyoshi, Akira Ushijima and Masato Nakazono. (2016). *Relationship between creek distribution density and irrigation form - Research on the distribution characteristics of the creek in Chikushi Plain (3)*, Proceedings of Chugoku Branch Research Workshop, Architectural Institute of Japan, 39: 741–744 (in Japanese).

Chapter 11

Effect of open dyke for flood disaster mitigation in Kyoto

Taisuke Ishigaki and Michiko Hayashi
Kansai University

Ryuji Kawanaka
Hydro Technology Institute Co. Ltd.

CONTENTS

11.1 INTRODUCTION

Traditional countermeasures for flood disasters in Japan are listed in Table 11.1. These are divided into structural and nonstructural measures, as shown in Table 11.1. Structure, person, and wisdom are keywords to understand each measure's effects. Nonstructural measures are based on self-help and mutual help actions concerning predicting and mitigating damages and the safe evacuation obtained by people's experiences. And structural measures are based on mutual and public help actions and have been worked up by trial and error. These structures are not irrefrangible but transformable during devastating floods and reconstruction after floods.

Table 11.1 Traditional countermeasures in Japan

Structural measures:
Dyke, open dyke, grove, flood plain, groin, weir
Nonstructural measures:
Forecast: legend, divine message, proverb
Protection for persons: village community
Land use control: keep area open
Floodproofing constructions: waterproofing house

DOI: 10.1201/9781003222736-13

The first river work for disaster mitigation was done by Emperor Nintoku in the 4th century, and river works were constructed until the 19th century. Some works are still functioning in Kyoto and other districts of Japan. These can mitigate flood damages but not prevent them perfectly; however, the natural environment around rivers is preserved as a result. Their function has been investigated qualitatively but not quantitatively. To study them quantitatively, it is necessary to study the function and availability using numerical or hydraulic models.

One of the structural measures, the open dyke system, is discussed here. Embankment construction technology began in the 4th century with the Manda Dike built along the Yodo River by the Emperor Nintoku. Later, warlords in the late Warring States period ingeniously flood control their territories by allowing river water to flow through breaks in the dikes when the water level was rising and return to the river through breaks in the dikes downstream when the water level was falling. It is said that the first person to build an open dyke system was Lord Takeda Shingen in the 16th century, and he built it as a unique dike for rivers flowing through the fan-shaped area of the Kofu Basin. It is said that these open dikes are used in steep-slope rivers, but the date of construction of the open dikes targeted in this chapter is unknown, but they are used in mild-slope rivers. Authors have studied the open dyke system in the study area before (Ishigaki & Kawanaka, 2007; Kawanaka et al., 2008); however, evacuation safety has not yet been studied. Inundation data by a two-dimensional shallow flow model is used in this chapter, and the flow condition and evacuation safety are discussed.

11.2 STRUCTURAL MEASURES IN KAMEOKA BASIN, KYOTO

Kameoka Basin is on the west side of Kyoto, and the area is about 40 km^2, as shown in Figure 11.1. This basin was under sea level 1 million years ago, so alluvial plains were formed there. The Katsura River runs through the northwest to the southeast basin and connects to a narrow valley called "Hozu Gorge." This gorge works as a constriction during floods, and the backwater extends over the basin. As the flooded water sometimes covered the lowland area, the village and ancient road are located on the terrace caused by a slip on a fault on the east-side mountain slope. A masonry embankment called "Iga-Bane" was built to mitigate flash flood damage on this mountain slope, as shown in Figure 11.2a (Ishigaki et al., 2020). The Katsura River was used as a

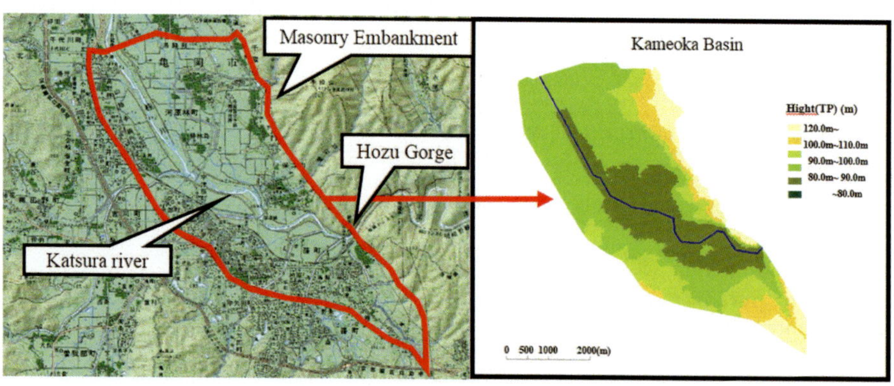

Figure 11.1 Study site in Kameoka Basin and Katsura River.

Figure 11.2 Masonry structures in Kameoka: (a) embankment, (b) attracting groin in "Hozu-Gorge."

water transportation system for woods and goods for the old capital of Japan, Kyoto, so some villages were developed along the river. Many attractive groins made of stones were set in the river to maintain the navigation channel, and some groins kept those shapes for more than 400 years (Figure 11.2(b)). The function of these groins was discussed in our previous works (Ishigaki et al., 2004, 2008).

The solid line in the figure shows the study area of the previous works (Ishigaki and Kawanaka, 2007; Kawanaka et al., 2008) to investigate the inundation process. The discontinuous bank of which upstream end is opened and it is connected to the downstream bank with a small angle of 10°–30° in this area. Then, the river water flows through open dykes to the surrounding flood plains and returns to the river through another open dyke after flooding. There are few houses near the river, a low-lying area inundated during floods.

The location of the studied discontinuous dyke is on the left bank of the Katsura River, as shown in Figure 11.3(a). This area is surrounded by the dyke of about 2.5 m

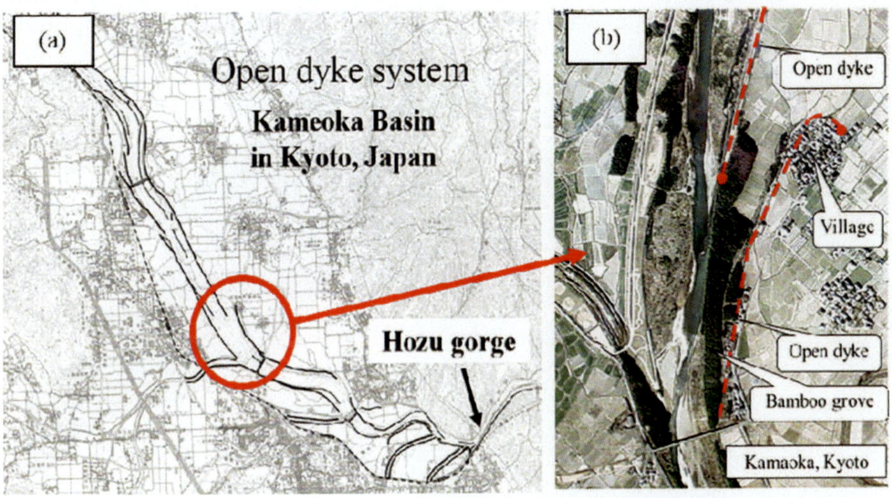

Figure 11.3 Open dyke system: (a) location along the Kizu River, (b) the studied open dyke.

Figure 11.4 Village surrounded by open dyke: (a) dyke, (b) raised-up house.

in height in Figure 11.4a, and some houses there are raised about 0.5 m to mitigate the damage of inundation in Figure 11.4b. Figure 11.3b shows that the open dyke is covered with a bamboo grove that absorbs the flood flow attack to the bank, and settles down the sand and rubbish in the flooded flow. This chapter discusses the effect of this open dyke on the adjacent villages.

11.3 METHOD OF INUNDATION CALCULATION

To calculate the inundation depth and velocity, a 2D shallow flow model was used with nonstructural mesh in the red line contour in Figure 11.5. The west side boundary was set along the right bank of the Katsura River, and the east boundary was set considering the ground elevation. A 5 m mesh data of ground elevation was used, as shown in Figure 11.5. The length of the triangle mesh is about 5 m, and the number of mesh is 179,520, of which Manning's n is 0.04 (natural river with trees). In villages A and B, the effect of the open dyke is thought about the difficulty of evacuation by using calculated flow depth and velocity.

The boundary condition at the upstream end is given with the river discharge obtained by the kinematic wave method. Figure 11.6 shows the hyetograph on

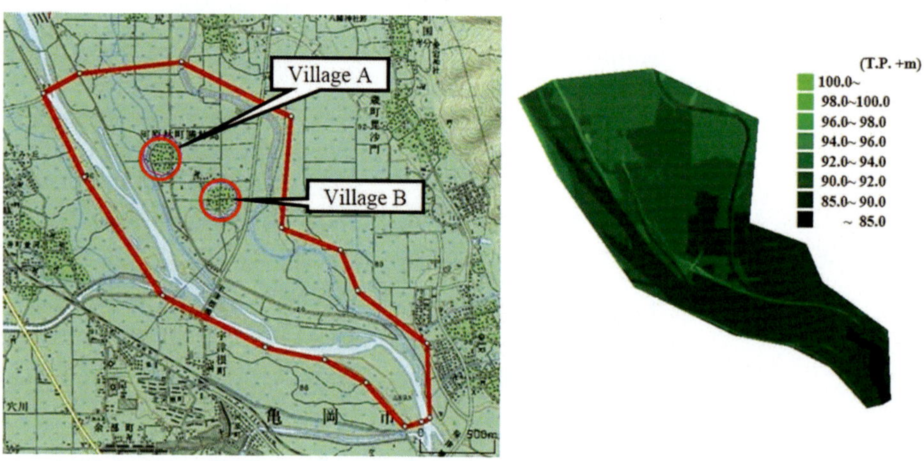

Figure 11.5 Calculating area and its elevation.

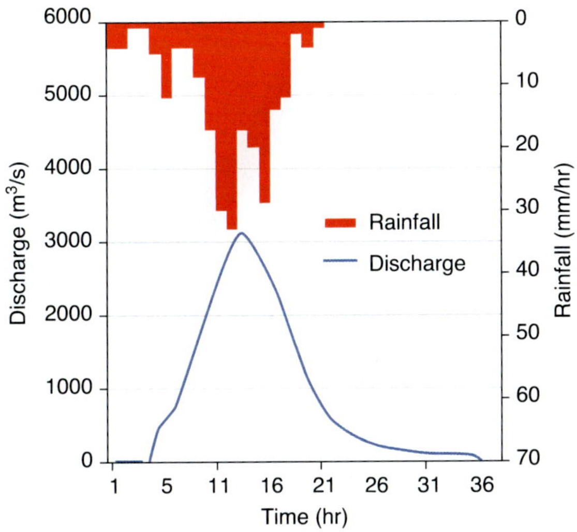

Figure 11.6 Rainfall and river discharge at the upstream boundary.

25 September 1953 and the calculated hydrograph. The rainfall caused the major flood disaster in this area. Total rainfall was 290 mm in 21 hours; the damage was four dead, 25 houses and 143 bridges washed away, 157 houses destroyed, 3,031 houses inundated, and 619 dykes destroyed. The hydrograph was calculated with 592 km^2 composed of 552 blocks using the kinematic wave method. The peak discharge is 3,100 m^3/s, almost equivalent to the design discharge of 3,500 m^3/s. Flow out discharge through the downstream end was set using Manning's formula with Manning's n of 0.04 and bed slope of 1/1,000.

11.4 EFFECT OF OPEN DYKE

The inundation process from the beginning of outflow to the maximum situation is shown in Figure 11.7. Flood flow runs through the open dyke, spreads to the flood plain, and flows over the small river. Some part of village A behind the open dyke is not inundated. All area of village B is inundated, but the inundation depth is not so deep. Figure 11.8 shows the distribution of velocity vectors at the beginning and the maximum situation. In village A, velocity is under 1.0 and 0.5 m/s on the flood plain.

By using these data, the criteria for safe evacuation can be estimated. The authors propose this criterion based on the evacuation test by using a real-size model of stairs and corridor installed in the Ujigawa Hydraulics Laboratory of Kyoto University (Ishigaki et al., 2010, Asai et al., 2010). Two kinds of evacuation tests were carried out. One test is a climbing test through a staircase model, and the other is a walking test through a corridor model. These tests' results show that evacuation safety during inundation is concerned with water depth and flow velocity. This means that the specific force per unit width, composed of drag force and hydrostatic pressure, can be used as the criterion for safe evacuation.

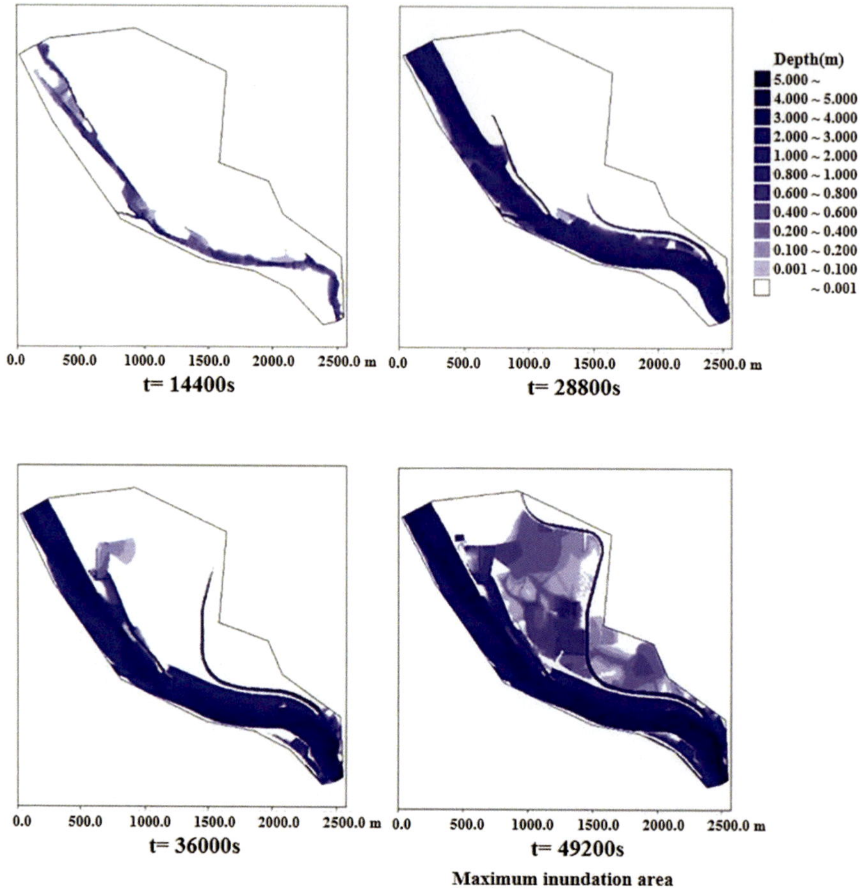

Figure 11.7 Calculated results of inundation depth and its process.

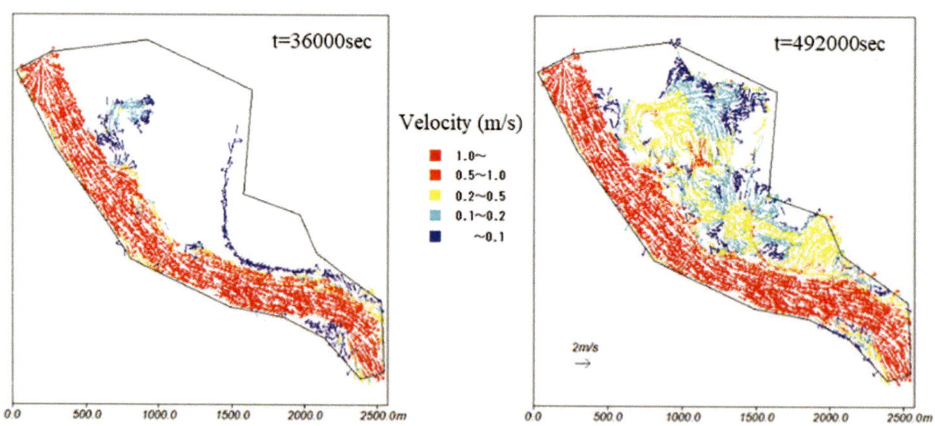

Figure 11.8 Distribution of velocity vectors.

Table 11.2 Criteria of safe evacuation presented by the specific force per unit width, M_0 (m^3/m)

	Limit of safe evacuation	*Difficult without any help*
Male	0.125	0.250
Elderly male	0.100	0.200
Female	0.100	0.200
Elderly female	0.080	0.160

The following equation calculates the specific force per unit width in Table 11.2, where V is velocity, h is the water depth, and g is the gravity acceleration.

$$M_0 = \frac{V^2}{gh} + \frac{h^2}{2} \tag{11.1}$$

The specific force is made up of two kinds of force. The first term in Eq. (11.1) is hydrodynamic force, hydrostatic force. The value of the limit of safe evacuation means the maximum magnitude that people can walk through flooded water without any help. This value depends on gender and age and is categorized into two situations. If the flow condition is over the value, people would sometimes be held up and swept away. Moreover, people cannot evacuate if the specific force per unit width exceeds the difficulty value without help in Table 11.2.

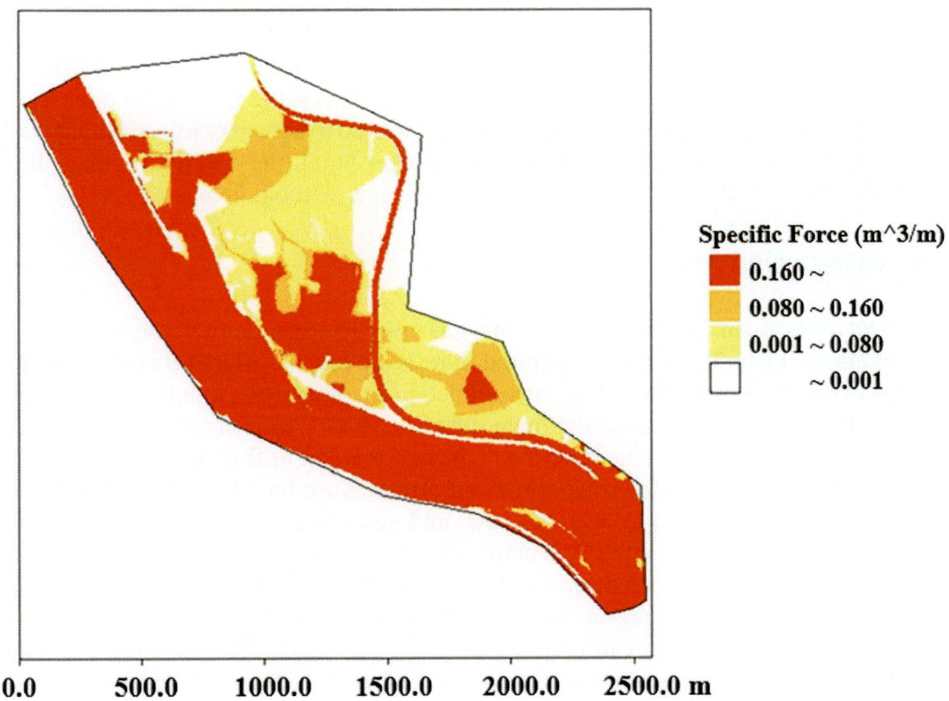

Figure 11.9 Distribution of the maximum criteria value of safe evacuation.

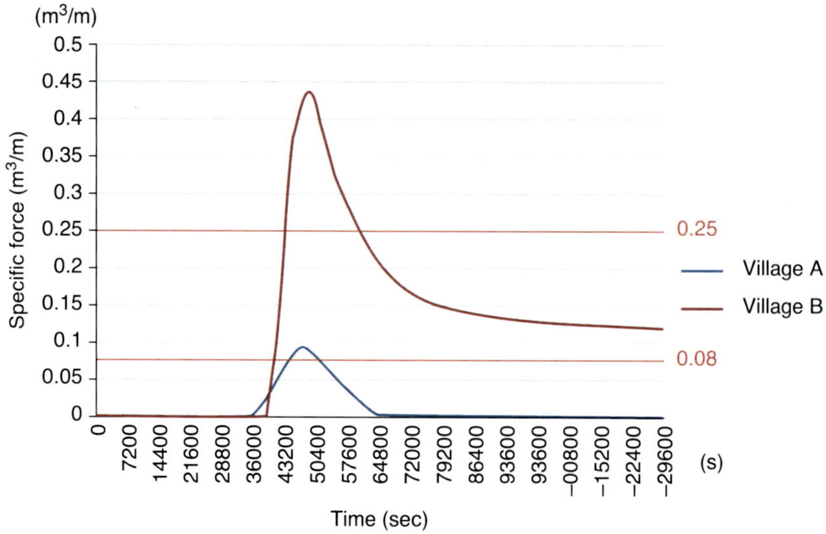

Figure 11.10 Time variation of safe evacuation criteria at villages A and B.

Figure 11.9 shows the distribution of the maximum value of the criteria. The minimum condition in which all inhabitants can make a safe evacuation is that the criteria should be under 0.080 for an elderly female. In most areas, the criteria are under 0.080, but in some places, the criteria are over 0.160. This means that there is a dangerous area in the flood plain. The time variation of criteria at the evacuation places in villages A and B is shown in Figure 11.10. At the evacuation place in village A, the criteria are just under 0.080, so all inhabitants can stay there. This is the effect of an open dyke system. This village is surrounded by the dyke, and the dyke works as a training wall and changes the direction of flood flow. On the other hand, the criteria in village B is over 0.25, which means that all people cannot evacuate without help.

11.5 CONCLUSIONS

The effect of the open dyke system in Kyoto was investigated quantitatively using a numerical model. From the results, it is found that the open dyke effectively mitigates the damage of inundation. In the village surrounded by an open dyke, inhabitants can stay in the evacuation place because the open dyke works as a training wall and changes the direction of flood flow. This function is the major effect of this open dyke system.

The present structural measures for flood disasters are based on prevention, mitigation, and resilience against flood disasters, and governments mainly conduct those. In contrast, traditional measures were mainly developed by inhabitants who lived in the flooded area, rebuilding them after the disaster. Traditional countermeasures consist of green infrastructure and preserving the natural environment around them, so these measures could be used if they were investigated quantitatively. These measures have a possibility that could be one method for ecosystem-based disaster risk reduction.

ACKNOWLEDGMENTS

The authors express their sincere appreciation for the support of the Kameoka City Office staff and the inhabitants who provided the data concerning the open dyke system.

NOTATION

g = gravity acceleration (m/s^2)
h = water depth (m)
M_0 = specific force per unit width (m^3/m)
V, u = flow velocity (m/s)

REFERENCES

Asai, Y., Ishigaki, T., Baba, Y. & Toda, K. (2010). Safety analysis of evacuation routes considering elderly persons during underground flooding, *Journal of Hydroscience and Hydraulic Engineering*, 28(2): 15–21.

Ishigaki, T. & Kawanaka, R. (2007). *Traditional flood management in the Kameoka Basin by open dyke system.* Pre-Conference Paper Volume of the International Conference on Water and Flood Management, ICWFM2007, Dhaka, Bangladesh, March, 2, pp. 479–485.

Ishigaki, T., Ueno, T., Rahman, M.M. & Khaleduzzaman, A.T.K. (2004, June). *Scouring and flow structure around an attracting groin.* Proceedings of 2nd International Conference on fluvial hydraulics, Naples, pp. 521–525.

Ishigaki, T., Asano, T. & Kawanaka, R. (2008). *Stability of old groins in the Katsura River during floods.* Proceeding of the International Conference on Scour and Erosion, Kyoto, Japan, November, pp. 310–315. (on CD-ROM)

Ishigaki, T., Asai, Y., Nakahata, Y., Shimada, H., Baba, Y. & Toda, K. (2010). Evacuation of aged persons from inundated underground space, *Water Science & Technology*, 62(8): 1807–1812, IWA Publishing.

Ishigaki, T., Kawanaka, R. & Hayashi, M. (2020). *Effect of masonry embankment located on mountain slope called as "Iga-Bane."* E-Proceeding of the 22nd IAHR-APD Congress 2020, September 7-1, Sapporo, Japan.

Kawanaka, R., Ishigaki, T., Kuroki, Y. & Ono, J. (2008). *1-D and 2-D simulation of the flow around discontinuous dyke.* Proceeding of 16th IAHR-APD Congress, Nanjing, pp. 993–998.

Chapter 12

Flood control strategy in Japan during the Edo period (the early 17th to mid-19th century)

Tadaharu Ishikawa
Tokyo Institute of Technology

Ryosuke Akoh
Okayama University

Hiroshi Senoo
TOKEN C. E. E. Consultants Co., Ltd.

CONTENTS

12.1 INTRODUCTION

At the beginning of the 17th century, Japan entered a stable era called the Edo period that was about 250 years long. Edo is the old name of Tokyo. Japan's population increased from about 10 million to 30 million due to the expansion of agricultural land and developing commerce and industry in cities. This social development was realized through the benefits of flood disaster mitigation works carried out on alluvial plains. Because civil engineers at the time could not build large enough embankments

DOI: 10.1201/9781003222736-14

to prevent river overflows, they designed levee systems to induce deliberate and safe river overflows onto floodplains when floods exceeded the channels' capacity to avoid catastrophic flooding.

Because hydraulic facility design in the Edo period depended highly on topography and land use regional characteristics, studies done thus far about them have mostly been carried out individually from a historical civil engineering technology viewpoint based on old documents. However, there were some strategies for flood control in those days that were too commonly used to be recognized as important and were thus not included in the old documents. Therefore, in this paper, common flood disaster prevention strategies in the Edo period are discussed based on the results of previously published numerical simulations of the hydraulic functioning of different types of flood control facilities in three regions. The three regions are shown in Figure 12.1, and details of the relevant flood control works can be found in the references (Section 12.2: Akoh et al. 2017, Section 12.3: Ishikawa and Akoh 2019, Section 12.4: Ishikawa and Akoh 2016, Section 12.5: Ishikawa and Senoo 2020, 2021).

12.2 METHODOLOGY

An ordinary two-dimensional shallow flow equation was converted into a finite difference equation using the finite volume method with an unstructured triangular mesh system. All of the equations are given in Ishikawa and Akoh (2019). Because of the absence of data on floodplain topography in the Edo period, the latest ground surface elevation map with a spatial resolution of 5 m (Geospatial Information Authority of Japan, 2015) was used. Large-scale topographic changes that occurred in the 20th century were removed by interpolation from the surroundings. Assumptions made for the flood hydrographs are described in the following sections.

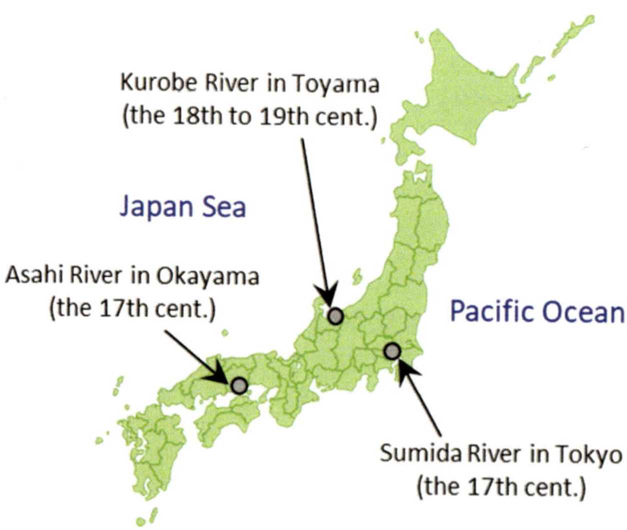

Figure 12.1 Study locations.

12.3 FLOOD CONTROL ON THE LOWLANDS OF EDO (TOKYO)

12.3.1 Overview of flood control facilities

In 1603, the political center of Japan was transferred from the Kansai Area, comprising Kyoto and Osaka, to Edo (now Tokyo), which was a new city built starting in 1590. After developing the waterway transportation network in the 17th century, Edo became Japan's largest economic, cultural, and consumption center. After the Edo administration disappeared due to political changes in 1868, Edo was renamed Tokyo, and in 1871 the new administration established Tokyo as the capital.

Figure 12.2a shows the first accurate map of Tokyo based on modern surveying, done in 1886 when the city's structure and land use had not yet changed substantially since the Edo period. Edo Castle and the ruling class dwellings were located on the diluvial terrace on the west side. The townspeople's residential areas were on the alluvial lowlands on the east side. Large-scale trading companies and warehouses were established on the banks of the Sumida River, which runs south toward Edo (Tokyo) Bay. To protect this area from flood damage, a large levee system was built in the upper reaches of the Sumida River, as shown in Figure 12.2b, the area of which is marked with a double rectangle in Figure 12.2a. Three levees were arranged in a funnel shape to dam up the floodwater from the upper reaches of the Sumida River, called the Ara River. The Nihon Levee, located apart from the river channel, was the most important among the three, and, therefore, the overall flood control works are called the "Nihon levee system" herein. The large, irregular plots upstream of the Nihon Levee represent paddy fields. The finely divided plots around the Asakusa area show the urban area at the northeastern end of Edo. The Nihon levee was built to a height of 3 m in 1621 and was removed in 1927.

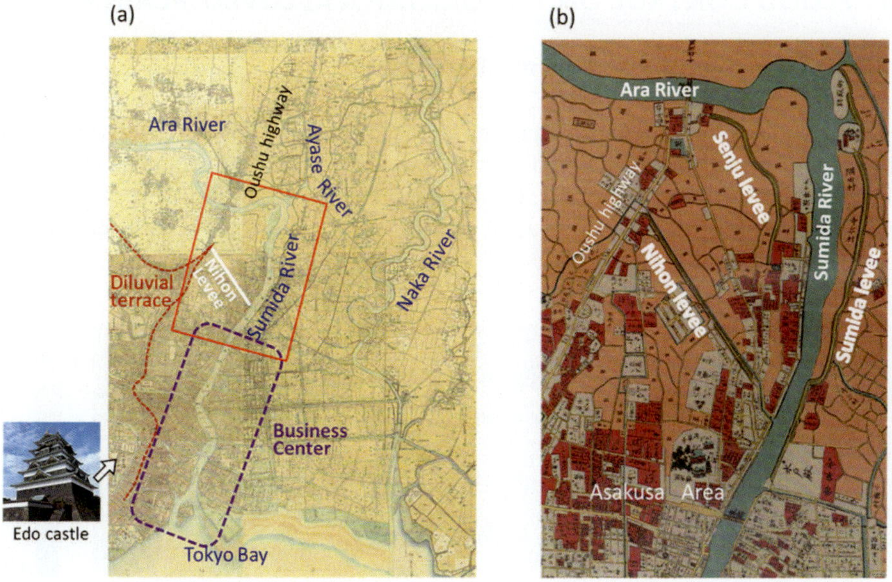

Figure 12.2 Sumida River flood control facilities.

It should be noted that the Ayase River joins the Sumida River on the upper right of Figure 12.2b. As shown in Figure 12.2a, a canal connects the Ayase River to the Naka River, flowing east toward Tokyo Bay. These waterway connections were mainly used for water transport but may have served as drainage channels during floods.

12.3.2 Results of numerical flow simulation

The earliest modern river plan, issued in 1911, when there were no continuous levees in the upper Ara River watershed, which had existed in the Edo period, assumed the flood discharge for levee design at 2,500 m^3/s. In the numerical simulation, therefore, the hydrograph of a large recently observed flood was used after it had been flattened to a peak discharge of 2,500 m^3/s, while keeping the total water volume unchanged.

Figure 12.3 shows the spatial distribution of inundation depth at three points in time. The locations of the levees and waterways and land use status are shown in the left figure. The flood covered the upstream floodplain between the Kumagaya levee and the diluvial terrace in the rising phase. In the flood peak phase, the inundated area expanded to the north and east side paddy fields through the Ayase River and the canal that connected it to the Naka River. Finally, the water that flooded the paddy fields was discharged from the Naka River to Edo (Tokyo) Bay in the receding phase. The flood flow dispersion was induced by the water level rise upstream from the Nihon levee system. The water level rise was only 2 m, but it was enough to change the flood flow direction because the ground in the area was almost flat. For later discussion, please note that flooding from the Ara River branched eastward following the ground slope before constructing the Sumida levee, which was built to reduce the inundation of the paddy fields in the Naka River Basin.

12.4 FLOOD CONTROL ON THE OKAYAMA ALLUVIAL PLAINS

12.4.1 Overview of flood control facilities

Figure 12.4 is a pictorial map drawn in the late 17th century showing an overview of the Okayama Alluvial Plain, which had been formed by the Asahi River. The terrain generally slopes toward the south, and, in old times, the Asahi River diverged toward the south. In the early 17th century, branch streams were collected into one channel near

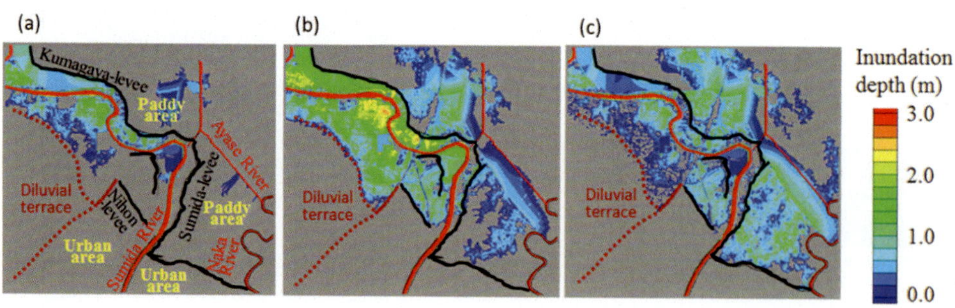

Figure 12.3 Spatial distributions of inundation depth during three phases.

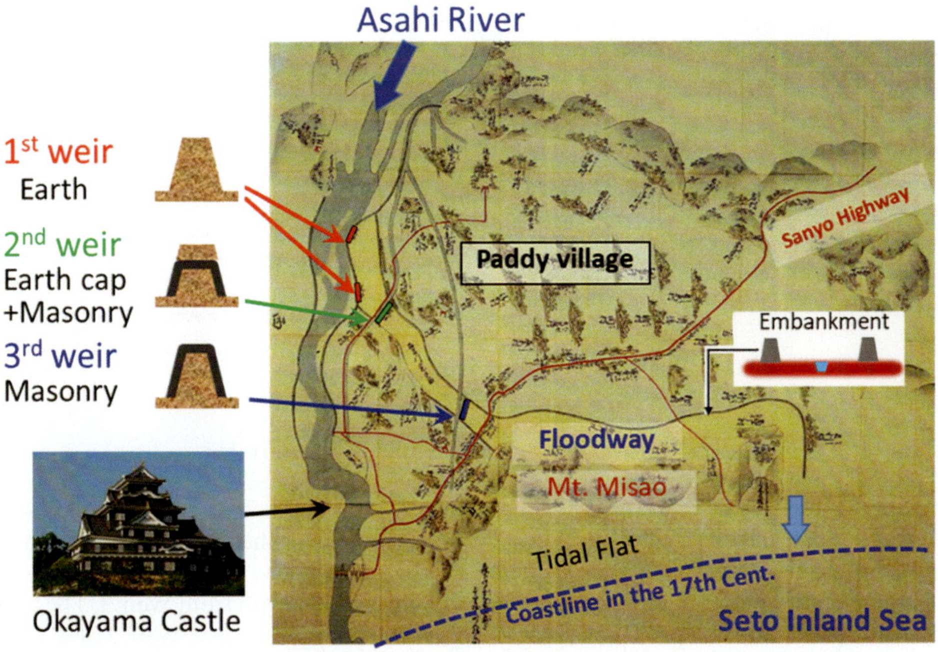

Figure 12.4 Floodway on the Okayama Alluvial Plain in the late 17th century.

Okayama Castle on the west side of the plain, and the old stream channels became used for irrigation. However, because frequent floods hindered the development of the castle town, the feudal clan government constructed a large-scale floodway toward the south, which discharged floodwaters to the Seto Island Sea from the east side of Mt. Misao.

Okayama River Management Office (1978) describe the process of floodway construction as follows: In 1669, two portions of the Asahi River embankment, being depicted as Weir 1 on the map, were lowered to induce river overflow. This earthen weir was to be washed down immediately after flood overtopping, but the backup levee behind Weir 1, on which Weir 2, a masonry structure with an earthen cap stably affixed at the location of overtopping, stopped flood dispersion. In 1670, a continuous embankment with a relative height of 2–3 m was constructed to form the outline of the floodway. Overflow water from Weir 2 was channeled into the floodway. The land in the floodway was generally used as paddy fields, and flashboard facilities were installed to close the levee openings to road traffic only during floods. Weir 3, constructed of masonry, was built in the floodway in the same year, but its function is not clear at present because of its destruction by a large flood in the 19th century.

12.4.2 Results of numerical flow simulation

After the construction of the flood bypass, while the frequency of flooding in the castle town decreased, it increased in the rural areas to the east. This "flood relocation" seems to have been intentional, as shown in Figure 12.5. Figure 12.5a shows the flow

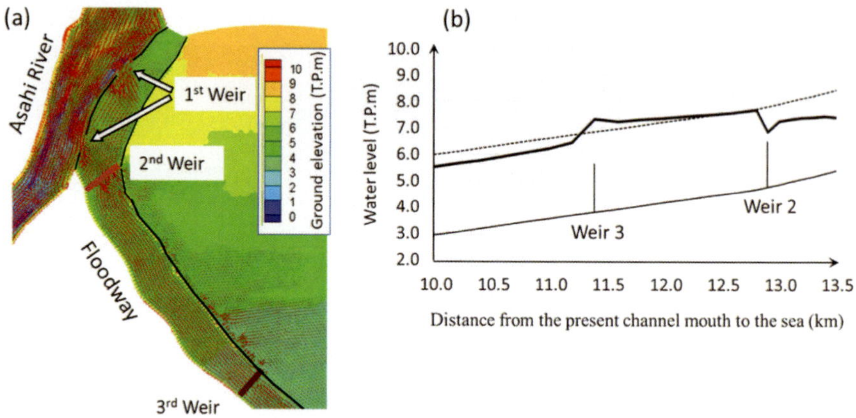

Figure 12.5 Flow field near the head of the floodway (Q = 3,200 m^3/s).

conditions near the head of the floodway in a large flood with a return period of about 12 years. The earthen Weir 1 had already washed out, and the flood intruded to the floodway through the two openings. The flow direction is rectified by Weir 2 toward the floodway. Note that levee overtopping was induced upstream from Weir 3. The longitudinal water surface profile plotted in Figure 12.5b suggests that Weir 2 and Weir 3 are arranged to intentionally confine the extent of possible levee breakage against floods that exceed the capacity of the floodway.

Results of 21 numerical simulations of various conditions showed that this branching to the floodway improved the safety of the castle town against flooding from a storm return period of 6.1 years to 8.7 years. But the results also suggested that flooding the eastern plain via intended floodway levee breakage resulted in further improvement from 8.7 to 12.2 years. In other words, intentionally flooding the eastside floodplain was an important part of the flood control measures for the castle town. It should be noted that the 17th century was a time when the weight of the economy in Japan shifted from agriculture to commerce and industry in the town.

Old documents indicate that farmers from the east villages occasionally petitioned the regional magistrate to reinforce the floodway levees. But the local government did not accept these entreaties but gave them tax remissions instead. This fact suggests that the regional administration gave comprehensive consideration to the farmers for the transfer of flood risk from the castle city to the farmland caused by the floodway system.

12.5 FLOOD CONTROL ON THE KUROBE ALLUVIAL FAN

12.5.1 Overview of flood control facilities

The Kurobe River is one of the fastest flowing rivers in Japan, running from Japan's central mountain range (3,000 m in elevation) to the Japan Sea via a deep canyon about 80 km long, carrying a large volume of sediment. The river forms a vast alluvial fan on the Japan Sea Coast, with a radius of 13 km, a slope of 1/100, and an apex angle of 60°.

Figure 12.6a shows the topography of the alluvial fan, on which the levee lines measured in 1894 and old river traces are also plotted in halftones. The levee lines were obtained from a modern survey carried out in 1894. The small numbers between the levee lines are the distance from the river mouth in km along the present river channel centerline. Figure 12.6b is a pictorial map drawn in 1785, the direction and scale of which is roughly adjusted to be the same as Figure 12.6a. However, a bit of disagreement was inevitable due to the geometric distortion of the pictorial map. The map contains many branch streams and discrete short levees, together with village names and highways.

The river improvement works on the alluvial fan were behind those on the alluvial lowland, because the river flow was very swift due to the steep terrain, and the main channel flow frequently changed among the many diverging streams. From the pictorial map, it appears that the branch streams were gradually cut off by short, discrete levees and became one dominant stream. The process of changing from the levees in Figure 12.6b to those in Figure 12.6a is not certain, but the levee construction was completed before the end of the Edo period.

12.5.2 Results of numerical flow simulation

Figure 12.7a shows the inundation area obtained from the numerical simulation, and Figure 12.7b shows the area estimated for the 1934 flood, the peak discharge of which was about 3,000 m^3/s (Teramura, 2008). Overall, the two results are similar: excess river water overflows from the upstream levee openings, down through the old river channel outside the left levee. The flow splits into two streams, one of which returns to the river channel through the downstream levee openings, while the other flows down toward the coast. The numerical simulations with different flow rates and river channel sandbar arrangements showed that the same flow pattern occurred on the right bank.

Figure 12.8 is a summary of the results of multiple numerical simulations. The excessive runoff from the canyon overflowed through upstream levee openings to old river traces adjacent to the river channel. The flow slowed there, and some floodwater returned to the main channel through the funnel-shaped levee openings in the middle zone and downstream. In other words, the old river traces, a little lower than their surroundings, were used as a long retention basin having exits downstream.

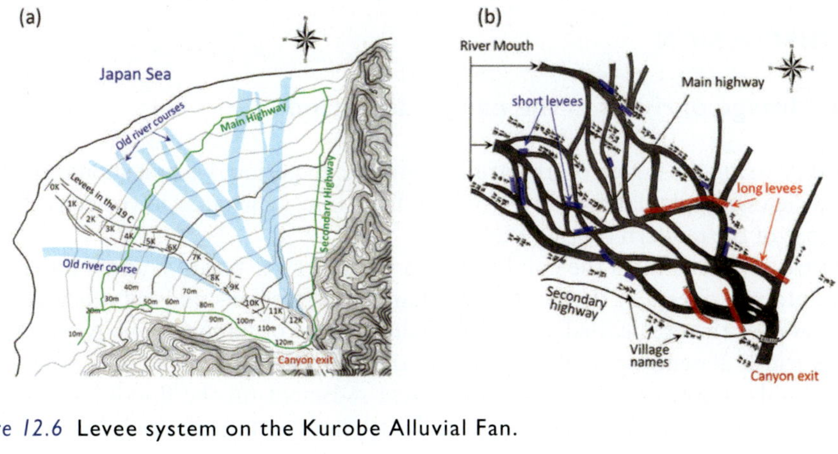

Figure 12.6 Levee system on the Kurobe Alluvial Fan.

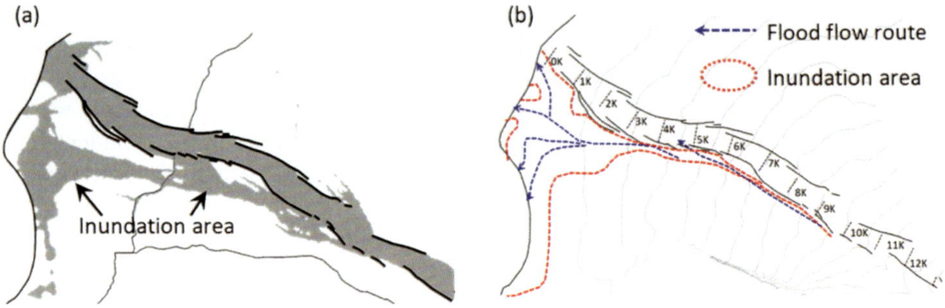

Figure 12.7 Inundation by 1934 flood ($Q = 3{,}000\ m^3/s$).

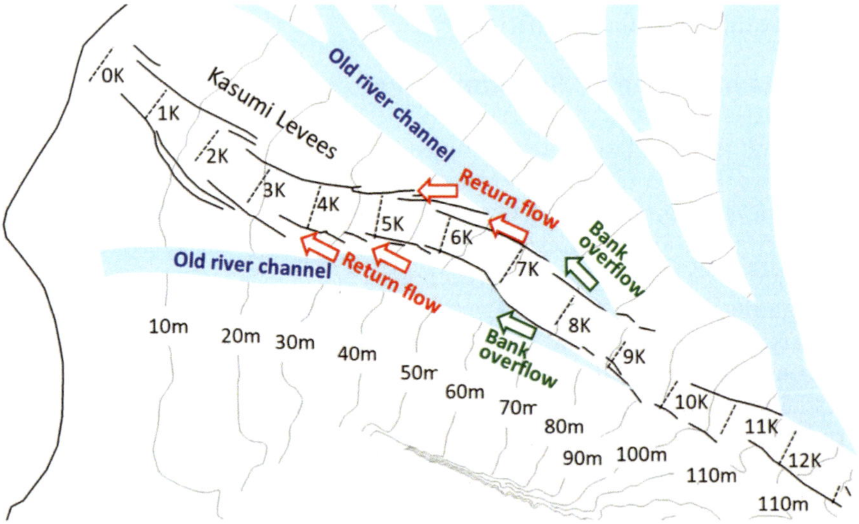

Figure 12.8 Function of levee openings on the Kurobe Alluvial Fan.

12.6 DISCUSSION

12.6.1 Image of rivers in the early Edo period

Before discussing the flood control strategy in the Edo period, we first should consider people's views of and relationship to rivers in that time, which may have been different from those of the modern times.

Figure 12.9 illustrates the condition of a natural river basin. Runoff water from the mountains collects in the main river channel via streams and tributaries, and it also flows over the area called a floodplain at the time of large runoff. A river's flow changes width depending on flow rate, and the floodplain becomes part of the river during floods. River overflow deposits coarse sediment on the floodplain's nearby riverbanks to develop natural levees and distributes the nutrient-rich fine sediment

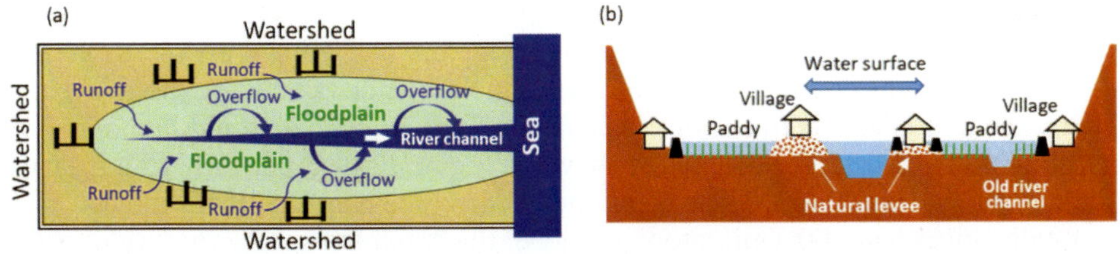

Figure 12.9 Images of rivers in early times.

over marshy lands behind. With the growth of human activity, villages were built on natural levees and the edges of floodplains. People created paddy fields in the marshy land and developed irrigation systems making use of old streams. In that sense, it can be said that they lived with or in part of the river. In addition, since transportation mainly depended on waterways at that time, commerce and industry in towns were also closely related to rivers. Therefore, civil engineers probably did not consider creating continuous embankments to separate rivers from the places people lived.

12.6.2 Common strategies for flood control in the Edo period

The flood control facilities described individually in Sections 12.3–12.5 seem quite different from each other. Still, eliminating the details, they all consisted of three parts as illustrated in Figure 12.10: (a) a river opening to divert flow toward the floodplain, (b) inundation space to disperse the flood, and (c) exits for drainage to the sea or river. Researchers in technological history often focus on (a) and (c), but the core of this strategy is the space (b) where floodwater is stored and flows out slowly. From this point of view, these facilities are a compound of retarding basins and flood bypasses, which are thought of as two different tools in the present flood control projects. Using the knowledge of modern hydraulics, these two kinds of facilities can be designed much easier than the compound facilities used in the Edo period. However, civil engineers in the Edo period did not even have the concept of flow rate, let alone knowledge of modern hydraulics. Therefore, the quantitative hydraulic design of these facilities was not possible. Perhaps their empirical knowledge was only the most basic principles of hydraulics, namely that water flows from high to low places, and that the strength of water's flow depends on the gradient of the water surface.

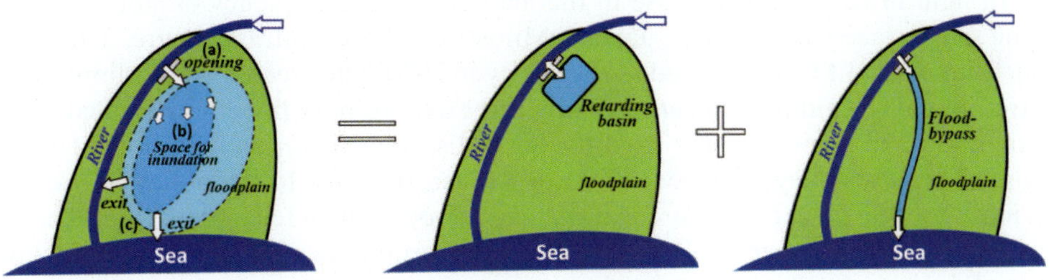

Figure 12.10 Images of compound flood control facilities.

Looking at this scheme from another perspective, it can be seen as an extension of the natural flood control shown in Figure 12.9a. In river basins in their natural condition, its width naturally expands toward the floodplain as the flow rate increases. The flood hydrograph flattens due to the difference in flow velocity between the river channel and the floodplain. If this simple mechanism were known, it would be possible to envision the effectiveness of the facility shown in Figure 12.10a, even only with the most fundamental hydraulic knowledge mentioned above. The people of that time may have thoroughly investigated the microtopography of the floodplains using the leveling technology available in the Edo period, correlated that information with past flood events, and designed the scheme in Figure 12.10a as a perturbation of those events. They could have assumed a path of branched flow basically similar to those that occurred in the past. To mitigate the damage, however, they would limit the inflow at the entrance and modify the flow on the floodplain a little using technology available at that time.

In the case of the Nihon levee system (Section 12.3), without the Kumagaya and Sumida levees, which were constructed to reduce the frequency of flooding in the paddy fields in the Ayase and Naka River basins, runoff from the upper Ara River would have flowed into the two basins following the terrain slope. So, means were devised to revert to the natural flood path only during large floods.

The design concept of the floodway on the Okayama Plain (Section 12.4) was similar. Before the Asahikawa embankment was constructed, a branched inundation flow flowed in the direction of Mt. Misao according to the topographical gradient. The floodway built in the 17th century was used to revive that natural flood path only in a major flood.

In the case of the Kurobe River (Section 12.5), the situation was more complicated than the other two areas, because the entire alluvial fan is covered with old river courses. The pictorial map of Figure 12.6b, drawn in 1785, shows that relatively long levees were concentrated in the upstream part. This fact suggests that they first prevented flood flow divergence at the uppermost section and then utilized only the old river courses adjacent to the main channel.

12.7 CONCLUSIONS

With westernization from the middle of the 19th century, flood control policies followed in the Edo period were replaced by the modern policy of surrounding the river channel with continuous levees. As a result, the frequency of river floods has decreased, but disasters caused by large outflows exceeding the designed channel capacity continue to occur and have been on the rise in the recent years.

In light of the recent increase in the magnitude and frequency of storms due to rising sea surface temperatures, Japan's Ministry of Land, Infrastructure, Transport and Tourism (MLIT) announced a new policy in 2020. They recognize the limitations of traditional river improvement works against extraordinary floods anticipated in the near future and emphasize the need for flood disaster mitigation measures in floodplains. The new policy of the MLIT is, in a sense, a return to the policy of the Edo period when the power of nature overmatched flood control technologies. Therefore, it is expected that further detailed analysis of more examples of flood control measures in the Edo period will give hints to the process of realizing the new policy.

ACKNOWLEDGMENTS

A part of this article is based on the paper titled "Hydraulic study on levee arrangement on Kurobe alluvial fan constructed in the early 19th Century" written by Ishikawa and Senoo and published in the proceedings of the 22nd IAHR-APD Congress, and its content has been expanded and modified to fit the IAHR monograph, by including the parts of three papers, Ishikawa and Akoh (2016), Ishikawa and Akoh (2019), and Ishikawa and Senoo (2021).

REFERENCES

Akoh, R., Ishikawa. T., Kojima, T., Tomaru, M., & Maeno, S. (2017). High-resolution modeling of tsunami run-up flooding: a case study of flooding in Kamaishi city, Japan, induced by the 2011 Tohoku tsunami. *Nat. Hazards Earth Syst. Sci.*, 17: 1871–1883.

Geospatial Information Authority of Japan. (2015). Available online: https://fgd.gsi.go.jp/download/menu.php

Ishikawa, T., & Akoh, R. (2016). Estimation of flood risk management in 17th century on Okayama Alluvial Plain, Japan, by numerical flow simulation. *Int. J. Saf. Secur. Eng.*, 6: 455–465.

Ishikawa, T., & Akoh, R. (2019). Assessment of flood risk management in lowland Tokyo areas in the seventeenth century by numerical flow simulation. *Environ. Fluid Mech.*, 19: 1295–1307.

Ishikawa, T., & Senoo, H. (2020). *Hydraulic study on levee arrangement on Kurobe alluvial fan constructed in the early 19th Century*. Proceedings of the 22nd IAHR-APD Congress 2020, Sapporo, Japan.

Ishikawa, T., & Senoo, H. (2021). Hydraulic evaluation of the levee system evolution on the Kurobe alluvial fan in the 18th and 19th centuries. *Energies*, 14: 4406. https://doi.org/10.3390/en14154406.

Okayama River Management Office (1978). *The history of Hyakken-gawa*, (edited by Taniguchi, S), Ministry of Construction (in Japanese).

Teramura, J. (2008). A study on function results of open levees when flood happened in Kurobe River. *Historical Studies in Civil Engineering, Japan*, 29: 43–50 (in Japanese).

Chapter 13

Changes in the historical river course and related flood risk in the Arakawa River basin in Japan and the role of still-existing secondary embankments in the recent 2019 flooding event

Norio Tanaka and Yoshiya Igarashi
Saitama University

CONTENTS

13.1 INTRODUCTION

Japan is an especially disaster-prone country, and flood disaster management practices have been continuously implemented. Matsuki (2012) classified three important periods of Japanese history according to their disaster management practices: (1) the Nara era (710–794), (2) the Edo era (1603–1867), and (3) since the Meiji Restoration (1868–) during which 'people started to suffer from flood disasters', 'local governments intervened in communities', and 'Japanese society was ruled by a central government,' respectively. In the Edo era, discontinuous open levees were installed in many rivers under the direction of local governments. Flood management techniques using a retarding basin (Ohgushi et al., 2016; Ishikawa and Akoh, 2018; Furuta and Shimatani, 2018) or a river course change were important countermeasures for the management of floods. Figure 13.1a shows examples of sites in which the retarding effects of discontinuous levees were recently re-evaluated (Okuma, 1987; Teramura and Okuma, 2005; Kawanaka et al., 2007; Nakashima et al., 2013; Furuta and Shimatani, 2018; Ishikawa

DOI: 10.1201/9781003222736-15

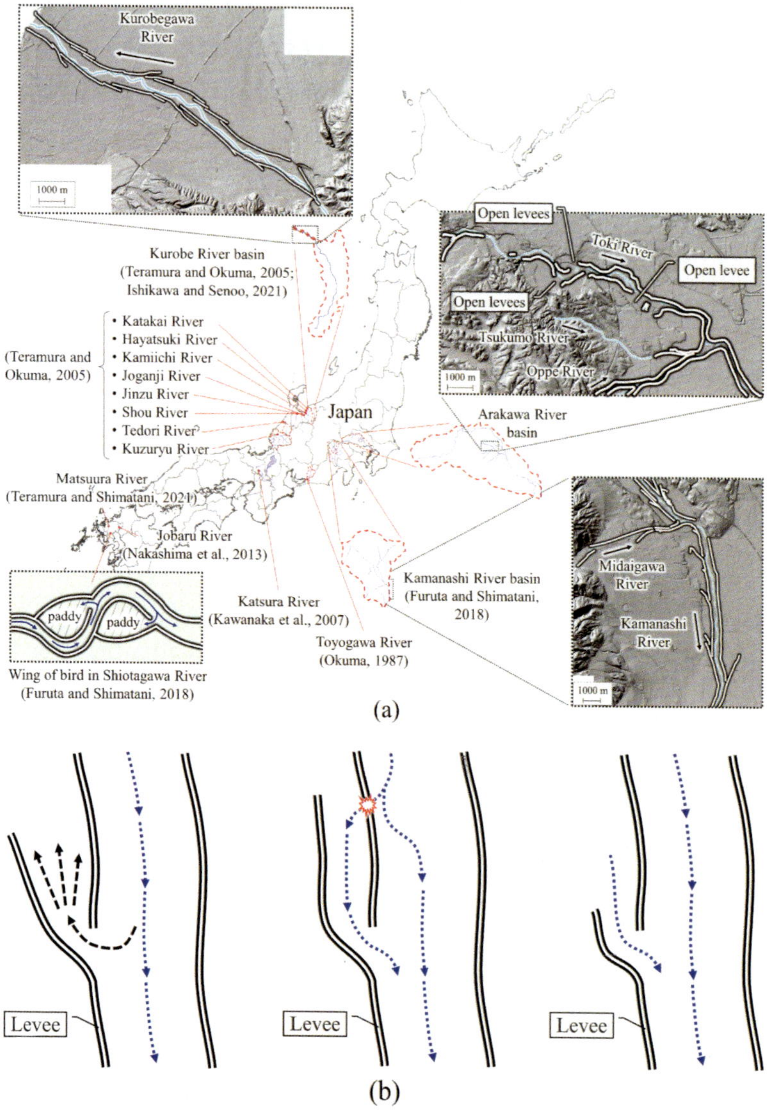

Figure 13.1 Discontinuous open levee system in Japan. (a) Examples of the retarding effects of discontinuous levees were recently re-evaluated, (b) schematic of the role of the discontinuous levee; left: flood storage and retardation, center: returning the overtopping flood water to the downstream river, and right: drainage of branch runoffs. (Modified from Teramura and Okuma, 2005.)

and Senoo, 2021; Teramura and Shimatani, 2021), including our target site, the Arakawa River. As shown in Figure 13.1b, discontinuous open levees not only store water during flood events but also return the water inundating the hinterland to the river. Levees were usually constructed manually in the Edo era, and they were not very high, around 1–3 m. Some of the open levees still exist, but their roles changed from being part of river levees to being detached secondary embankments after the construction of high continuous levees along the river.

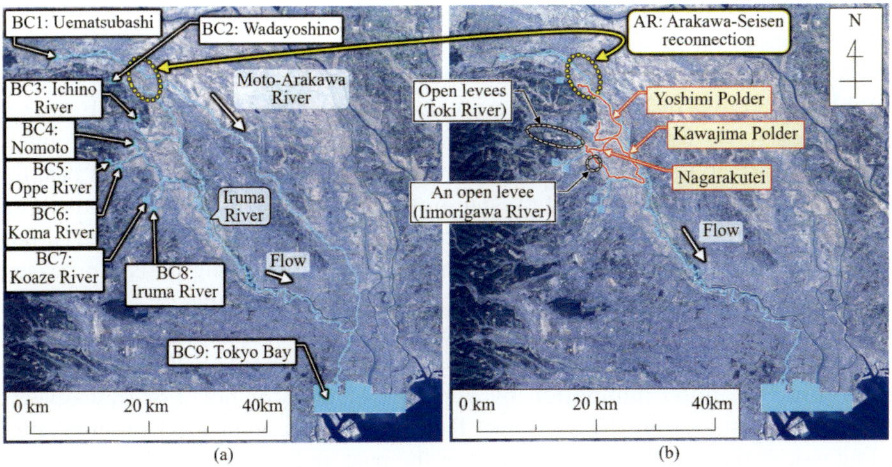

Figure 13.2 River network before and after the river course change in 1629 (Arakawa-seisen Reconnection: AR). (a) Before AR, (b) after AR.

Figure 13.2 shows the change in the river course in 1629, from the Moto-Arakawa River to the present Arakawa River in Japan (Arakawa-seisen Reconnection, hereafter AR) in the basin where Metropolitan Tokyo now exists. After the AR, flood inundation frequently occurred in the middle stream watershed of the Arakawa River basin, especially in the branch rivers shown in Figure 13.2a, at a frequency of around once in ten years, till around 100 years ago. Many secondary embankments for controlling the flood flow in the hinterland of river embankments were constructed. In the Yoshimi and Kawajima areas especially, polders (ring levees), shown in Figure 13.2b, were newly constructed and strengthened by heightening the levees, respectively, after the AR. Many river confluences are on the west side of the Kawajima Polder, where a discontinuous open levee system mainly manages floods. In the open levee area, including the area in between the Kawajima and Yoshimi Polders, many retarding ponds can be assumed to have appeared during flood events. After the Meiji Restoration, with progress in flood management, most of the open levees in this area disappeared. It is now changed to a continuous levee system, including 'gate' or 'gate and pump station' systems. However, upstream of the Toki River—the open levee system still exists. In addition, old embankments of the Oppe River and a part of the Kawajima Polder still exist as a secondary embankment, along the Iimori Embankment and Nagarakutei Embankment, respectively, even after the installation of continuous levees along the Oppe and Ichino rivers. The locations of the three important old levees are shown in Figure 13.2 as a dotted line or an arrow.

After large floods in 1910 and 1913, diversion channels, high embankments, and dams (the Futase, Urayama, Kakkaku, and Takizawa Dams) were constructed upstream on the main Arakawa River. In the last 100 years, till 2019, no large flood breaching the embankment of the main Arakawa River occurred. However, a flooding risk still exists when high-level floods occur. Although many secondary embankments still exist, their roles are not clear (Tanaka et al., 2019). Tanaka et al. (2019) concluded that a flooding risk area against the recurrent flood period of 200 years is located in a similar area. However, the present levee is quite high (6–8 m), compared with those (around 3 m) 100–400 years ago. The study demonstrated that the historical secondary embankments inside the Yoshimi Polder and Nagarakutei Embankments still can store the inundation flow and delay flooding in the downstream region for around 1–8 hours under the simulated flood conditions.

Typhoon Hagibis in 2019 caused severe damage to many large rivers in East Japan. The Arakawa River was one of the large rivers affected. Breaching of river embankments also occurred in the basins of Arakawa's branches. Re-evaluation of past flood management works, which did not prevent but still mitigated floods, can contribute to developing new approaches to managing extreme floods in the near future (Ishikawa and Akoh, 2018, Tanaka et al., 2020). Global climate change might also increase the frequency of flood discharges that exceed the channel flow capacity. Therefore, the objective of this study was to clarify whether historical secondary embankments, also known as the Kasumitei banks or open levees (hereafter noted as OL), played a role in controlling the flood inundation by Typhoon Hagibis or not. For that objective, a numerical simulation of the Arakawa River basin was conducted to elucidate the flooding risk in the Arakawa River basin in the past and compare the inundated area to the flooding situation in the 2019 Typhoon Hagibis.

13.2 MATERIAL AND METHODS

This study adopted a two-dimensional nonlinear depth-averaged flow model in the Arakawa region (continuity equation: Eq. (13.1), momentum equations: Eqs. (13.2)–(13.5), as used in Tanaka et al. (2019).

$$\theta \frac{\partial \eta}{\partial t} + \frac{\partial Q_x}{\partial x} + \frac{\partial Q_y}{\partial y} = 0. \tag{13.1}$$

$$\frac{\partial Q_x}{\partial t} + \frac{\partial}{\partial x}\left(\frac{Q_x^2}{\theta(\eta+h)}\right) + \frac{\partial}{\partial y}\left(\frac{Q_x Q_y}{\theta(\eta+h)}\right) + \theta g(\eta+h)\frac{\partial \eta}{\partial x} + \theta \frac{\tau_{bx}}{\rho} = 0 \tag{13.2}$$

$$\frac{\partial Q_y}{\partial t} + \frac{\partial}{\partial x}\left(\frac{Q_x Q_y}{\theta(\eta+h)}\right) + \frac{\partial}{\partial y}\left(\frac{Q_y^2}{\theta(\eta+h)}\right) + \theta g(\eta+h)\frac{\partial \eta}{\partial y} + \theta \frac{\tau_{by}}{\rho} = 0 \tag{13.3}$$

$$\tau_{bx} = \frac{\rho g n^2}{(\eta+h)^{7/2}} Q_x \sqrt{Q_x^2 + Q_y^2} \tag{13.4}$$

$$\tau_{by} = \frac{\rho g n^2}{(\eta+h)^{7/2}} Q_y \sqrt{Q_x^2 + Q_y^2} \tag{13.5}$$

where x and y are the horizontal coordinates in a Cartesian coordinate system; subscript x or y means the direction of the respective value; Q the discharge; τ_b the bed shear stress; t the time; $\eta + h$ the total water depth; h the local still water depth (on land, the negative height of the ground surface); η the water surface elevation; g the gravitational acceleration; ρ the density of water; and n the Manning roughness coefficient.

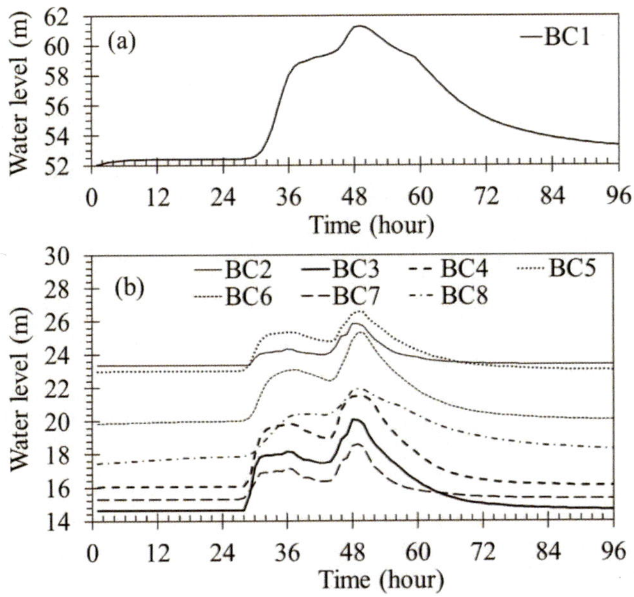

Figure 13.3 Water levels used for boundary conditions in simulations by a two-dimensional depth-averaged model. (Locations of boundaries are shown in Figure 13.2.)

The elevation and linkages of the Arakawa main river and its branches, shown in Figure 13.2, were modeled in a two-dimensional nonlinear depth-averaged flow model. Boundary conditions were applied in the locations of BC1–BC9 in Figure 13.2a. After the AR, boundary conditions were also set in the same location. A set of model equations was solved by the finite-difference method using a staggered leap-frog scheme. An upwind scheme was used for nonlinear convective terms to maintain numerical stability. A semi-Crank–Nicholson scheme was used for the bed friction and drag terms. The discharge hydrographs were hypothesized to be similar to those of the 1947 typhoon, which caused severe damage to the Arakawa River (hereafter, 47TF). For the flood risk in old times, a ten-year return period of flooding (Figure 13.3) was selected, because the embankment height was around 1–3 m around 400 years ago, and the flood frequency was about once in ten years in the Edo era. The peak values were assumed to be not very large, because flooding occurred upstream of the boundaries in many locations.

To compare the flood risks in the Edo era and now, field investigations were conducted after the 2019 Typhoon Hagibis, which caused severe damage in Japan. In our study site, breaching occurred at six points in the Iruma River and one in the Ichino River basin.

13.3 REPRODUCTION OF FLOODING RISK IN THE EDO ERA BEFORE AND AFTER THE RIVER COURSE CHANGE (ARAKAWA-SEISEN: AR)

Figure 13.4 shows the flood inundation pattern before and after the AR at 47TF (The recurrent period of floods is ten years.). Before the AR, flooding occurred radially in

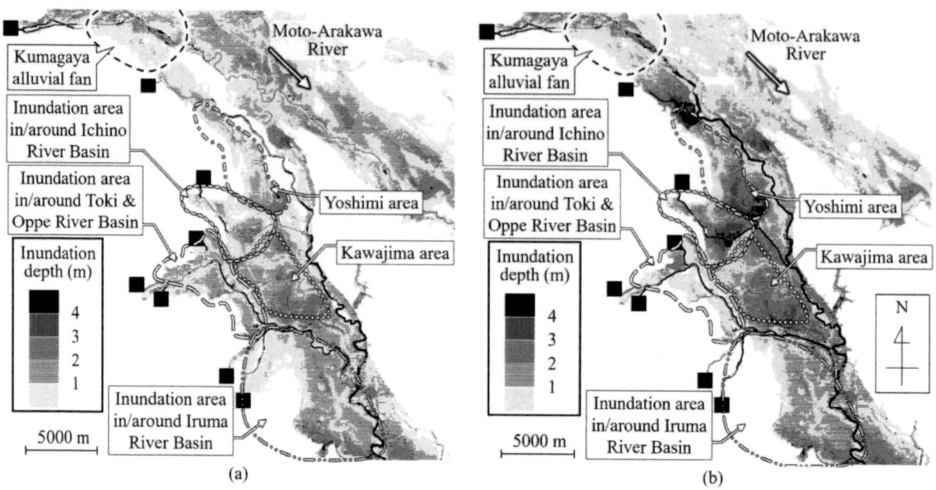

Figure 13.4 Flood inundation area under the scenario of 1947-type floods with ten-year return periods (47TF) when there was no embankment or the embankment height was low. (a) Before AR (no embankment), (b) after AR (discontinuous levee system).

the Kumagaya alluvial fan. There were no embankments at that time. On the left-hand side of the Arakawa River (Moto-Arakawa River area), the flood inundation is more severe in Figure 13.4a than that shown in Figure 13.4b. Compared with the left side, the inundation was minor on the right side of the alluvial fan. Some areas were not inundated. However, inundation became severe on the right-hand side of the Arakawa River after the AR. Especially, the inundation depths became larger in the Yoshimi and Kawajima areas in which the polder was reportedly newly constructed, or the polder height was raised, respectively, after the AR. At that time, because the ring levee heights were around 3 m, floods with a ten-year return period could overtop the embankments. In the Toki and Oppe rivers and the Iruma River basin, inundation occurred from the open parts of the discontinuous levees even after the AR. Although the inundation depth and area had declined compared with that before the AR, flooding risk existed in the same area even after installing the open levee system.

Tanaka et al. (2019) discussed the present flooding risk area for a 1947-type flood (Here, the recurrent period of floods was not ten years but 200 years.) and confirmed that the simulated inundated area was similar to that in the historic 1910 flood event. However, the flood hydrographs were not the same. Historically, two polders (Yoshimi Polder and Kawajima Polder) were constructed in these areas with banks around 3–4 m in height. Even though the embankment height has been increased (now around 5–8 m in height), the flooding occurs in a similar area when the recurrent flood period is around 200 years. After that report (Tanaka et al., 2019), flooding occurred in the 2019 Typhoon Hagibis in the branch areas (mainly the Ichino River basin and Toki and Oppe River basin in Figure 13.4).

These flooding risk locations were generated by the river course change around 400 years ago, and these locations were severely attacked by floods around once in ten years, 100–400 years ago. Thus, many secondary embankments and open levees were constructed in the area. Although the embankment height was raised and the flow capacity largely increased, the flooding risk was supposed to exist in the region when the flood level exceeded the river's capacity.

13.4 FLOOD INUNDATION AREA IN THE 2019 TYPHOON HAGIBIS EVENT

In the 2019 Typhoon Hagibis, the area shown in Figure 13.5 was inundated. Similar flood inundations for the Ichino River Region and the Toki and Oppe Regions shown in Figure 13.4 were observed. The risk in the main Arakawa River did not appear, because the embankment of the main Arakawa River is high, and flood discharge at the 2019 typhoon was relatively lower than the design level in the main Arakawa watershed compared with the watersheds in branches of the Iruma River.

Using the flood inundation area map of the Ministry of Land, Infrastructure, Transport and Tourism of Japan (MLIT), the areas around the Toki River inundated by the 2019 Typhoon Hagibis were estimated. Using the inundation map and the land elevation data (5 m grid data) measured using Laser Profiler (MLIT), the maximum inundated water depths obtained in the field survey by our research team, the inundated volumes were estimated and are shown in Figure 13.5. The flooding from inland water occurred mainly in Iimori, Kuzu, and Tsukumo areas. The breaching of the levee greatly affected the inundation volume in other areas. B4 and B6 affected the volume

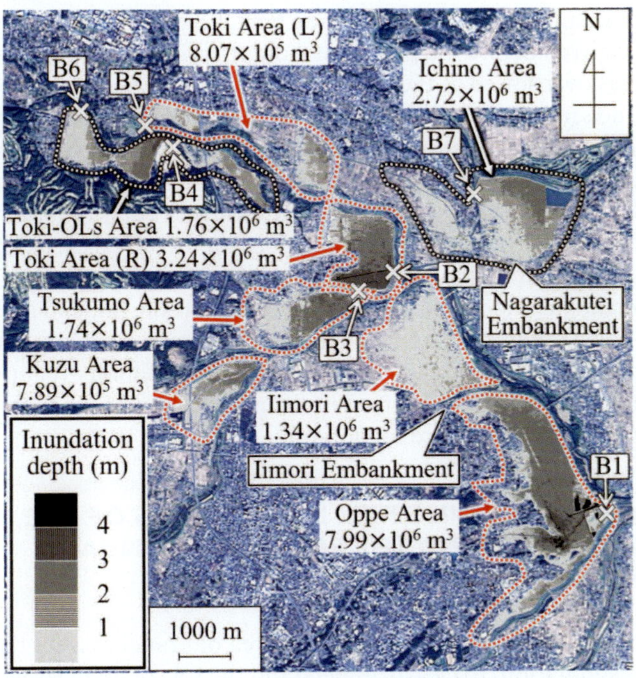

Figure 13.5 Flood inundation area and the inundation volume of the 2019 flood. (B1–B7 show the locations of breaches in the embankment.)

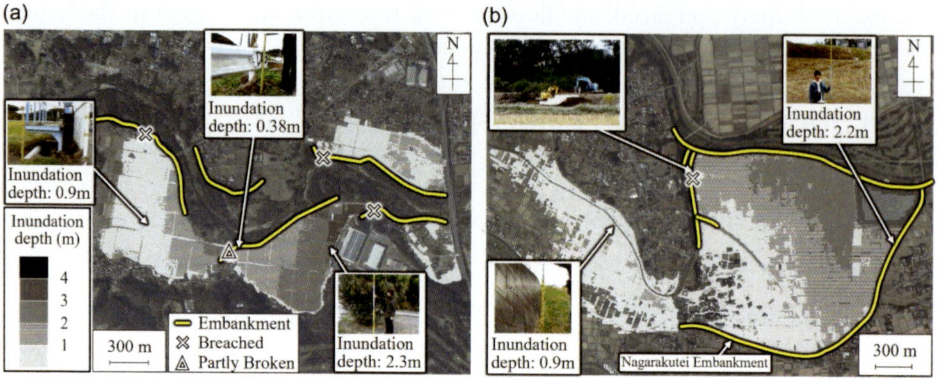

Figure 13.6 Flood inundation area and inundation depths at the 2019 flood (B4–B7 show the locations of breaches of the embankment.). (a) At open levees (OLs) in the Toki River, (b) north part of the Nagarakutei Embankment.

in the 'Toki open levees (OLs) Area,' B2 and B3 affected the 'Toki-R Area,' B5 affected the 'Toki-L Area,' B5 and B7 affected the 'Ichino Area,' and B1 affected the 'Oppe Area.' A large amount of floodwater was stored in this region. From the small river that runs from the 'Toki River (L)' to the 'Ichino River,' the inundation volume in the 'Ichino River' area comes not only from B7 but also from B5 (Figure 13.6). The still-existing open levees in the Toki River stored water in the 2019 flood, but overtopping of the levee occurred, and the embankment was breached. The storage mechanism from the open part occurred before the peak, but storage cannot be expected near the peak, because the levees overflowed at the 2019 flood level.

If there are no secondary embankments (especially the Nagarakutei Embankment as shown in Fig.13.6b), inundation from the Toki River and a branch of the Ichino River is not stopped in the 'Ichino Area', but also is assumed to occur in Kawajima Town (inside the present Kawajima Polder). The Nagarakutei Embankment is a part of a polder constructed around 400 years ago. It is not directly used as an embankment for the Ichino River, a branch of the Arakawa River, but the embankment still has a role to play in managing floods in this area, and to limit the flooded area.

Although the Arakawa River has a large flood retention lake, storing around 39,000,000 m^3 of water, it was almost filled by the 2019 typhoon event (35,000,000 m^3). If the flood inundation had not occurred in the middle area of the Iruma River basin, it could have affected the downstream flooding.

13.5 CHANGE IN FLOODING RISK IN BRANCHES OF THE ARAKAWA RIVER AND THE ROLE OF SECONDARY EMBANKMENTS

In Figure 13.7, the flood inundation area in the area of branches after the AR is shown in detail. In this simulation, to clearly understand the role of open levees, the locations of the boundaries, BC4, BC5 and BC6 in Figure 13.2, were set slightly upstream of the OLs area in the Toki River, Oppe River and Koma River, respectively. After the AR and before the Meiji Restoration, the inundation mechanism was different from that

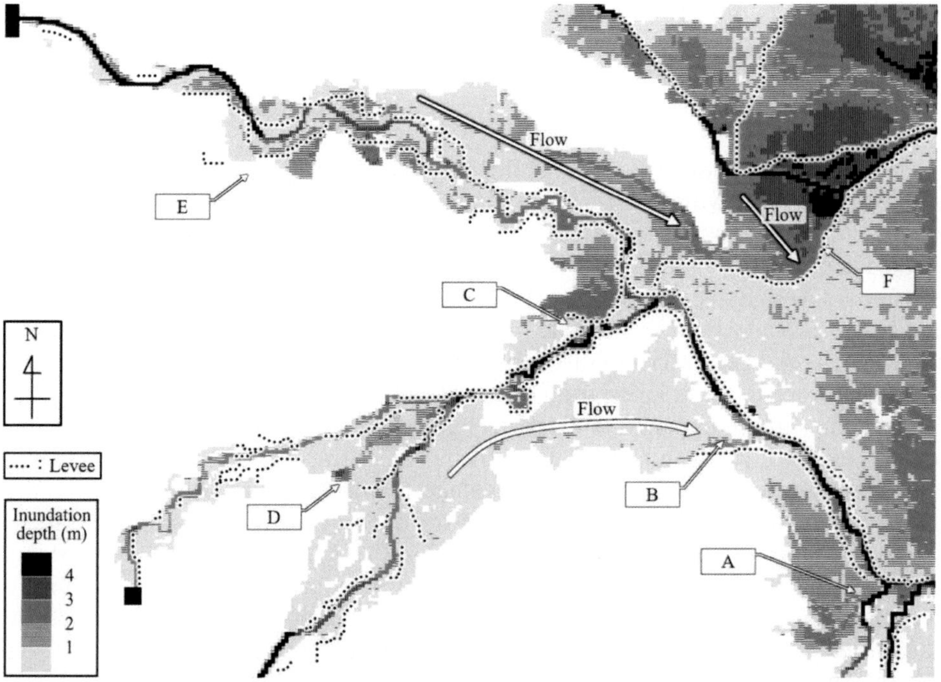

Figure 13.7 Flood inundation patterns when discontinuous levees were installed after the AR. (Similar areas were inundated in the 2019 flood, Figure 13.5.)

of the 2019 flood. For example, inundation occurred due to the overflow from the open levees of the Koma and Oppe Rivers in area D. Just downstream of and along the open levees of the Koma and Oppe Rivers, residential areas existed and direct hitting by the inundated flow was avoided. Area D acted like a retarding basin. In the 2019 flood, Area D was also inundated not by the open levee system, because the levees are now continuous, but by the small rivers between the Oppe and Koma Rivers after the gate was shut, because of the high-water levels of the two rivers (inland-water flooding). Area A was inundated in the 2019 flood by the breaching, followed by overflow from the Oppe River. After the AR, the area was inundated through the open part of the OLs. In between areas A and B, the Iimori Embankment still exists. This old levee remained in the area even after constructing the continuous levee system (an open part was closed, and a gate-pump system was constructed). The Iimori Embankment stopped the inundation current from breaching A's location (also shown as B1 in Figure 13.5). Thus, only inland-water flooding occurred in the Iimori River basin in the 2019 flood, and the inundation depth in Area B was lower than that in Area A. The open part of the levee inundated Area C after the AR. In the 2019 flood, however, the inland water also inundated when the gate was shut down because of the high-water level of the Oppe River. This is similar to the situation of Area D. After the gate was closed, however, breaching of the left-hand side of the levee occurred, and the inundation current also went to the northern part of C. In the 2019 flood, the flow from the levee-breach section (B5) and the overflow from the left-hand side of the Toki River inundated the Toki (L) and Ichino Areas, and the flow from B7 was stopped by the Nagarakutei Embankment (Figure 13.5). The inundated water ran along the river and accumulated on the north side of the Kawajima Polder (Nagarakutei Embankment). This is almost the same pattern as seen in the simulation after the AR in Figure 13.7. The Nagarakutei Embankment still played a role in decreasing the inundated area in the Kawajima Polder area.

Open levees and polders are an old flood mitigation system constructed mainly in the Edo era together with traditional countermeasures like elevated housing for flood defense, called *mizuya* or *mizuka* in Japanese (Siyanee et al., 2009), and flood guard moats (Kamaebori) (Tanaka and Hasemi, 2018) to protect settlements and houses from floods. At that time, flooding occurred more frequently, but people in the high-flood risk area knew how to evacuate or survive. Senoo and Ishikawa (2018) demonstrated the function of a series of discontinuous levees located in the early 19th century along the Kurobe River in Japan from a similar numerical simulation in our study. The levee openings in the upper river reach diverged the flow to old river channels. After the flood receded, the inundated water returned from the openings in the lower river reach. Our simulation in Figure 13.7 also suggests that the inundating flow from the right-hand side of the Koma River flowed to the opening of the Iimori Embankment (B) and returned to the Oppe River after the flood receded. The open levee systems shown in Figure 13.7 can be assumed to have functioned as a system like the Kurobe River. Ishikawa and Akoh (2018) also discussed the flood-retarding function of the Nihon levee system built in the 17th century and located upstream of Edo, which was the capital at that time.

However, floodplains and low-lying land, which are at high risk in water-related disasters, have been rapidly urbanized in Japan to support the sharp population increase and economic growth following World War II. Countermeasures to increase flow capacity and strengthen disaster prevention structures were systematically implemented or developed together with the reconstruction efforts after disasters (Inoue and Kamogawa, 2012). To decrease high-flood risk areas and develop residential areas, many discontinuous levees systems (open levees) were closed, and the recurrent flooding area increased in many places. However, this study shows that, although the critical levels of the flood discharge have been raised and the flooding mechanism is changed, similar areas can be inundated in a large flood event when the flood level exceeds the disaster prevention capacity. The lessons of the 2019 Typhoon Hagibis show that the area needs (1) flood retention ponds for large flood events, (2) non-structural measures to reduce the discharge itself by keeping or providing retention ponds like paddy fields, and (3) land-use planning such as living outside the inundated areas or changing part of the area to a flood-retarding basin. After the 2019 flood event, the MLIT in Japan and Saitama Prefecture are now planning to make five-side-overflow ponds along the rivers in the affected area. Side-overflow ponds are already installed in many places, but they need to be increased due to the heavy rainfall caused by climate change. The side-overflow ponds are more effective than the open levees, because they can store the water volume near the peak of the flood flow by appropriately planning the side-wall height.

Faisal et al. (1999) also discussed the importance of non-structural measures for flood management, especially preservation of retention ponds and flood zoning. They recommended a well-coordinated combination of both structural and non-structural measures for the long-term flood mitigation strategy of Dhaka City, Bangladesh. As one of the non-structural measures, using a flood hazard map (Shidawara, 1999) for helping evacuation has spread widely in Japan. However, land-use planning or flood zoning for the high-risk areas and methods to increase the flood retention using paddy fields to decrease the flood discharge quantitatively should be discussed in the future.

13.6 CONCLUSION

This study investigated that the difference between patterns before and after the course of the Arakawa River was changed around 400 years ago, and the flooding area was compared with the recent 2019 flood area of the Typhoon Hagibis. The role of the historical secondary embankments remaining in the investigated area was clarified by comparing the flood risk area in numerical simulations and the actual area inundated in 2019.

Although the flood inundation pattern differs a little, the Iimori Embankment (a part of the remaining old discontinuous embankment along the Iimori River) and Nagarakutei Embankment (a part of the Kawajima Polder, which is the secondary embankment for the Ichino River now) have been demonstrated to still play a role to reduce the flood inundation area. Therefore, the historical flood management system should be kept or strengthened to mitigate future flood inundations. Recently, not only in Japan, rainfall magnitudes and the total amount of water in a flood event are increasing under the effects of climate change. As one of the measures for adaptation to climate change, the system of discontinuous open levees and secondary embankments may reduce the inundation area in the hinterland and reduce the peak discharge into the downstream river. Discussions of total water management should include 'water retardation' and 'inundated current management' by using or further improving the old wisdom in association with land-use management of a high-flood risk area.

ACKNOWLEDGMENTS

The authors would also like to acknowledge Mr. Hayato Kajitani and Mr. Kengo Fushimi for helping with the data preparation in numerical simulations and the data collection in the field investigation. Further analysis based on Tanaka et al. (2020), which was presented in the 22nd IAHR-APD Congress, Sapporo, Japan, was conducted in this study.

FUNDING

This work is partly supported by the River Foundation (No. 2019-5211-037) and KAKENHI (Typhoon19).

REFERENCES

Faisal, I. M., Kabir, M. R., & Nishat, A. (1999). Non-structural flood mitigation measures for Dhaka City. *Urban Water*, 1(2): 145–153.

Furuta, N., & Shimatani, Y. (2018). Integrating ecological perspectives into engineering practices – Perspectives and lessons from Japan. *International Journal of Disaster Risk Reduction*, 32: 87–94. https://doi.org/10.1016/j.ijdrr.2017.12.003

Inoue, M., & Kamogawa, M. (2012). The need to change the concept of water-related disaster prevention. Quarterly Review of NISTEP Science & Technology Foresight Center, No. 44.

Ishikawa, T., & Akoh, R. (2018). Assessment of flood risk management in lowland Tokyo areas in the seventeenth century by numerical flow simulations. *Environmental Fluid Mechanics*, published online https://doi.org/10.1007/s10652-018-9616-6

Ishikawa, T., & Senoo, H. (2021). Hydraulic evaluation of the levee system evolution on the Kurobe alluvial fan in the 18th and 19th centuries. *Energies*, 14(15): 4406.

Kawanaka, R., Ishigaki, T., & Shimada, H. (2007). Hydraulic experiments on the flow around open dyke (in Japanese with English abstract). *Proceedings of the Hydraulic Engineering*, 51: 745–750.

Matsuki, H. (2012). Tripod scheme in flood disaster management in Japan. *Journal of Disaster Research*, 7: 582–589.

Nakashima, H., Ohgushi, K., & Hino, T. (2013). Numerical simulations of flood and inundation for evaluating the effects of Nokoshi and Open levee in Jobaru River (in Japanese with English abstract). *Journal of Japan Society of Civil Engineers, Ser. B1 (Hydraulic Engineering)*, 69(4), I_1537-I_1542.

Ohgushi, K., Nakashima, H., Hino, T., Morita, T., & Jansen, T. (2016). A study on Jobaru River basin management by numerical simulations of flooding and sediment deposition with field survey. *Lowland Technology International*, 18(1): 23–30.

Okuma, T. (1987). *A study on the function and etymology of open levee (in Japanese)*, Proceedings of the 7th Conference of Historical Studies in Civil Engineering in Japan, 259–266.

Senoo, H., & Ishikawa, T. (2018). *Hydraulic function of the Kasumi levee system on the Kurobe Alluvial Fan of the 19th century*. E3S Web of Conferences, Vol. 40, p. 06032.

Shidawara, M. (1999). Flood hazard map distribution. *Urban Water*, 1: 125–129.

Siyanee, H., Piyapong, J., Toyoda, Y., Mizuta, T., & Kanegae, H. (2009). An influence of social network on knowledge transferring in flood mitigation and preparedness: A case study of Waju Area, Ogaki City, Gifu Prefecture, Disaster Mitigation of Cultural Heritage and Historic Cities, 3: 275–282.

Tanaka, N., Fushimi, K., Kajitani, H., & Igarashi, Y. (2019). *Effects of existing historical second embankments in the floodplain for river flood inundation control*. E-Proceedings of the 38th IAHR World Congress, Panama City, Panama.

Tanaka, N., & Hasemi, Y. (2018). *Flume experiments on the effective shape of a historical flood guard moat, Kamaebori, around a protective mound and structure, Mizuka, in frequent flood occurrence area*. Proceedings of the 21st IAHR-APD Congress 2018, pp. 1239–1246, 2–5 September 2018, Yogyakarta, Indonesia.

Tanaka, N., Kajitani, H., & Fushimi, K. (2020). *Historical river course changes and paddy fields developments in the Arakawa River Basin, Japan and the role of second embankments in the recent 2019 flooding event*. Ext-abstract of the 22nd IAHR-APD Congress, Sapporo, Japan.

Teramura, J., & Okuma, T. (2005). A study on evolution and role of open levees on alluvial-fan rivers in the Hokuriku District: from the viewpoint of decentralization of river-engineering decision making (in Japanese with English abstract). *Historical Studies in Civil Engineering*, 24: 161–171.

Teramura, J., & Shimatani, Y. (2021). Advantages of the open levee (Kasumi-Tei), a traditional Japanese river technology on the Matsuura River, from an ecosystem-based disaster risk reduction perspective. *Water*, 13(4): 480.

Chapter 14

Teizan Canal

History and its effectiveness for tsunami energy reduction

Hitoshi Tanaka and Nguyen Xuan Tinh
Tohoku University

Kiyoshi Hashimoto
Miyagi Prefecture

CONTENTS

14.1 INTRODUCTION

The Great East Japan Earthquake and Tsunami of 2011 caused massive damage to people and infrastructures in the northeast area of Japan. This large-scale disaster provided valuable lessons on the huge consequences and potential impact of mega-earthquakes and mega-tsunamis. There have been many studies on the effects of structural solutions, such as coastal embankments and breakwaters, and nonstructural solutions such as pine tree forests, mangrove forests, and coastal dunes for reducing tsunami energy. In the case of mega-tsunamis, however, combining multiple solutions for tsunami disaster mitigation is highly necessary, instead of applying a single countermeasure. As illustrated in Figure 14.1 proposed for the Sendai Plane area, a typical combined system consists of (1) resilient coastal embankments, (2) shore-parallel canal, (3)

DOI: 10.1201/9781003222736-16

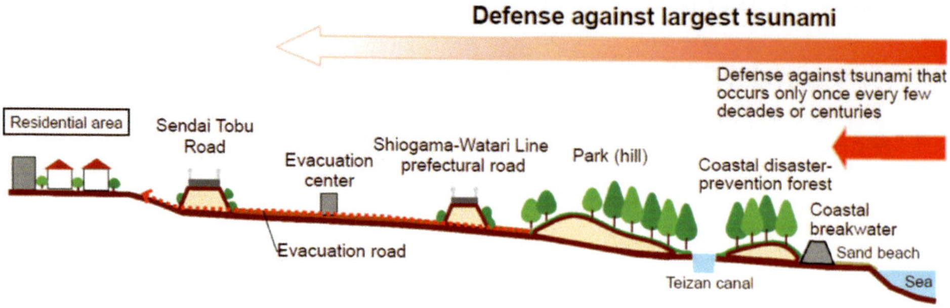

Figure 14.1 Combined multiple solutions against a mega-tsunami (Sendai City, 2021).

tsunami-mitigative coastal vegetation zone, (4) elevated road, and (5) evacuation facilities, etc. (Sendai City, 2021). The main objective of this research is to investigate the effectiveness of a historical shore-parallel canal on tsunami mitigation in the Sendai Plane area using a numerical model.

14.2 TEIZAN CANAL

As illustrated in Figure 14.2, the canal system on the Sendai Bay Coast consists of the Teizan Canal (Kobiki Bori, Shinbori, Ofuneiri Bori), Tona Canal, and Kitakami Canal. The name "Teizan Canal" is a general term for the entire canal system from the Abukuma River mouth to the Old Kitakami River mouth. It was initially constructed in the 1620s by Kawamura Magobei (1575–1648, Figure 14.3) by the feudal lord Masamune Date (1567–1636, Figure 14.4) for navigation during the Edo era. The name "Teizan" is a posthumous name of Masamune Date. Subsequently, the shore-parallel canal has been extended further into the northeast direction, as shown in Figure 14.2. It remains with additional functions such as historical water environment, flood water drainage in the low land area, and irrigation.

The Teizan Canal is approximately 24–45 m wide with the maximum water depth of about 1.3 m (Adityawan & Tanaka, 2016). Figure 14.5 shows a comparison of coastal features in 1701 and 2021. The old map indicates the Teizan Canal between the Abukuma and Natori River mouths. In contrast, on the right-hand side of the Natori River mouth, a canal is not drawn on the old map, as this canal was constructed from 1870 to 1872, as shown in Figure 14.2. Figure 14.6 illustrates paintings drawn on the wall along Ofuneiribori (see Figure 14.2), showing canal navigation and rice storage in a granary facing the canal during Edo Era.

The river embankment along the whole of the canal was seriously damaged during the 2011 Tohoku Tsunami. However, it has been reported that the canal had a disaster reduction effect against the Tsunami. Hence, Miyagi Prefectural Government (2013) has drawn up a "Vision for Restoration and Reconstruction of Teizan Canal" symbol of prefectural restoration after the earthquake and tsunami disaster. In this report, the tsunami mitigation effect has been added to the multiple functions of a canal as a part of the tsunami mitigation plan, besides environmental, historical, and flood management functions, and has stated the necessity of proper management and maintenance of the canal.

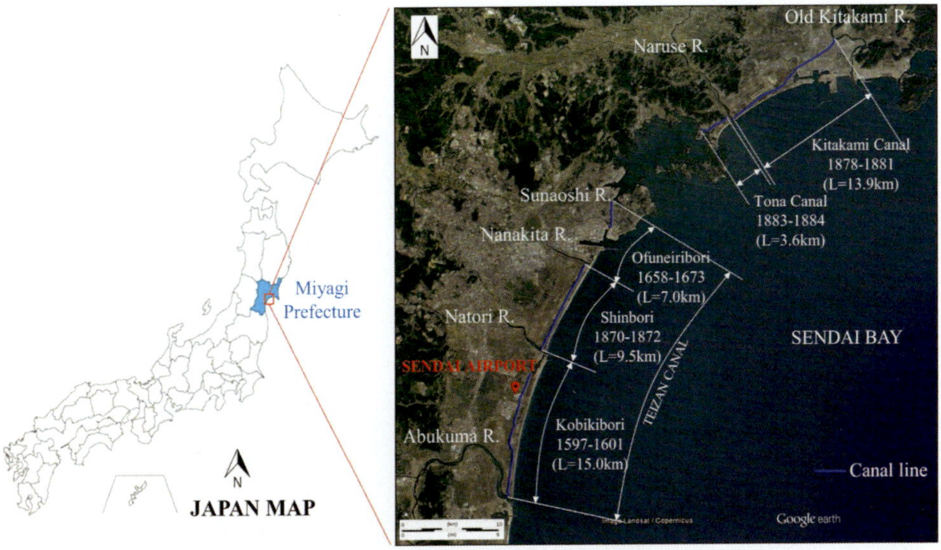

Figure 14.2 Location of the shore-paralleled canal along Sendai and Ishinomaki Coast.

Figure 14.3 Magobei Kawamura (1575–1648).

Figure 14.4 Masamune Date (1567–1636).

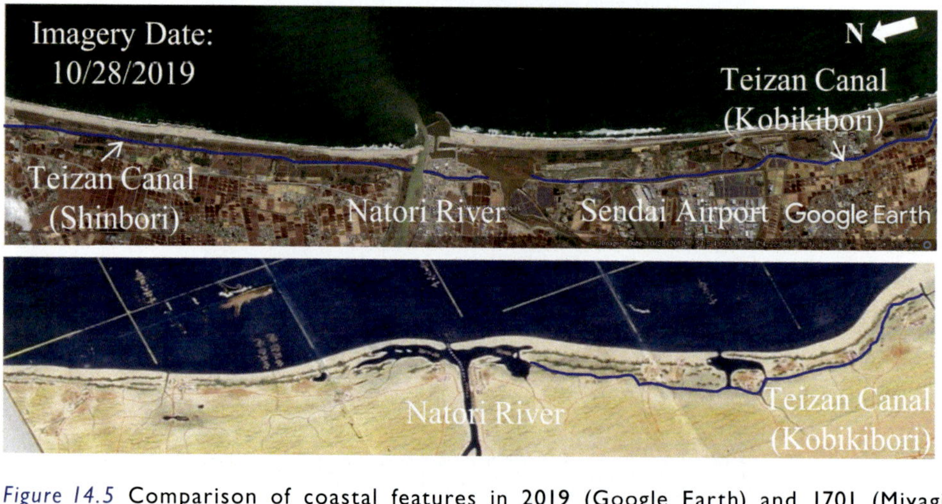

Figure 14.5 Comparison of coastal features in 2019 (Google Earth) and 1701 (Miyagi Prefectural Archives).

(a)

Figure 14.6 Wall paintings along Teizan Canal. (a) Wall paintings along the canal. (b) Navigation through the canal. (c) Rice granary along the canal.

(Continued)

Figure 14.6 (Continued) Wall paintings along Teizan Canal. (a) Wall paintings along the canal. (b) Navigation through the canal. (c) Rice granary along the canal.

Figure 14.7 schematically indicates tsunami mitigation effects of a shore-parallel canal. During the tsunami run-up period, it may cause a delay in tsunami arrival; meanwhile, during the tsunami run-down period, it enables collecting tsunami return flow, resulting in the reduction of the inundated area, which eases earlier restoration in the Tsunami affected area (Tanaka et al., 2014).

In order to apply a shore-parallel canal as a part of multiple defense systems against tsunamis, as depicted in Figure 14.1, it is highly necessary to understand its effectiveness quantitatively. Many researchers have investigated the effects of pine trees forest, artificial and natural structures, and coastal dunes, such as Tanaka (2009) and Nandasena et al. (2012). These studies have confirmed that these canals had tsunami

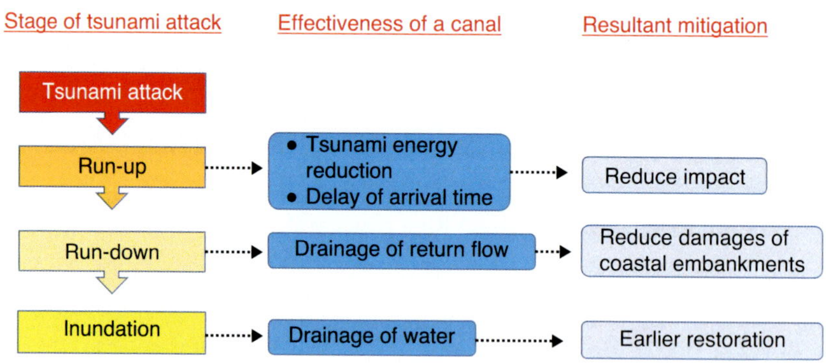

Figure 14.7 Tsunami mitigation effects of a shore-parallel canal.

disaster reduction effects, such as delay of tsunami arrival time and efficiently collecting tsunami return flow during the run-down tsunami process. However, research on the relationship between tsunami disasters and shore-parallel canals in tsunami mitigation has not been widely carried out. Recently, to clarify the tsunami disaster reduction effect of a canal on Sendai Bay Coast, Hashimoto and Tanaka (2018) and Tanaka et al. (2018) carried out numerical investigations on the tsunami disaster reduction effect of the existing canal by using a two-dimensional numerical model. The present study aims to evaluate the effectiveness of a shore-parallel canal on tsunami mitigation along the Ishinomaki Coast, which is closer to the 2011 earthquake epicenter than the Sendai Coast.

14.3 NUMERICAL MODEL

Numerical model is presently a valuable tool for assessing tsunami risks. There have been several numerical studies dealing with tsunami mitigation effects of a canal, such as Niimi et al. (2013), Dao et al. (2013), Oshiro et al. (2015), Watanabe et al. (2016), and Rahman et al. (2017). However, these investigations have been limited to vertical two-dimensional phenomena, including Tsunami plunging into a canal. Hence, it is difficult to utilize their results in the actual planning of tsunami mitigation. In this study, a 2D numerical model has been developed to evaluate a canal's tsunami mitigation effect concerning various hydrodynamic quantities in the land area, including tsunami arrival time, water level, and flow velocity.

14.3.1 Governing equations and computational cases

The governing equations of the numerical model are nonlinear long-wave equations (shallow water equations) given by Eqs. (14.1–14.3) as follows.

$$\left[\text{Continuity equation}\right]$$

$$\frac{\partial \eta}{\partial t} + \frac{\partial M}{\partial x} + \frac{\partial N}{\partial y} = 0 \tag{14.1}$$

$$\left[\text{Equation of motion}\right]$$

$$\frac{\partial M}{\partial t} + \frac{\partial}{\partial x}\left(\frac{M^2}{D}\right) + \frac{\partial}{\partial y}\left(\frac{MN}{D}\right) + gD\frac{\partial \eta}{\partial x} + \gamma_b^2 \frac{M\sqrt{M^2+N^2}}{D^2} = 0 \tag{14.2}$$

$$\frac{\partial N}{\partial t} + \frac{\partial}{\partial x}\left(\frac{MN}{D}\right) + \frac{\partial}{\partial y}\left(\frac{N^2}{D}\right) + gD\frac{\partial \eta}{\partial y} + \gamma_b^2 \frac{N\sqrt{M^2+N^2}}{D^2} = 0 \tag{14.3}$$

where (x, y): the horizontal coordinate system, t: the time, η: the water level, (M, N): the water discharge per unit width in x- and y-direction, respectively, and γ_b^2: the friction coefficient ($= g\, n^2\, \mathrm{D}^{-1/3}$, n: Manning's roughness coefficient, D: the total water depth).

In the numerical analysis, a finite difference method with the Leap-frog technique is applied to achieve stable computation in the whole region from the tsunami source

Table 14.1 Computational conditions

Item	*Setup condition*
Computational domain	From the Pacific Ocean to Miyagi Prefecture Coast
Grid spacing	450 m: Off Sanriku Coast 150 m: Off Southern Sanriku Coast 50 m: Off Miyagi Prefecture Coast 10 m: Miyagi coastal area
Governing eqs.	Nonlinear shallow water equations
Numerical method	Finite difference (Leap-flog method)
Tsunami source model	Central Disaster Management Council
Tidal condition	T.P. + 0.023 m (MSL in Sendai Bay)
Computation duration	5 hours. $\Delta t = 0.1$ second
Others	Roughness coefficient: Kotani et al. (1998) Coastal structure: no damage River discharge: no discharge

area to the Tsunami affected land area. The details of the computational conditions are summarized in Table 14.1.

14.3.2 Cases of numerical simulation

The present numerical study investigates three cases with different canal cross-sections, as in Table 14.2. Case 1 is a simple assumption without the canal. Case 2 corresponds to a typical cross-section under the present condition. In addition to these two fundamental cases, Case 3 is added to investigate the effect of canal shape by elevating embankment height to T.P. +4.5m. The shape of each cross-section is illustrated in Figure 14.8.

14.4 RESULTS AND DISCUSSION

In order to evaluate the effectiveness of the canal on Tsunami mitigation in the land area, five computation output points are set up, which are located before, inside, and behind the canal, as illustrated in Figure 14.9. The model results of tsunami arrival time, inundation depth, and the flow velocity are extracted at these five computation points to make a comparison among three canal conditions defined in Table 14.2 and Figure 14.8.

Table 14.2 Numerical simulation cases

Case	*Canal width (m)*	*Embankment height (T.P.m)*
Case 1 (No canal)	–	–
Case 2 (Present situation)	22	+2.25
Case 3	22	+4.5

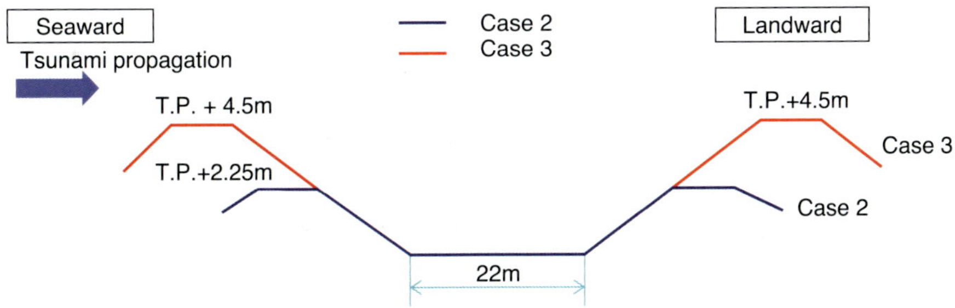

Figure 14.8 Cross-section of construction canal for test cases.

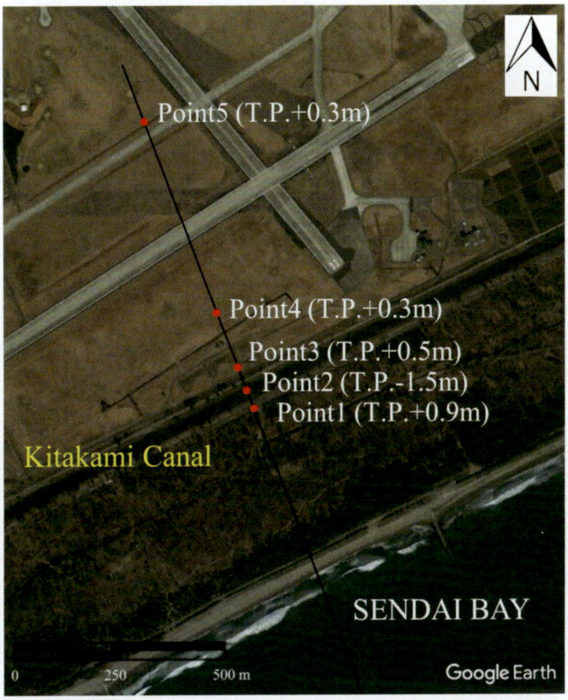

Figure 14.9 Locations of model result output points for comparison.

14.4.1 Effect on tsunami arrival time

Figure 14.10 shows the tsunami arrival time at five extracted points for all three cases. The difference in tsunami arrival time increases with the distance from the shoreline. In addition, the model results indicate that the tsunami wave takes longer to reach the inland location by introducing a canal than in a case without a canal (Case 1). It is interesting to note that by elevating the canal embankment from T.P. +2.25 m (Case 2) to T.P. +4.5 m (Case 3), the tsunami arrival time can be significantly delayed by 72 seconds at Point 5.

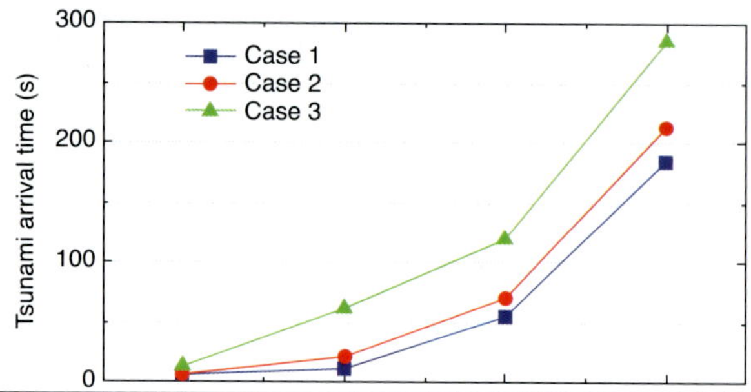

	Point2	Point3	Point4	Point5
Case 1	6	12	56	185
Case 2	6	22	71	214
Case 3	15	62	120	286

Figure 14.10 Effect on tsunami arrival time.

14.4.2 Effect on the maximum water level

According to Koshimura et al. (2009), tsunami inundation depth is the most influential factor in destroying houses and structures. Figure 14.11 compares the maximum water depth from Point 1 to Point 5 for all the three cases. With a shore-parallel canal in Case 2 and Case 3, the waterline's spatial gradient changes significantly compared to a case without a canal (Case 1). The water level behind the canal at Points 4 and 5 in Case 3 was significantly reduced by 0.39 and 0.44 m, respectively, compared with Case 1.

14.4.3 Effect on the maximum tsunami flow velocity

Figure 14.12 shows spatial variation of the maximum flow velocity under three canal conditions. At Point 2 inside of the canal, a remarkable velocity reduction can be observed by 2.18 and 2.87 m/s in Case 2 and Case 3, respectively.

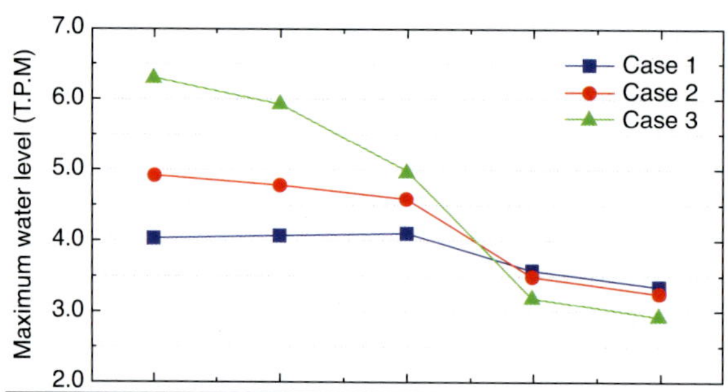

	Point1	Point2	Point3	Point4	Point5
Case 1	4.01	4.05	4.09	3.56	3.35
Case 2	4.91	4.77	4.58	3.46	3.23
Case 3	6.28	5.93	4.95	3.17	2.91

Figure 14.11 Effect on tsunami water level.

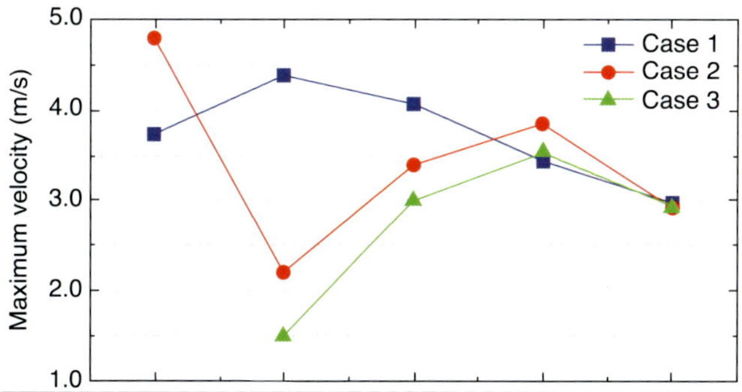

	Point1	Point2	Point3	Point4	Point5
Case 1	3.73	4.38	4.06	3.45	2.98
Case 2	4.79	2.20	3.40	3.85	2.92
Case 3		1.51	3.01	3.54	2.93

Figure 14.12 Effect on tsunami flow velocity.

14.5 CONCLUSIONS

In this study, a detailed investigation of the effectiveness of a shore-parallel canal along the Ishinomaki Coast is numerically carried out regarding tsunami disaster mitigation. In contrast to the previous numerical studies, the present investigation highlights a two-dimensional hydrodynamic behavior of the 2011 Tsunami in Sendai plain. Three numerical experiments are carried out with different canal cross-sections to investigate the effectiveness of a canal shape for reducing tsunami impact. Among various factors characterizing tsunami impact, tsunami arrival time, tsunami water level, and tsunami-induced velocity are quantified for the different cross-sections of a canal. As a result, elevating canal embankment causes high effectiveness for various hydrodynamic characteristics. Therefore, a canal system can be utilized as a part of a multilayered defense system in an area expecting a devastating tsunami disaster in the near future.

ACKNOWLEDGMENTS

The authors would like to express their thanks for financial support from the Tohoku–Tsinghua University Research Fund (2020–2021).

REFERENCES

Adityawan, M. B. & Tanaka, H. (2016). *Investigating the effect of old river mouth and the Teizan Canal in Sendai Coast to the 2011 Tsunami*. In: V. Santiago-Fandino, H. Tanaka, M. Spiske (Eds), *Tsunamis and Earthquakes in Coastal Environments: Significance and Restoration*, Springer, pp. 125–136

Dao, N. X., Adityawan, M. B., Tanaka, H., & Lin, P. (2013). *Effectiveness of a shore-parallel canal to reduce tsunami impact*. Proceedings of 35th IAHR Congress, Chengdu, China.

Hashimoto, K. & Tanaka, H. (2018). Reduction effect of tsunami disaster by canals along Sendai Bay Coast. *Journal of JSCE Ser.B3 (Ocean Engineering)*, 74(2): 157–162 (in Japanese).

Koshimura, S., Oie, T., Yanagisawa, H., & Imamura, F. (2009). Developing fragility functions for tsunami damage estimation using numerical model and post-tsunami data from Banda Aceh, Indonesia. *Coastal Engineering Journal*, 51(03): 243–273.

Kotani, M., Imamura, F., & Shuto, N. (1998). Tsunami run-up simulation and damage estimation by using GIS. *Proceedings of Coastal Engineering, JSCE*, 45: 356–360 (in Japanese).

Miyagi Prefectural Government (2013). *Vision for Restoration and Reconstruction of the Teizan Canal*, 35p (in Japanese).

Nandasena, N. A. K., Sasaki, Y., & Tanaka, N. (2012). Modeling field observations of the 2011 Great East Japan tsunami: Efficacy of artificial and natural structures on tsunami mitigation. *Coastal Engineering*, 67: 1–13.

Niimi, T., Kawasaki, K., Mabuchi, Y., Nagayama, K., Tsuji, T., Oie, T., & Matsuda, K. (2013). Numerical examination on tsunami mitigation effect of Teizan Canal. *Journal of Japan Society of Civil Engineers, Ser.B2 (Coastal Engineering)*, 69(2): 211–215 (in Japanese).

Oshiro, T., Nakaza, E., Inagaki, K., Rahman, M. M., & Egashira, S. (2015). Effect of canal and vegetation on tsunami disaster mitigation. *Journal of Japan Society of Civil Engineers, Ser. B2 (Coastal Engineering)*, 71(2): 187–192 (in Japanese).

Rahman, M. M., Schaab, C., & Nakaza, E. (2017). Experimental and numerical modeling of tsunami mitigation by canal. *Journal of Waterway, Port, Coastal, and Ocean Engineering*, 143(1): 4016012.

Sendai City (2021). Multiple Defenses to Minimize Tsunami Damage. https://sendai-resilience.jp/en/efforts/government/development/ (accessed August 24, 2021).

Tanaka, N. (2009). Vegetation bioshields for tsunami mitigation: Review of effectiveness, limitations, constructions, and sustainable management. *Journal of Landscape Ecology Engineering*, 5, 71–79.

Tanaka, H., Adityawan, M. B., Udo, K., & Mano, A. (2014). *Breaching and tsunami water drainage at old river mouth locations during the 2011 tsunami*. Proceedings of the 34th International Conference of Coastal Engineering, Seoul, Korea.

Tanaka, H., Hashimoto, K., & Tinh, N. X. (2018). *Effectiveness of a shore-parallel canal for reducing tsunami impact*. Proceedings of 21st Congress of the Asia and Pacific Division of the International Association of Hydraulic Engineering and Research (IAHR-APD), Yogyakarta, Indonesia.

Watanabe, S., Mikami, T., & Shibayama, T. (2016). *Laboratory study on tsunami reduction effect of Teizan Canal*. Proceedings of 6th International Conference on Application of Physical Modelling in Coastal and Port Engineering and Science, Ottawa, Canada.

Chapter 15

Major restorations in main channels and the inverted siphon of Tatsumi Aqueduct

Nobuyuki Tamai
The University of Tokyo
NPO for Wise Learning on Tatsumi Aqueduct

Masaru Kitaura and Toshikazu Ikemoto
Kanazawa University
NPO for Wise Learning on Tatsumi Aqueduct

Haruhiko Todo
Kanazawa City

CONTENTS

DOI: 10.1201/9781003222736-17

15.1 INTRODUCTION

15.1.1 Social features

The first lord of the Kaga Clan, Maeda Toshiie, moved his government to Kanazawa in 1583. The stonewalls and mounds of Honmaru, Ninomaru, and Sannomaru fortresses were constructed over three years, and the keep tower, palace, and watchtowers were constructed in 1586. Kanazawa Castle was constantly developed in response to politics in its early stages. It was often destroyed by fires, for example, in 1620 and 1631, which destroyed large areas of a castle town and a part of the castle (Kagahan Shiryo, 1980a).

The third lord, Maeda Toshitsune, ordered in 1632 to construct Tatsumi Aqueduct to obtain sufficient water resources for the castle for fire prevention. The aqueduct was completed in the same year. There are many remaining questions on the aqueduct because no recorded material was written during the construction time. One exception is Kagahan Shiryo. However, the statement itself is pretty short and simple, as shown in the following sentence. In a section of the year 1632 (in Kan'ei 9 in the Japanese calendar), the statement is written only by "In this year water of Saigawa river reached in the castle. This is called Tatsumi Aqueduct" (Kagahan Shiryo, 1980b).

There are three most famous books that refer to Tatsumi Aqueduct. They are "*Etsutoga Sansyushi* (abbreviated by EST)," "*Kinjo Shinpiroku* (abbreviated by KS)," and "*Mitsubo Kikigaki* (abbreviated by MK)." Although "MK" is the oldest book; it was published in 1704 through 1711, about 70 years after the construction. In this paper, the contents of three books (EST, KS, and MK) were collected through explanations in Kaga Tatsumi Aqueduct. In Kagahan Shiryo, the editor added explanations extracted from the three books stated above to the original Maeda family documents.

15.1.2 Technical and historical features

The aqueduct conveys water to Kanazawa Castle through an inverted siphon crossing a wide moat of the castle. The upper part of the aqueduct, starting from the intake, has been constructed by an unlined tunnel whose length amounts to approximately 3.3 km. The difference in elevation of the pipeline of the inverted siphon reaches 17 and 21 m during the history of Tatsumi Aqueduct, which is a quite large value in the Edo Period in Japan. The objective of the construction work was explained as fire prevention in all three books (EST, KS, and MK) (Kitamura, 1983a). The chief engineer of civil engineering design and construction works in situ was Itaya Heishiro (in KS and MK). EST mentions an additional name. Because the editor of Kagahan Shiryo adopted the idea that the construction was led by Itaya Heishiro alone, this idea has become the prevailing one (Kitamura, 1983b).

During Edo Period, the aqueduct was owned and operated by the Kaga feudal government under the Maeda Clan, who was named the ruler of the Kaga domain by the Tokugawa Shogunate. Tatsumi Aqueduct supplied water to wide fronts, such as fire prevention, moats of the castle, gardens, and for a better castle environment, Kenrokuen garden, irrigation, and waterpower for the gunpowder mills of the Kaga feudal government.

By Meiji Restoration, political power moved from Tokugawa Shogunate to a new government under Emperor in 1868. Japan experienced turmoil due to the transition of political power over several years. The major function of Tatsumi Aqueduct changed to irrigation, and a union of farmers was organized for the management. The inverted siphon stopped supplying water to Kanazawa Castle in Meiji Period because the Kaga feudal government in the castle was disassembled.

It is reported that Iwazeki Aqueduct in the Akita feudal domain (current Akita Prefecture) was completed in 1631 using drainage techniques in mines. The sectional shape and alignment of the tunnel were similar to those of Tatsumi Aqueduct (Aoki, 1989). The Akita feudal government owned eminent domestic silver mines in those days. It is left for the record that many prospectors from the Kaga feudal domain earned their living by working in the silver mine (Yamaguchi, 2008). From the fact that 11 mines were in the Kaga feudal domain at that time (Ogawa, 1962), it can be thought that the mining technology was applied to Tatsumi Aqueduct tunnel construction method, and a large number of miners working in mines in the Kaga feudal domain were mobilized to dig a tunnel. However, neither historical nor old documents on the mobilization of miners in the Kaga feudal domain to Tatsumi Aqueduct project exist in the exchange of technical information on Iwazeki Aqueduct.

This paper expanded the paper contents presented at IAHR-APD Congress in Yogyakarta (Tamai et al., 2018). Constraints of the discharge of the inverted siphon for firefighting and the very late establishment of a fire brigade system imply that the castle's Tatsumi Aqueduct for fire prevention was not as effective as the Kaga feudal government had hoped.

Creative civil engineering technologies with excellent construction management have brought several awards to Tatsumi Aqueduct. Tatsumi Aqueduct was designated as a national historical site by the Ministry of Education, Culture, Sports, Science, and Technology of Japan in 2010. In 2018 the whole waterway of Tatsumi Aqueduct between the intake and Kanazawa Castle park, including the remains of the inverted siphon, was certified as Civil Engineering Heritage by the Japan Society of Civil Engineers. Tatsumi Aqueduct was also endowed IAHR-APD Hydro-Environment Heritage Award in 2020.

15.2 CHARACTERISTICS OF TATSUMI AQUEDUCT PROJECT

15.2.1 Topographical features of Tatsumi Aqueduct

The city of Kanazawa is located in the center of Ishikawa Prefecture, as shown in Figure 15.1. The city center is divided into three by Saigawariver and Asanogawa rivers and is surrounded by Utatsuyama hills, Kodatsuno terrace, and Teramachi terrace, which are listed from the north. The Kanazawa Castle is located on the edge of the Kodatsuno terrace, as shown in Figure 15.2.

The aqueduct, approximately 11 km in length, starts from the Higashi-iwa intake on the right bank of the upper Saigawa river. The aqueduct flows through a tunnel approximately 3.3 km long, connects to an open channel part on a west slope of Kodatsuno terrace, and finally reaches Kenrokuen garden. The average gradient from 'Kiji' (the original intake) to the Kenrokuen garden is approximately 4‰. However, it is approximately 1‰ to the Nishikimachi town on the Kodatsuno terrace in the upper part of Tatsumi Aqueduct, mainly composed of tunnels.

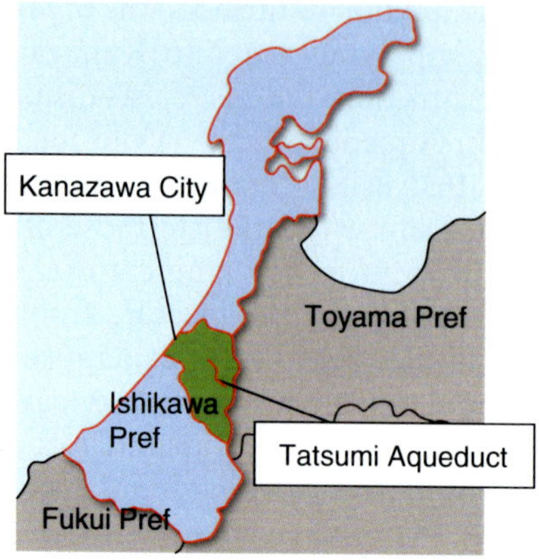

Figure 15.1 Location map of Tatsumi Aqueduct.

Figure 15.2 Geography around Kanazawa Castle.

15.2.2 A unique feature in the tunneling technique

Tatsumi Aqueduct was constructed at the beginning of the 17th century. A long tunnel in the upstream part and an inverted siphon with high water pressure in the downstream part are symbols of very advanced technologies at the time. The construction took only nine months to complete the aqueduct, including the inverted siphon conveyed from Kenrokuen garden to Kanazawa Castle.

There are several aqueducts in Japan where tunnels were dug in the 17th century, but a unique construction method was adopted in Tatsumi Aqueduct. As shown in Figures 15.3 and 15.4, there were many adits toward the tunnel's terraced cliff at the wall about 20–30 m apart. Such adits were called "windows" and used for ventilation, daylighting, and carrying out spoil generated by digging the tunnel. The adit system also made simultaneous tunnel excavation possible for higher efficiency.

In addition, since the height of the entrance of adit almost accords with the riverbed height of Saigawa river in the late Pleistocene of Quaternary, it is estimated that the bed level of the aqueduct was designed to trace the old riverbed of Saigawa river at the time when Kodatsuno terrace was formed about 40 thousand years ago (Aoki, 1983a).

The excavation process dug two pilot tunnels from the upstream and downstream sides. When the pilot tunnel was joined, the section of the pilot tunnel was enlarged. Small niches were dug on the wall surface of the tunnel inside, of which an oil-filled plate with a wick was placed to improve the work efficiency by illuminating surrounding areas.

Figure 15.3 Main tunnel (left) and an adit (right).

Figure 15.4 Illustration of Tatsumi Aqueduct in 1809 (Bunka Rokunen Tatsumi Yosui Ezu [1]: In this study, old documents and drawings are listed by number in the rear part of journals and books in a modern system.).

Central axes of pilot tunnels excavated from upstream and downstream were set to form a zig-zag shape on a horizontal plane to prevent pilot tunnels from passing through in parallel. The digging technique to prevent failure completed the project in only nine months.

Tatsumi Aqueduct installed an inverted siphon to deliver water to Kanazawa Castle to cross the Hyakkenbori moat from Kenrokuen garden. The inverted siphon overcame a large head difference of more than 10m (see Figure 15.13). The inverted siphon of Tatsumi Aqueduct was an outstanding innovative hydraulic structure at the time.

15.2.3 Restoration from severe damage by earthquake

Several drawings wrote the whole of Tatsumi Aqueduct during the period of the Kaga feudal domain. The oldest drawing among these was written after the 1799 big earthquake. It can be confirmed that several sections were replaced with tunnels after the earthquake, and Sandan Ishigaki (the three-tier masonry wall of about 260m in length) was constructed in the Mekuradani valley, which was considered to be the most vulnerable geology (Figures 15.5 and 15.6). At the relatively downstream location in the section where the tunnels appear in succession, it was confirmed that many open channel sections were converted to tunnels after the earthquake, and the intake had been extended.

Remodeling after the earthquake prevented the open channel from being buried by a cliff collapse. By 1834, the length of the tunnel part had increased by about 1 km and extended to 4.5 km in total after the earthquake in 1799 (Aoki, 1983b).

Figure 15.5 Sandan Ishigaki.

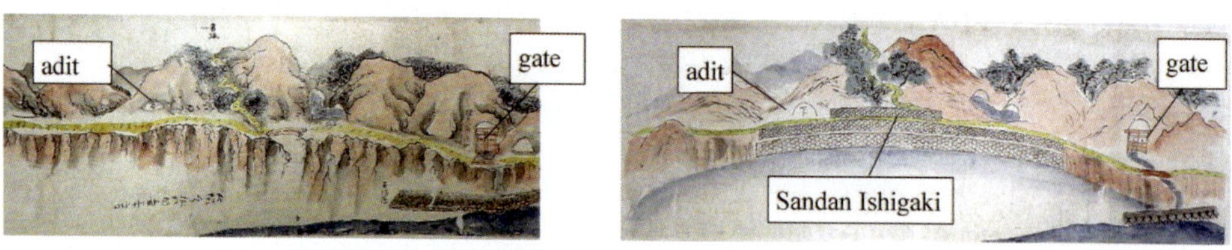

Figure 15.6 Change of views around Mekuradani valley by the 1799 earthquake: (a) before the earthquake; (b) after the earthquake (Tatsumi Jyosui Esujino Ezu [2]).

Figure 15.7 Tunnel by piling stones with an arch type.

In the Meiji period, the management of the main constituent of Tatsumi Aqueduct turned into a private sector from the Kaga feudal government. At the end of the Meiji Period, an inspection of the tunnel part of the aqueduct revealed that a large-scale rehabilitation to prevent the collapse of the deteriorated tunnel was required. Then a maintenance construction with an arch type was carried out (Figure 15.7). As a result, most of the tunnels of the upstream part retained their original form. At the same time, many sections of the aqueduct constructed in the Edo period have been renovated without detailed records.

15.3 HISTORICAL CHANGE OF A ROUTE OF A PRESSURIZED SECTION OF AN INVERTED SIPHON

15.3.1 Outline of land-use change in the current Kenrokuen garden area

It is necessary to know the change of land use of the current Kenrokuen garden area because an upstream well of the inverted siphon is located in this area called Chitosedai. At first, in the 17th century, the Kenrokuen garden area was a district of samurai warriors where the chief vassals of the Kaga Clan were stationed with subordinates. From the topographic point of view, Kanazawa Castle needed a strong defensive base against the enemy's attack from the Kodatsuno terrace, which extends in the southeastern direction.

In 1676, the fifth lord Tsunanori made the Renchitei garden in the slant place on the underside of the samurai district. He gradually relocated the group of samurai warriors to create a vacant lot owned by the Kaga Clan in 1697. This policy signified the end of the Warring States Period and symbolized the stabilization of the foundation of the Edo Shogunate. From this point on, the Kaga Clan focused on developing art and culture. In 1792, the eleventh lord Harunaga constructed the Meirindo school for higher education for young samurais in Chitosedai. In 1822, the twelfth lord Narinaga removed the school, built a magnificent mansion for his retirement life, and named the garden "Kenrokuen." The thirteenth lord Nariyasu removed the mansion and changed the whole area to a Japanese-style garden.

15.3.2 First period (the year 1632–1634)

Mitsubo Kikigaki [3] explains that Tatsumi Aqueduct changes to a buried water pipe on the north side of the residence of the Okumura family (shown in Figure 15.8 (Enpo Kanazawazu [4]). In this study, we call this location point A hereafter, and a red circle in Figure 15.8 shows the point.

By mapping Figure 15.8 with a current map, point A is located at the top of the slope of the Kodatsuno terrace (48 m above sea level). When point A is taken as an inlet of the inverted siphon to the castle, it is higher than that of the Sannomaru fortress of the castle (44.6 m above sea level) but lower than the Ninomaru fortress of the castle (50 m above sea level). Therefore, Tatsumi Aqueduct could supply water up to the Sannomaru fortress at the time of completion in 1632 (Aoki, 1983c).

15.3.3 Second period (the year 1634 to late 18th century)

To supply water to the Ninomaru fortress, it was necessary to move the inlet further upstream and to extend buried water pipes to make the inlet position higher than the Ninomaru fortress. When excavating the bottom of the aqueduct at the time of road construction in 1981, a wooden water pipe was discovered in the section about 300 m long (Refer to Figure 15.9). As a result of carbon analysis, the pipe was estimated to have been built in the 17th century. The historical book "Etsutoga Sansyushi" in the early 19th century stated the total distance of the buried pipes from the intake to Renchitei garden. However, the position of the gate was unclear (Aoki, 1983d).

The oldest drawing—a whole route of Tatsumi Aqueduct was discovered recently. This drawing [2] makes the gate position clear, since the distance and content written on this drawing are consistent with the contents of several historical books. When we take a reference point at a step boundary of the Kodatsuno terrace, it is clarified that the upstream edge of the wooden water pipe starts from just a downstream point of identified location of Ishibiki Water Gate. A long pressurized wooden pipe

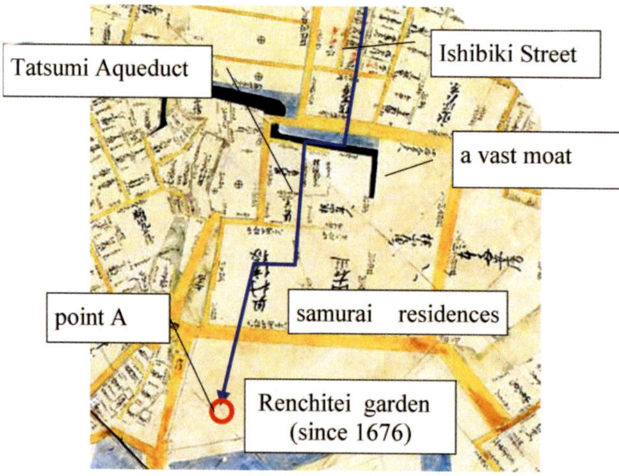

Figure 15.8 Kenrokuen area in the 17th century (Enpo Kanazawazu [4]).

Figure 15.9 Buried wooden pipes at a construction site. (Photo by M. Azechi, 1981.)

was installed for two years until 1634 between point A in Figure 15.8 and point B in Figure 15.10, exceeding 1 km. The difference in elevation of the pipeline of the inverted siphon reached 21 m.

15.3.4 Third period (between late 18th century and mid-19th century)

It was difficult to manage this long inverted siphon, and it was necessary to shorten the length as much as possible. Then we investigate how the inlet was moved from point B to the downstream side and still possible to supply water to the Ninomaru fortress.

In "Kanazawajyo Gotenzu" [5], a castle map drawn around the 1750s, the inlet is drawn in the Renchitei garden. In addition, in " Kanazawajyochu Mizunote no hizu" [6], which describes the route of the inverted siphon in the castle in 1807, the distance from the inlet to the Ishikawamon gate was written. While measuring the distance from the Ishikawamon gate back to the Renchitei garden in the current map, the position of the inlet of the inverted siphon (point C in Figure 15.11) is determined. Furthermore, the altitude at point C is 51–52 m, and it was confirmed that it is possible to supply water to Ninomarufortress.

The inlet was moved to point C because it could change the aqueduct's route after the samurai residence changed to a vacant lot. There is a record that the aqueduct was flowing in the Renchitei garden in 1730 [7]. The record indicates that the aqueduct was moved to draw water into the garden.

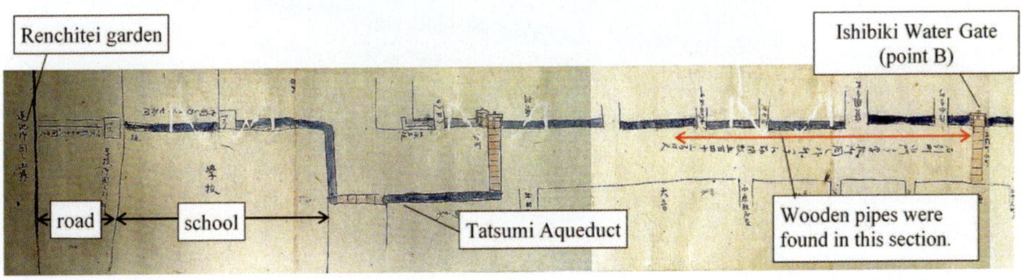

Figure 15.10 Position of the Ishibiki Water Gate in Tatsumi Jyosui Esujino Ezu [2].

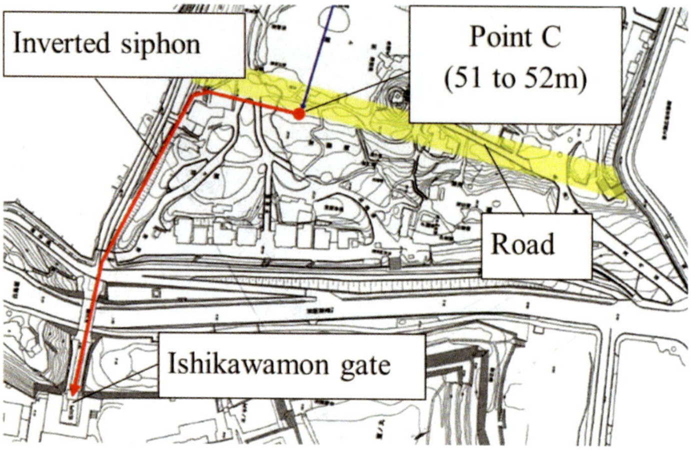

Figure 15.11 Location of point C.

15.3.5 Fourth period (between middle 19th century and early Meiji period) and summary

After the magnificent mansion for the lord's retirement life (the 13th Lord Maeda Narinaga) was built, the integral utilization of these sites began. A meandering stream in the mansion appeared, and the garden was given the name "Kenrokuen." The next phase changed to remove the mansion and expand the Kenrokuen garden. It can be confirmed that the inlet approaches the present position, and the whole area changes to a Japanese-style garden based on several drawings (Nagayama, 2006). In addition, wooden pipes used for the inverted siphon were replaced by stone pipes in the middle of the 19th century, enabling a more stable water supply (Figure 15.12). A conceptual diagram between the inlet of the inverted siphon and Kanazawa Castle is shown in Figure 15.13.

Figure 15.12 Remains of stone pipes near the final point of the inlet of the inverted siphon in Kenrokuen garden.

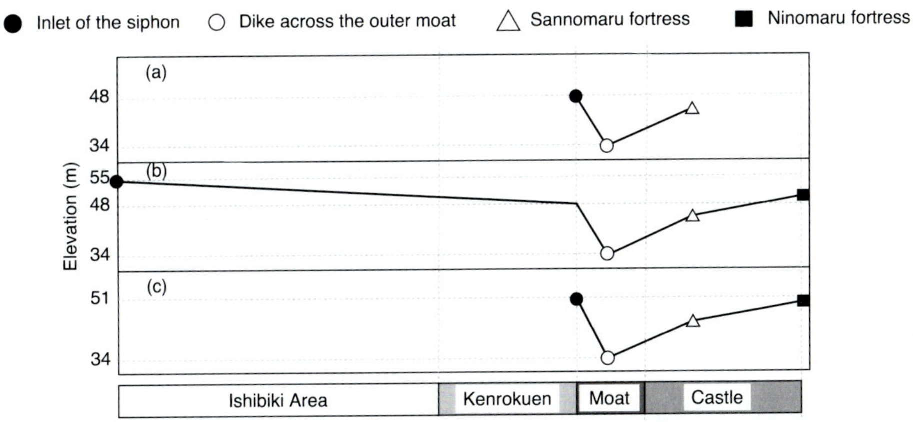

Figure 15.13 Schematic diagram of the volition of the inverted siphon over time, based on the position and altitude of the upstream wells : (a) year 1632; (b) year 1634; (c) year about 1730.

15.4 THE DISCHARGE RATE OF THE INVERTED SIPHON AND CONSIDERATION OF FIRE PREVENTION FOR THE CASTLE

The discharge rate by the inverted siphon from point A shown in Figure 15.8 to Sannomaru fortress in 1632 calculated for the siphon arrangement shown in Table 15.1 was $0.0039\,m^3/s$. In two years, the upstream well of the inverted siphon was reinstalled to a higher location upstream (point B in Figure 15.10). In 1634, the inverted siphon reached Ninomaru fortress, conveying $0.0020\,m^3/s$ (Tamai and Yamada, 2006; Aoki, 2001).

To extinguish a fire, it is necessary to quickly spray a large amount of water on a burning house. Firefighting tools with hoses were not available in the 1630s in Japan. It is considered that spraying $0.004\,m^3/s$ of water is not enough to extinguish the fire in the case of a wooden house effectively.

Big fires in the castle town of Kanazawa, including castle areas, were recorded in 1620, 1631, 1635, and 1644 during the early half of the 17th century (Kagahan Shiryo, 1980c). The Kaga Clan introduced several countermeasures after these big fires. In the case of the 1631 fire, which destroyed even half of the castle area, a general law stipulated the behavior of samurai and townspeople within the same year. In December of the same year, six groups were established to make fire patrol and firefighting for the central part of the town (Kagahan Shiryo, 1980d). However, the castle was an off-limit zone even under firefighting operations. In 1639, fire patrol and firefighting members

Table 15.1 Elevation of key wells and the pipe length of the siphon

	Point B Figure 15.10	*Kodatsuno entrance, Kenrokuen*	*Point A Figure 15.8*	*Sannomaru fortress*	*Ninomaru fortress*
Elevation (m)	55	53.5	48	44.6	50.2
Segment length (m)	0	1,020	350	340	300

were reinforced by increasing members of high-ranking samurai families called Hitomochi–Gumi members in the Kaga feudal government (Hitomochi–Gumi members compose the front lines of the military line of Kaga Clan on a battlefield.) (Kagahan Shiryo, 1980e). In 1661, the first fire brigade was organized by two units, each consisting of four groups. Several family names of Hitomochi–Gumi members were seen among the head of each group (Kagahan Shiryo, 1980f).

When we see the process of development of the fire prevention system of the Kaga Clan, the highest impact seems to be a big fire that occurred every several years, not the completion of Tatsumi Aqueduct. If the significant purpose of Tatsumi Aqueduct is fire prevention, it is natural that a fire brigade should be organized immediately. This large time gap indicates that fire prevention was not a top objective. Therefore, a traditional understanding is more suitable to be changed.

15.5 CONCLUSIONS

This historical water project signifies a high level of understanding of the formation of the terrain, surveying, and advanced construction techniques and management in terms of the historical water technology in Japan.

In this study, the authors extended reviews on the unique characteristics of Tatsumi Aqueduct from the standpoint of sustainability of the aqueduct. A new finding is obtained by a detailed inspection of a drawing called "Tatsumi Jyosui Esujino Ezu"—the oldest drawing to show the whole route of the aqueduct and was drawn after a 1799 big earthquake in the Kanazawa area. It is edited by the chief engineer of the Kaga feudal government with memos on damage and restoration projects. To these memos, the length of the tunnel part was increased to prevent damage to open channel sections by slope failure.

The inlet of the inverted siphon, which made the water supply by Tatsumi Aqueduct possible to Ninomaru fortress, is written in old documents as Ishibiki Water Gate. However, the exact location of the gate has been unknown. This study identified the location of Ishibiki Water Gate on the oldest traditional drawing, which shows the detailed plan of Tatsumi Aqueduct and was discovered recently.

It is clarified that the pressurized pipe length for the inverted siphon in the Kan-ei period (around 1634) exceeded 1 km, considering the identified location of Ishibiki Water Gate. The chronological change of the location of the inlet of the inverted siphon is summarized, considering the land-use change in the current Kenrokuen garden area.

Old documents of the Kaga feudal government in the first half of the 17th century were studied on fire prevention and fighting. A big fire broke out in Kanazawa in 1635, but no article on the use of Tatsumi Aqueduct for fire prevention and fighting was found. The patrol system was strengthened four years later than this fire. It is also clarified that the first fire brigade was established in 1661, 29 years later than the completion of the Tatsumi Aqueduct, casting doubt on the standard explanation that the main purpose of Tatsumi Aqueduct was fire prevention.

In this study, old documents and drawings are listed below.

[1] Bunka Rokunen Tatsumi Yosui Ezu, Owned by the Ishikawa Prefectural Museum of History.

[2] Tatsumi Jyosui Esujino Ezu, Owned by the Ishikawa Prefectural Museum of History.

[3] Mitsubo Kikigaki, Owned by Kanagawa City Public Library Tamagawa Branch, History of Kaga feudal domain before Manji 1658.
[4] Enpo Kanazawazu, Owned by Kanazawa City Public Library Tamagawa Branch, 1673 to 1681 (map of Kanazawa Castle town in the 1670s).
[5] Kanazawa Goten Ezu, Owned by Kanazawa City Public Library Tamagawa Branch.
[6] Kanazawajyochu Mizunoteno Hizu, Owned by Kanazawa City Public Library Tamagawa Branch.
[7] Kanazawa Castle Fushin Sakuji Historical Records 2, Ishikawa Prefectural Research Institute of Kanazawa Castle, 12 pp.

ACKNOWLEDGMENTS

This article is based on the conference paper titled "Changes of a route and an upstream well for an inverted siphon of Tatsumi Canal in the City of Kanazawa" in the Proc. of the 21st IAHR-APD Congress 2018, Yogyakarta, Indonesia, and has been improved and modified to fit the book.

FUNDING

The authors would like to express their sincere thanks for financial support from the Hokuriku Regional Management Service Association in 2015, River Fund in the category of Fund for Groups for Rivers in 2016 and 2017, and the Chubu Regional Division of the Japan Society of Civil Engineers for Research Fund to the third author.

REFERENCES

Aoki, H. (1983a). *Kaga Tatsumi Aqueduct* (ed. Takabori, K.), Research Group of Cultural Assets related to Tatsumi Dam, 368–369 (In Japanese).
Aoki, H. (1983b). *Kaga Tatsumi Aqueduct* (ed. Takabori, K.), Research Group of Cultural Assets related to Tatsumi Dam, 448 (In Japanese).
Aoki, H. (1983c). *Kaga Tatsumi Aqueduct* (ed. Takabori, K.), Research Group of Cultural Assets related to Tatsumi Dam, 380–384 (In Japanese).
Aoki, H. (1983d). *Kaga Tatsumi Aqueduct* (ed. Takabori, K.), Research Group of Cultural Assets related to Tatsumi Dam, 385–389 (In Japanese).
Aoki, H. (1989). *Kaga Tatsumi Aqueduct Higashiiwa Tunnel and surrounding Environment-A Record related with Tatsumi Dam Construction* (ed. Takabori, K.), Research Group of Kaga Tatsumi Aqueduct Higashi-iwa Tunnel, 37 (In Japanese).
Aoki, H. (2001). *Advanced technology in the early modern period in the Tatsumi Aqueduct*, Ph.D. thesis submitted to Kanazawa University, 130pp (In Japanese).
Kagahan Shiryo (1980a). *Maeda Ikutoku Kai* (Copyright Holder), Reprinted Edition, Seibundo Publication, 2: 460–463 and 641–646 (In Japanese).
Kagahan Shiryo (1980b). *Maeda Ikutoku Kai* (Copyright Holder), Reprinted Edition, Seibundo Publication, 2: 692–696 (In Japanese).
Kagahan Shiryo (1980c). *Maeda Ikutoku Kai* (Copyright Holder), Reprinted Edition, Seibundo Publication, Annex. Chronological Table, 2: 999–1006 and 3: 1018 (In Japanese).
Kagahan Shiryo (1980d). *Maeda Ikutoku Kai* (Copyright Holder), Reprinted Edition, Seibundo Publication, 2: 664 (In Japanese).

Kagahan Shiryo (1980e). *Maeda Ikutoku Kai* (Copyright Holder), Reprinted Edition, Seibundo Publication, 2, Kan'ei 16: 905–906 (In Japanese).

Kagahan Shiryo (1980f). *Maeda Ikutoku Kai* (Copyright Holder), Reprinted Edition, Seibundo Publication, 3: 912 (In Japanese).

Kitamura, G. (1983a). *Kaga Tatsumi Aqueduct* (ed. Takabori, K.), Research Group of Cultural Assets related to Tatsumi Dam, 243–250 (In Japanese).

Kitamura, G. (1983b). *Kaga Tatsumi Aqueduct* (ed. Takabori, K.), Research Group of Cultural Assets related to Tatsumi Dam, 250–263 (In Japanese).

Nagayama, N. (2006). *Comprehensive Story of Kenrokuen Garden-History and Utilization*, Katsura Publisher, 307 pp (In Japanese).

Ogawa, K. (1962). Mines of Kanazawa feudal domain, *Journal of Ecchu Local History*, No. 24, 36 (In Japanese).

Tamai, N. and Yamada, T. (2006, June). *Purpose of the Tatsumi Aqueduct examined through the capacity of reverse-siphon, Historical studies in civil engineering*, Proceedings of the Annual Conf. of JSCE, 26: 153–159 (In Japanese).

Tamai, N., Todo, H. and Ikemoto, T. (2018, September). *Changes of a route and an upstream well for an inverted siphon of Tatsumi Canal in the city of Kanazawa*, Proceedings of the 21st IAHR-APD Congress, 1145–1152.

Yamaguchi, K. (2008). A mining town in Akita feudal domain in the early Edo period, 233–250, *Collected Works of Keiji Yamaguchi*, Vol.2, Azekura Publisher (In Japanese).

Part III

Historical water projects and traditional water technologies in Korea

Chugugi, invented in the 15th century in Korea, the instrumental rain gauge first in the world

The Korean peninsula has been affected by a monsoon climate with heavy rainfall in summer and dry seasons other than summer. Typhoon has also been frequently set on, sometimes causing catastrophic disasters, while at the same time inducing positive effects on rice cultivation badly affected by frequent droughts. Inland navigation had been common in Korea since the Koryo dynasty in the 10th century until the early 20th century, while they used the river channels themselves without any large canalization activities.

This part introduces the kingdom-age reservoirs in Korea, first built in the 4th century AD, and registered as the World Heritage Irrigation Structure by

DOI: 10.1201/9781003222736-18

International Commission on Irrigation and Drainage. This part also introduces the old practice and technologies of stream improvement works, especially in urban streams, such as the Cheonggyecheon stream in Seoul and the Yongsan-gang river in Damyang. In those river works, we can find the wisdom of the old generations using live plants for reveting and reinforcing the levees. It is the old version of the close-to-nature river works, which may be a good example of Confucius's philosophy of "Gaining new knowledge by reviewing old things." Last, this part introduces the ancient rainfall gauge—*Chugugi*, invented in the 15th century AD, which is recognized as the world's first instrumental rain gauge.

Chapter 16

Kingdom age irrigation for paddy farming under monsoon in Korea

Jin-Yong Choi
College of Agriculture and Life Sciences, Seoul National University

CONTENTS

16.1 INTRODUCTION

The Korean Peninsula is located in the Far East of Asia. South Korea occupies the southern portion of the Korean Peninsula, extending some 1,100 km from the Asian mainland. The Yellow Sea flanks this mountainous peninsula to the West and the East Sea to the East. Its southern tip lies on the Korea Strait and the East China Sea.

South Korea has a humid continental and subtropical climate. It is affected by the East Asian monsoon, with precipitation heavier in summer during a short rainy season, which begins at the end of June through July. Winters can be frigid, with the minimum temperature dropping below −20°C in the inland region of the country. In Seoul, the average January temperature range is −7°C–1°C, and the average August temperature range is 22°C–30°C. Winter temperatures are higher along the southern coast and considerably lower in the mountainous interior. In most parts of the country, summer can be uncomfortably hot and humid, with temperatures exceeding 30°C in most parts of the country. South Korea has four distinct seasons; spring, summer, autumn, and winter. Spring usually lasts from late March to early May, summer from mid-May to early September, autumn from mid-September to early November, and winter from mid-November to mid-March (Choi, 2016).

Rainfall is concentrated in the summer months of June through September. The southern coast is subject to late summer typhoons that bring strong winds and heavy rains. The average annual precipitation varies from 1,370 mm in Seoul to 1,470 mm in Busan. There are occasional typhoons that bring high winds and floods (Wikipedia.org, Climate of South Korea).

Korea has a monsoon called "Jang-ma" from June to July. It provides abundant rainfall for paddy cultivation under a comparably sound condition for agriculture with

DOI: 10.1201/9781003222736-19

four vivid seasons and moderate temperatures. Therefore, the wet and dry seasons repeat every year, with seasonal variations of precipitation requiring irrigation and drainage systems for regular agricultural activities. Usually, June through August is the wet season, while most of the yearly rainfall occurs during this period, and the other nine months have about 30% of the annual rainfall. Therefore, most crops are cultivated from March to October, except for protected farming and winter crops. It is the reason to store the water for paddy farming before the coming monsoon and why a reservoir is required to cultivate paddy in Korea. Figure 16.1 demonstrates the locations of the two representative kingdom age reservoirs in Korea that will be introduced.

Figure 16.1 Map of South Korea and the location of two representative kingdom age reservoirs.

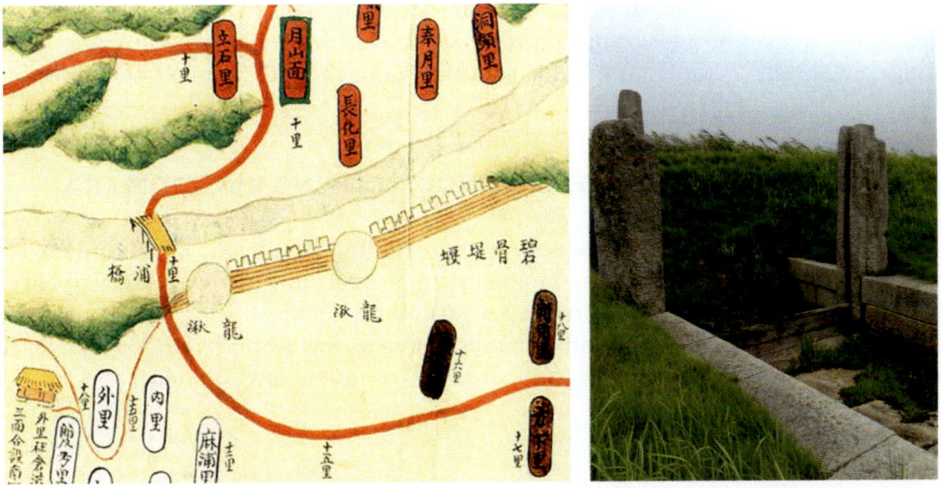

(a) (b)

Figure 16.2 Byeogolje map and the second watergate. (a) Map of Byeokgolje in 1872 (Source: Gyujangkak, Seoul National University). (b) The second watergate Jangsaenggeo of Byeokgolje. (Author's photograph.)

Rice has been the essential grain as the staple food of Koreans since rice was introduced. It was reported from excavated relics that rice cultivation started about 1000 BC. Even though rice cultivation in Korea seemed to have started around 1000 BC, records of structural construction as the national projects for irrigation and drainage systems can be found from the 60s BC, when the ancient countries were established. The mainstream of irrigation and drainage history in Korea is concerned with struggling with floods and droughts to obtain a stable food supply by relying on the ideology that farmers are the fundament of heaven and earth in Korean 'Nong-ja-cheon-ha-ji-dae-bon'. (Choi, 2016)

Therefore, enhancing agriculture and irrigation of paddy fields have been a principal responsibility governing the country in Korean history. Furthermore, when the country is faced with hardship accompanying destabilizing social movements or starvation due to natural disasters, such as droughts and floods, the government has attempted to alleviate these disasters by constructing new irrigation systems or rehabilitating existing ones.

This chapter introduces four representative reservoirs constructed during the kingdom era in Korea, registered as the World Heritage Irrigation Structure (WHIS) by International Commission on Irrigation and Drainage (ICID), to demonstrate the Korean historical efforts for paddy farming under the weather variation in monsoon areas. It includes Byeokgolje (Figure 16.2) and Hapdeokje, constructed from AD 330 to AD 1400s.

16.2 HISTORY OF IRRIGATION PROJECT DURING KINGDOM ERA

Endless Korean efforts had continued to overcome the water shortage since they settled down in the Korean Peninsula. From the records and relics, dam construction crossing the river for storing water can be found from early BC and continued during the Dynasty. Before the modern period, agriculture was the most crucial industry to sustain the country. Therefore, the Dynasty made efforts to keep the agricultural productivity while confronting weather variation, including drought. This history of agricultural policies and irrigation and drainage projects implemented during the kingdom age during four different kingdom periods is explained in the KRC and KCID (2014) in detail.

16.3 TWO REPRESENTATIVE RESERVOIRS DURING KINGDOM ERA

In this chapter, two representative reservoirs, including Byeokgolje and Hapdeokje, which were constructed during the kingdom era, are introduced. Two reservoirs are historical heritages of engineered irrigation structures demonstrating the efforts and skills of Korean ancestors, and both were registered as the World Irrigation Heritage Structure of ICID (International Commission on Irrigation and Drainage).

16.3.1 Byeokgolje

Byeokgolje is located in Gimje City, Chollabukdo province, famous as the rice bowl of Korea. Byeokgolje was the first and the most significant reservoir in Korea. In Samguksagi—the historical book of the ancient three kingdoms, it is written that the reservoir was built in the 21st year of Shilla King Heulhae's reign (AD 330). Then, King Wonseong of the Unified Shilla Kingdom rebuilt it in the 6th year of his ruling period (AD 790). Later, King Hyeonjong and King Injong of the Goryeo Dynasty reconstructed it. And then, it was remodeled by King Taejong of the Joseon Dynasty (AD 1415) in the 15th year of his ruling. But, it was flooded away in the reign era of the 2nd year (AD 1420) of King Sejong's reign after heavy rain. Afterward, Dongjin Farmland Reform Association converted the bank into a waterway for framing irrigation in 1925. As a result, the reservoir lost most of its original form and remains. (KCID, 2016)

Byeokgolje installed five floodgates at construction, but now 3 km long bank, the 2nd floodgate, and the 5th floodgate remain. The second watergate is called Jangsaenggeo, and the fifth watergate, Gyeongjanggeo, appears to the south for about 2 km along the bank.

In the agricultural age, flood control was a matter of survival. Therefore, the Byeokgolje had been damaged over the years due to not enough capacity for floods. It is estimated that at least 320 thousand people per year were mobilized only to build a bank. So many people gathered to repair the bank when the bank experienced a flood. The repairing works continued until the facility lost its function as an irrigation structure in the 1900s. The large-scale irrigation facility was the key industry for founding a country. When we consider the social structure at that time, it can be guessed that the construction, maintenance, and repair of Byeokgolje were important national projects.

Byeokgolje marks an epoch in the history of Korean Science, supposing that it shows us that the ancient kingdom already had such high-developed engineering skills to build the great reservoir with soil and stoned watergates. The dam and watergate materials obtained from vicinity nature demonstrate how they utilize clay and finely honed stones to reserve and release irrigation water. Therefore, Byeokgolje and its repairing monument in the Joseon Dynasty have been designated as the 111th Historical Relics of Korea. (KCID, 2016) and registered as a World Irrigation Heritage Structure of ICID in 2016.

16.3.2 Hapdeokje

The Hapdeokje (Figure 16.3) is located about 90 km south of Seoul, Korea's capital city, and in Dangjin City in the northern part of Chungcheongnam-do Province. The Hapdeokje is a representative reservoir constructed during the Chosun Dynasty, demonstrating Chosun engineers' ability to build a dam for longer than 8 km and design the dam in an arc shape to resist the water pressure. They constructed the dam layered with clay and wooden leaves to stabilize the dam and prevent the water flow through it, which was found during an archaeological study in 2002 (Seong, 2002). The year of initial construction is not known, but it was rehabilitated during the Joseon Dynasty. After rehabilitation, it had 1,771 m of dam length, 4 m of dam height, 6.5 m of dam crest width, 11.0–18.0 m of dam bottom width, 8–9 km of circumference, 102 ha of water surface area, and 720 ha of irrigated area (KCID, 2017). Hapdeokje had eight watergates to take out irrigation water, but it is unclear that particular gate was used

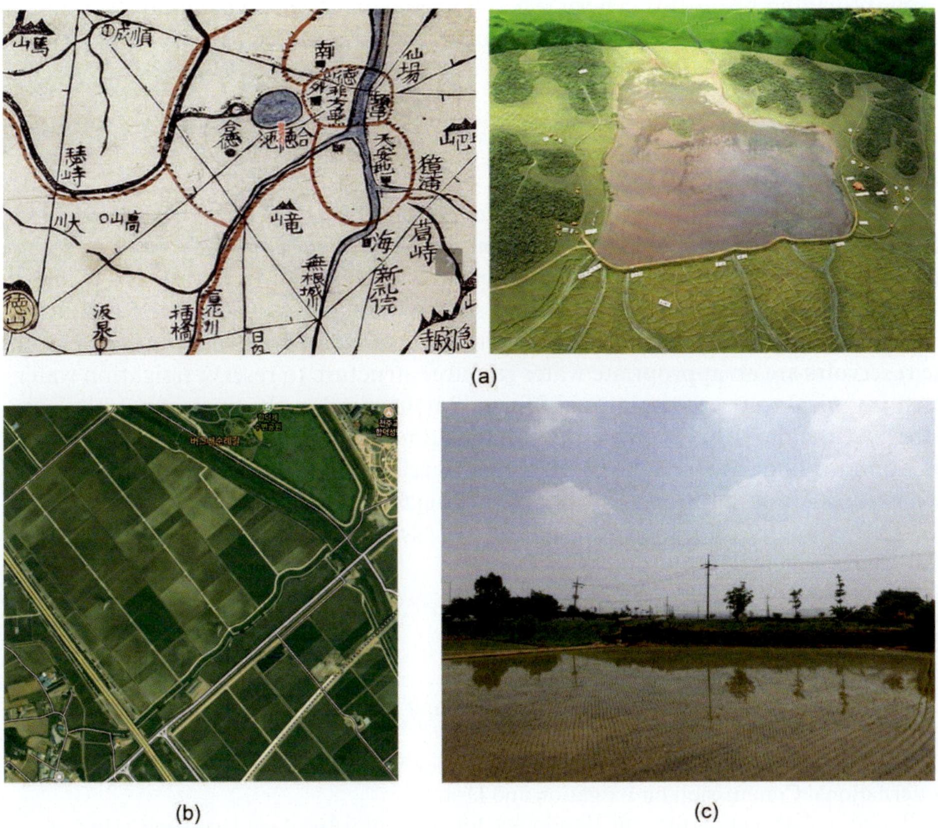

Figure 16.3 Maps, aerial photographs, and watergate of Hapdeokje. (a-1) Map of Hapdeokje in 1861 (Daedongyeojido). (Source: Gyujangkak, Seoul National Univ.) (a-2) Hapdeokje during Chosun Dynasty. (Author's photograph from Hapdeokje Museum.) (b) Present aerial view of Hapdeokje. (c) A stone-made watergate and part of embankment remained of Hapdeokje. (Author's photograph.)

to release water as a spillway. One stone-made gate (Waemok gate) remains, showing the original structure, and it can be used as a flood release spillway.

It was damaged by the floods in 1473 (the 3rd year of King Seongjong) and rehabilitated in 1474. In 1506, the dam lost its function but was rehabilitated and excavated in 1768 (the 44th year of King Youngjo) to mobilize 12,000 human resources from the Hongju area. In 1778 (2nd year of King Jeongjo), 8,053 people (4,553 from Hongju county and 3,500 from neighboring counties) worked for two broken dam sections. In 1792 (the 16th year of King Jeongjo), 6,500 people (3,000 from Hongju county and 3,500 from neighboring counties) were mobilized for rehabilitation (KCID, 2017). Also, it was rehabilitated again in 1800 (the 24th year of King Jeongjo), the inside slope of the dam was reformed with a vertical concrete wall in 1913, and the spillway was repaired with concrete structures. In the 1960s, Yedang reservoir was built 16 km south of the Hapdeokje. All the Hapdeokje and the Hapdeokje's irrigation area was included in the benefited area of the Yedang reservoir. The Hapdeokje was famous during the Joseon Dynasty for blooming lotus flowers in the reservoir (KCID, 2017). Lotus covers most of the water surface due to shallow water depth and good soil during the blooming season, providing beautiful scenery. The Hapdeokje area was rehabilitated as a museum and park complex recently. The Hapdeokje was registered as a World Irrigation Heritage Structure of ICID in 2017.

16.4 CONCLUSIONS

In this article, the kingdom age irrigation effort of Korean was reviewed. Paddy rice is a staple food of Korean, and they have put their measures for producing rice stable under the challenging weather since rice was introduced and settled down as a primary food. Korean efforts to produce food and secure water for irrigation have continued since more than 2,000 years ago, constructing dams for storing water in a reservoir. The reservoirs are an appropriate water resource structure to reserve irrigation water, which is typical to overcome drought for paddy farming in the monsoon area. They had the wisdom to utilize consolidated clay and natural materials for dam construction and finely honed stones for watergates to release the irrigation water. Through the measures for obtaining water during the kingdom era, we can learn about Korean ancestors' willingness for food and water—two essential resources for human sustainability and challenges to nature.

REFERENCES

Choi, J.-Y. (2016). Irrigation and drainage in Korea and ICT applications, *Irrigation and Drainage*, 65, 157–164.

KCID. (2016). Application form of Byeokgolje for the World Heritage Irrigation Structure in International Commission on Irrigation and Drainage.

KCID. (2017). Application form of Hapdoekje for the World Heritage Irrigation Structure in International Commission on Irrigation and Drainage.

KRC and KCID. (2014). Agricultural development in Yeongsan-gang River Basin, Korea, from poverty to prosperity, Korea Rural Community Corporation (KRC).

Seong, J.-Y. (2002). Dangjin Hapdeokje, Chungnam National University Museum Series, 24, Chungnam National University Museum.

Chapter 17

Reconstruction of the 1855 extreme flood and historical flood mitigation projects in the capital of Joseon Dynasty, Korea

Hyeonjun Kim and Cheolhee Jang
Korea Institute of Civil Engineering and Building Technology

CONTENTS

17.1 INTRODUCTION

Flood is one of the most severe natural disasters. In Asia, 3,454 disasters were recorded from 1970 to 2019, with 975,622 lives lost and USD 1.2 trillion reported in economic damages. Most of these disasters were associated with floods (45%) and storms (36%) (WMO, 2021). Global climate change predictions suggest a more frequent hydrometeorological extreme, including floods (IPCC, 2007).

Recent extreme flood events recorded at the end of the last and the beginning of the 21st century in many parts of the world have drawn attention (Benito et al., 2004). In Prague, the extreme flood of 1784 was reconstructed (Elleder, 2010). A 1,000-year history of typhoon landfalls in Guangdong, southern China, was reconstructed from Chinese historical documents records (Liu et al., 2001).

The ancient extreme floods in the urban stream were evaluated using instrumental rainfall data and flood damage records from historical documents. The hydrologic and hydraulic simulation models evaluated flood peaks, levels, and areas. Also, the estimated flooding area was compared to the flood damage records from the historical document—the *Veritable Records of the Joseon Dynasty*. River works and flood records were investigated through the Korean historical documents of the Joseon Dynasty (1392–1910) (Kim, 1998, 1999; KICT, 2005; Kim, D. et al., 2007; Kim, K. et al., 2007; Yoo et al., 2015); these include the Joseon Wangjo Silok (*Veritable Records of the*

DOI: 10.1201/9781003222736-20

Joseon Dynasty), which covers the entire period of the dynasty, and contains detailed records on river maintenance and improvement. The Annals of the Joseon Dynasty were registered in the UNESCO Memory of the World Register (UNESCO, 1997).

From the beginning of the Joseon Dynasty, the stream banks overflowed every time there were floods. The stream was usually polluted due to a lack of streamflow in the dry period. In 1760, during King Yeongjo, 200,000 people were recruited to widen the stream and build up the stone embankments, and the waterways were straightened to their present condition (Kim, 1998; Seoul Metropolitan Government, 2003). This study introduces a brief history of systematic urban river management and the 500 years of flood mitigation projects in ancient Korea.

17.2 LOCATION

The Cheonggyecheon (originally called *Gaecheon*, 開川, or "opened waterway") is a partially urban stream centered in the old downtown of Seoul, Korea, now known as a famous case of successful urban river restoration (Seoul Metropolitan Government, 2003). The Cheonggyecheon is the second tributary of the Hangang River, originating from the mountains Bukaksan and Inwangsan. The Cheonggyecheon catchment is 50.96 km^2 in size and 13.75 km long (Figure 17.1). The average annual air temperature is 12.8°C, and the average annual precipitation is 1,455 mm (the rainy season from June to August is 960 mm, 66% of the total) (Noh et al., 2005).

17.3 THE EXTREME FLOOD ON JULY 16, 1885

The flood damages were investigated through historical documents listed in Table 17.1. From 1770 to 1910, the daily rainfall, which is over 200 mm, was classified. The rainfall data measured by *Chugugi* in Seoul was used for evaluation (KICT, 2005; Yoo et al.,

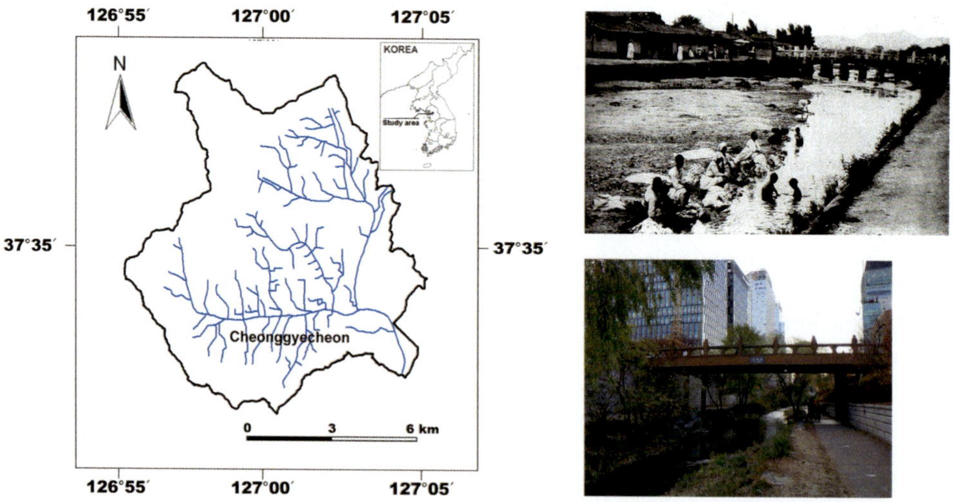

Figure 17.1 Boundary and stream networks of the Cheonggyecheon catchment and landscapes of stream in the early 1900s (upper right) and 2005 (bottom right).

Table 17.1 The extreme daily rainfall measured by *Chugugi* and flood damage records in the capital from the *Veritable Records of the Joseon Dynasty*

Flood event	*Daily rainfall (mm)*	*Flood records*
July 16, 1885	392.0	520 houses washed away
July 19, 1832	348.4	3,166 houses and 64 deaths
July 5–July 31, 1832	1,395.3	10,836 houses and 357 deaths (nationwide)
August 30, 1865	345.4	1,322 houses and many deaths
August 17, 1833	337.5	1,218 houses
September 5, 1851	281.1	Many houses washed way
July 10, 1816	240.1	–
July 1, 1792	236.8	Many houses washed way
July 11, 1816	224.0	3,688 houses (July 11, 1816)
July 8, 1888	216.8	No record
August 28, 1875	214.5	Many houses washed way

2007), and flood records were investigated for the entire period of the Joseon Dynasty (Kim, 1999; KICT, 2005). The ancient river geomorphology is assumed from the excavation report (Seoul Museum of History, 2006) and the Cheoggyecheon restoration project (Seoul Metropolitan Government, 2003; KWRA, 2003). The largest daily rainfall, 394 mm, was measured on July 16, 1885, and 520 houses were swept away in a day. The flood was most severe on July 19, 1832, with 348 mm of rainfall. Over 3,100 houses were washed out, and 64 persons were drowned.

The Hydrologic Modeling System (HEC-HMS) was applied to calculate flood hydrograph using the selected daily rainfall measured by the *Chugugi*. The Huff distribution Random Cascade model was considered to distribute daily rainfall to hourly rainfall. After computing flood peaks for each extreme rainfall, the River Analysis System (HEC-RAS) was used to estimate high flood levels and flooding depth with DEM. Then, the estimated flood area was compared to the flood damage records from the historical document (KICT, 2005).

From the applications for the several extreme floods and rainfall distributions, when the flood flow rate is over 190 m^3/s, the flood is overflowed, and flooding occurs near the riverside and low land area. When the flood peak is 342 m^3/s, the flooding area expands to the tributary area (Figure 17.2).

To analyze spatial flood damage, the flood level is overlaid on DEM (Digital Elevation Model). We assumed that the flooding damage occurred with a flood depth is over 0.5 m because of the traditional Korean house's *Daecheongmaru* (a kind of floor of the raised main hall of wooden floor). In the case of the mainstream, the flooding area is 1.96 km^2 for flood peak 194.6 m^3/s, and flood depth is over 0.9 m. Let us consider the area of flood depth over 0.8 m. The flooding area is 5.69 km^2 for the flood peak of 342 m^3/s. The damaged houses could be estimated at 1,040. It is comparable to 530 houses damaged in the *Veritable Records of the Joseon Dynasty* (Table 17.2).

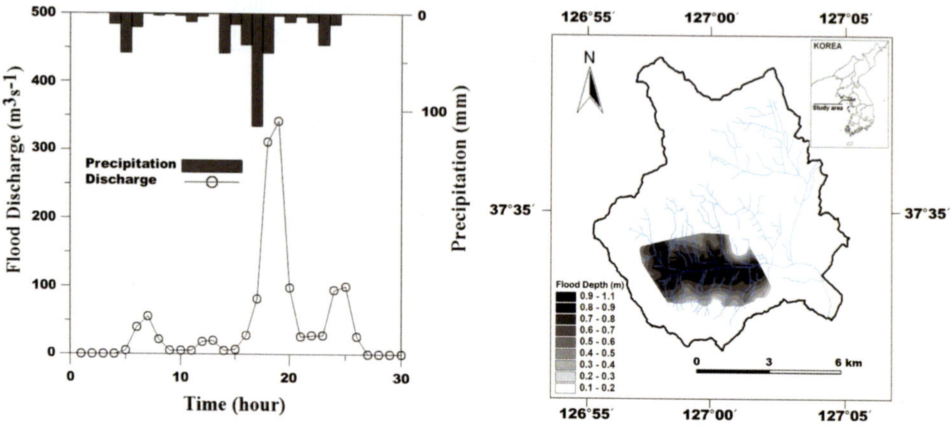

Figure 17.2 Estimated flood hydrograph and flooding area of mainstream and tributaries of Cheonggyecheon stream on July 16, 1885.

Table 17.2 Results of simulated flooding depth and area estimated house damages according to the extreme daily rainfall of 392 mm and flood peak 342 m^3/s

Flood depth (m)	*Flooding area (km^2)*	*Estimated house damage*
0.6–0.7	1.39	180
0.7–0.8	0.61	190
0.8–0.9	3.43	590
0.9–1.0	0.26	80

17.4 RIVER MANAGEMENT FOR FLOOD MITIGATION

After establishing the Kingdom of Joseon in 1392, the capital was relocated from *Gaesung* to *Hansung* (now called Seoul). The first King Taejo (1335–1408) recognized that city walls and river works were essential for the safe settlement of the new kingdom. In 1396, the Seoul City Wall construction took 49 days by 118,070 mobilized workers from local provinces. Earthen walls were built in the flat valley areas, and stone walls were constructed in the hillside regions, but the water gates in the main stream crossing the city were not completed because of cold weather and a short construction period (Kim, 1998).

After big floods hit the new capital in 1407 and 1410, river works were undertaken. For systematic plan and construction, the *Gaegeodogam* (開渠都監, river works agency) was temporally established in 1412, and 52,800 soldiers were re-mobilized from three local provinces for 31 days, from January 15 to February 15, despite the winter conditions. Stone embankments reinforced the streams near the palace, while wood and soil were used downstream. Also, several stone bridges, including *Gwantonggyo,* were built to protect against floods. As shown in old hand-drawn maps of the capital in the 19th century, the mainstream was artificially straightened through river maintenance works (Kim, 1998; Seoul Development Institute, 2010). Ancient tributaries are in Figure 17.3.

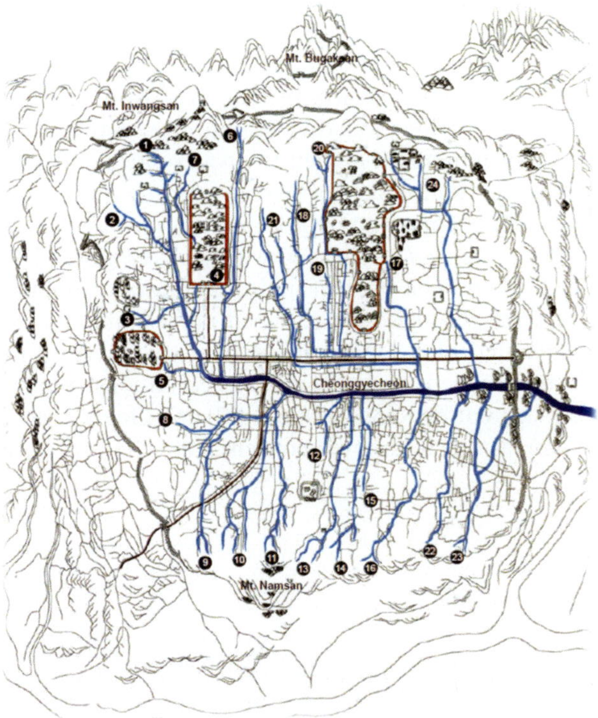

Figure 17.3 The tributaries of Cheonggyecheon (Seoul Development Institute, 2010).

Many houses were washed away caused by a severe flood in 1421, and the water depth on the main roads in the capital city exceeded 0.6m on August 8 (Kim, 1999). The capital's mayor recommended plans to enhance river maintenance, implemented by King Sejong (1397–1450). He conducted the dredging and expansion of the mainstream and its tributaries, built two more water gates for flood discharge, and replaced wooden bridges across the mainstream with stone. It was a landmark river engineering project to improve the capital's infrastructure. He completed the river improvement, including small tributaries. During the early Joseon Dynasty, infrastructure construction flourished, but there were few river maintenance records in the capital for 200 years after the 16th century (1537–1735). As a result, the riverbed had risen nearly to the level of the bridges (up to 2.5 m high) due to sediment deposition from the upland erosion of surrounding mountains (Kim, 1998).

After several famines in the 17th century, the capital population increased from 80,000 in 1657 to 200,000 in 1669. The rapidly increasing population caused serious environmental and social problems. A city slum along the Cheonggyecheon was built, and the unregulated deforestation of surrounding mountains produced severe landslides, soil erosion, and river flooding. The maintenance of the stream became a national issue again. Starting in 1752, the 21st King Yeongjo (1694–1776) spent seven years collecting public opinions and convincing the people before finally deciding to initiate a large-scale dredging project.

The *Juncheonsa* (濬川司, dredging work agency) was established to plan and carry out the project and river maintenance. The *Juncheonsajeolmok* (濬川司節目, code of dredging work) was officially announced in October of 1759. The dredging project lasted from February 18 to April 15, 1760. Two hundred thousand workers were recruited, including 50,000 paid laborers and 150,000 citizens mobilized from five local provinces (Kim, 1998). The mainstream corridor was divided into five sections from Songgigyo to Yeongdogyo across Ogansumun and arched five water gates under the capital wall to reduce the construction period. After the river works, the total length reached 3,000 walk steps (one walk step is approximately 1.4 m), the width upstream was over 20 walk steps, and the width of *Yeongdogyo* was 52 walk steps (Figure 17.4).

The project was completed in 57 days. The beds of mainstream and tributaries were dredged to reduce flood water levels, bridges were repaired, stone embankments were renewed, willow trees were planted along the embankment to prevent bank erosion, and the waterways were straightened up to the present conditions. Further large-scale dredging works were carried out by Kings Sunjo (1833) and Gojong (1873), demonstrating that dredging the stream was a national concern until the end of the Joseon Dynasty.

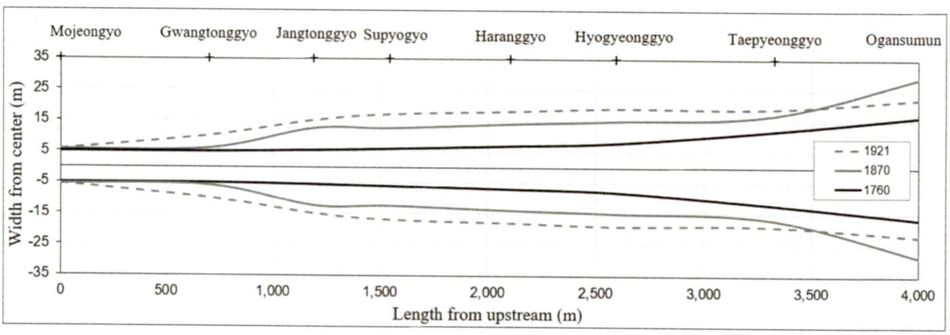

Figure 17.4 Channel width changes of Cheonggyecheon mainstream after 1760s dredging work; the channel width was expanded in 1870 and 1921 to prevent flood.

Figure 17.5 Photos of stone-retaining walls in the tributary in a postcard around the early 1900s (left, Kwon, 2005) and mainstream in the excavation of Cheonggyecheon for the restoration after 2003 (right, Seoul Museum of History, 2006).

The roughly scabbled granite was employed for the stone-retaining wall with two types of shapes in Figures 17.5 and 17.6. The normal shape (left) comprises large basement stone layers, and the wedge shape (right) has a trapezoidal, rectangular stone and large basement to enhance stability.

The substructures of the stone bridge were also enhanced to prevent scouring and ground subsidence. The wooden piles were installed, and the riverbed under the bridge was covered with ripraps and a large, flat stone basement (Figure 17.7).

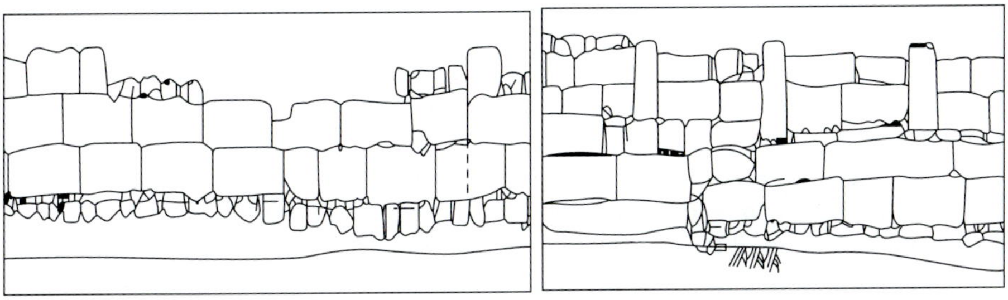

Figure 17.6 Types of stone-retaining walls (Seoul Museum of History, 2006).

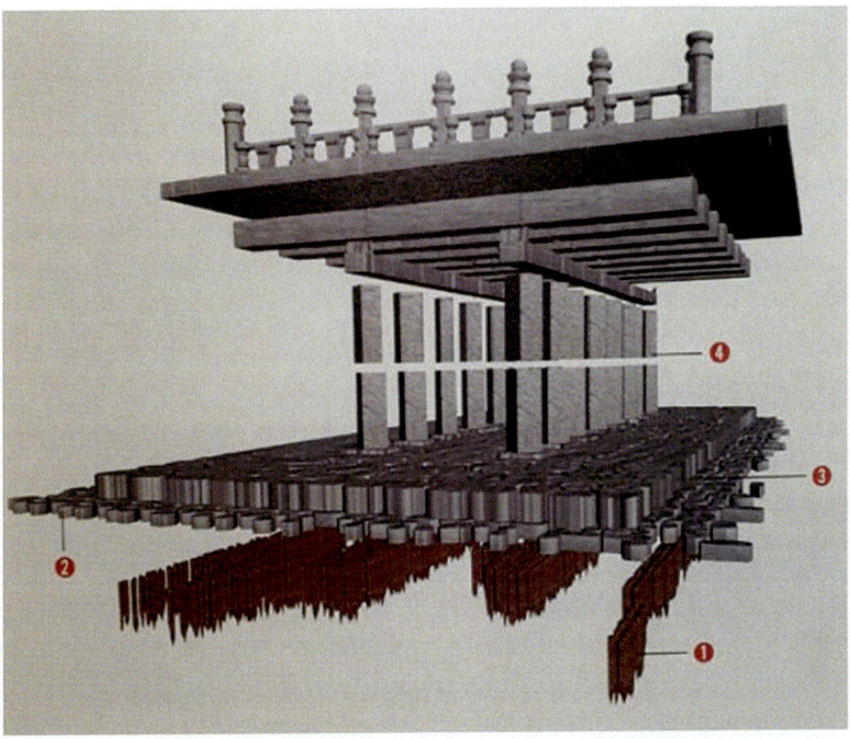

Figure 17.7 Substructures of Gwangtonggyo (Seoul Museum of History, 2006); the substructures of the bridge were constructed with (1) wooden piles to prevent ground subsidence, (2) ripraps with stones and rubbles, (3) stone basement against scouring, and (4) piers of the bridge.

17.5 CONSTRUCTION REPORT AND GUIDELINES FOR DREDGING WORKS

To guide future dredging and river management, King Yeongjo ordered the recording of all details regarding the project, printing them in a book, *Juncheonsasil* (濬川事實, a construction report of 1760s dredging project) (Seoul Historiography Institute, 2001). The information recorded included expenses, laborers, the number of horses used, etc. This book became a reference for later dredging works and composed of the king's preface, the *Juncheonsasil* (detailed contents of dredging work), and the *Juncehonsaje-olmok* (a code for dredging work). The book's preface was written by King Yeongjo, in which he described the dredging work's importance, and the purpose was for the safety of the people, not the civil work itself. In Figure 17.8, the painting *Juncheonsisayeol-mudo* (濬川試射閱武圖) illustrates the construction process vividly.

Some words were inscribed on the piers of stone bridges, *Gwangtonggyo* (broad-way bridge) and *Supyogyo* (watermark bridge), including "*Gyeongjinjipyeong* (庚辰地平)," which means riverbed level mark in the year of 1760 (Figure 17.9a and b). After that year, every time the carved words were covered by sediment, it was understood

Figure 17.8 *Juncheonsisayeolmudo*—a hand-drawn painting to explain dredging works in 1760 and the landscape of the Ogansumun (Seoul Development Institute, 2010); On the top of the Ogansumun (five water gates), the king sits in a chair under a large sunshade, and high officers stand around the king. The workers dig in the riverbed using spades and shovels, and remove material from the stream. Along the stream embankment, willow trees were being planted. The willow is commonly used in river bank enhancement, because it inhabits waterfront areas and produces extensive root systems to strengthen the soil bank.

(a) (b) (c) (d)

Figure 17.9 Ground level marks for dredging. After the dredging works in 1760, 1883, and 1869; The ground level marks were inscribed on the piers of *Gwangtonggyo* (Br.) and *Supyogyo* (Br.), From left *Gyeongjinjipyeong* (1760) on Supyogyo (a), *Gyeongjinjipyeong* (1760) (b), *Gyesagyeongjun* (癸巳庚濬, 1833) (c), and *Gisadae-jun* (己巳大濬, 1869) (d) on *Gwangtonggyo*. (Photos by Hyeonjun Kim.)

that dredging work should be carried out. As a result, floods in the capital's downtown have been somewhat controlled, although every subsequent king of the Joseon Dynasty has worried about the stream's burial by sediment.

17.6 CONCLUSIONS

The extreme floods in the capital, Cheonggyecheon urban river, were selected to analyze flood quantity and level using modern hydrologic and hydraulic simulation models. We applied the HEC-HMS model for the design flood using rainfall data measured by the *Chugugi*. We calculated the flood levels and inundated areas by the HEC-RAS flood simulation model. Ten extreme flood events after 1770 and the damage records were compared with rainfall records. On July 16, 1885, the daily rainfall was 392 mm, estimated to be more than 50 years' flood frequency in modern flood design criteria. The flood peak was calculated with 342 m^3/s, flooding depth was over 0.7 m, and flooding area was larger than 3.4 km^2. In the historical document, 520 houses were swept away caused by the flood on July 16, 1885. The historical document and rainfall data measured by the *Chugugi* can be utilized to understand the ancient floods.

River maintenance is crucial to reducing flood risk in ancient Korea and modern nations. The evolution of Cheonggyecheon—the capital of the Joseon Dynasty for 500 years shows the essence of the river works, including dredging, improvements, policy, and institution.

The primary river management stage was started in the early 15th century to establish the capital city's safety against floods. After the big floods in 1407 and 1410, the river works agency was established in 1412, the natural stream corridor was straightened, and a stone embankment was constructed near the palace area. The completing stage after the 18th century was started in 1760. King Yeoungjo had listened carefully to the public opinions and prepared the great river work, including enhancement of embankment and dredging the sediment deposited over 200 years. He prepared the dredging project very systematically. First of all, the code and agency for dredging work were enacted. Moreover, 50,000 paid workers were employed to support the relief of the poor after the severe famines. The retaining stone wall was expanded downstream, and the standard guideline for dredging work was proposed and kept until the early 20th century.

The stone embankment was planned for the steep upstream area, and the earthen river bank with trees was designed for the mild downstream. The dredging was based on human labor and a plow pulled by oxen. The dredged sand was removed to a remote area from the stream to prevent re-deposition. The code systematically managed the Cheonggyecheon, *Juncheonsajeolmok* in 1759, river management agency, *Gaegeodogam* in 1412, *Juncheonsa* in 1760, and periodical dredging works. Occasionally, nationwide mobilization and volunteering with citizens are necessary for river management. After the grand dredging works in 1760, the details of the whole project were recorded in a project report, *Juncheonsasil*, which was used as a guidebook for river management.

Moreover, the reference mark "*Gyeongjinjipyeong* (庚辰地平)" for riverbed management was inscribed on the pier of the bridge. The Joseon Dynasty's dredging work was evaluated as a project to protect people from flooding and ensure safety in their life. As a result, floods in the capital's downtown had been somewhat controlled. Since then, every king of the Joseon Dynasty deeply considered river maintenance, including dredging and embankment against sedimentation and flooding. Through the history of dredging works, we can understand the traditional technology for river management and the attitude of ancient society against natural hazards like a flood.

ACKNOWLEDGMENTS

This article is based mainly on the "A review on ancient urban stream management for flood mitigation on the capital of the Joseon Dynasty, Korea," authored by us and published in the *Journal of Hydro-environment Research* (Kim & Jang, 2019) and the conference article, which was presented in 21st IAHR-APD Congress (Kim et al., 2018). It has been rewritten fitting this monograph, focusing on reconstructing the extreme flood and systematic river management policy and mitigation projects by reinforcing and improving graphs and tables.

FUNDING

This research was initiated with the support of the 2005 SOC Project (05-GIBANGU-CHUK-D03-01) through the Design Criteria Research Center for Abnormal Weather-Disaster Prevention (DCRC-AWDP) in KICTTEP of MOCT. This research was also supported by the Basic Research Program (20160160-001) of Korea Institute of Civil Engineering and Building Technology.

REFERENCES

Benito, G., Lang, M., Barriendos, M., Llasat, M.C., Frances, F., Ouarda, T., Thorndycraft, V.R., Enzel, Y., Bardossy, A., Coeur, D., & Bobee, B. (2004). Use of systematic, palaeoflood, and historical data for the improvement of flood risk estimation. Review of scientific methods. *Natural Hazards*, 31(3), 623–643.

Elleder, E. (2010). Reconstruction of the 1784 flood hydrograph for the Vltava River in Prague, Czech Republic. *Global and Planetary Change*, 70, 117–124.

IPCC (2007). IPCC Fourth Assessment Report. http://www.ipcc.ch

KICT (2005). Improvement of Design Criteria of Hydrologic Structures for Defense against Abnormal Floods (in Korean).

Kim, D., Yoo, C., & Kim, H. (2007). Evaluation of major storm events measured by Chukwooki and recoded in Annals of Chosun Dynasty: 2. Quantitative approach. *Journal of Korea Water Resources Association*, 40(7), 545–554 (in Korean).

Kim, H. (1998). Survey of River Works in the Joseon Dynasty, Korea Institute of Construction Technology (in Korean).

Kim, H. (1999). Investigation Report on Flood Records of the Joseon Dynasty. Korea Institute of Construction Technology (in Korean).

Kim, H., & Jang, C. (2019). A review on ancient urban stream management for flood mitigation on the capital of the Joseon Dynasty, Korea. *Journal of Hydro-Environment Research,* 22, 14–18.

Kim, H., Jang, C., & Hong, I.P. (2018). Analysis of ancient extreme floods in Chenggecheon urban river based of Chugugi rainfall data and historical documents during 19th century in Seoul, Korea, *Proceedings of the 21st IAHR-APD Cong. 2018, Yogjakarta, Indonesia*, Special Session 1: 1–4.

Kim, K., Yoo, C., Park, M., & Kim, H. (2007). Evaluation for the usefulness of Chukwookee data in rainfall frequency analysis. *Journal of Korea Water Resources Association*, 40(11), 851–859 (in Korean).

Kwon, H.H. (2005). Postcards from Joseon, Minumsa, Seoul, Republic of Korea (in Korean).

KWRA. (2003). Report on hydraulic and numerical experiment for Cheonggyecheon restoration (in Korean).

Liu, K., Shen, C., & Louie, K. (2001). A 1,000-year history of Typhoon Landfalls in Guangdong, Southern China, Reconstructed from Chinese Historical Documentary Records. *Annals of the Association of American Geographers*, 91(3), 453–464.

Noh, S., Kim, H., & Jang, C. (2005). Application of WEP model to the Cheonggyecheon watershed. *Journal of Korea Water Resources Association*, 38(8), 645–653 (in Korean).

Seoul Development Institute (2010). A City and Its Stream, Seoul (in Korean).

Seoul Historiography Institute (2001). Juncheonsasil and Jugyojinam (in Korean).

Seoul Metropolitan Government (2003). Cheonggyecheon Restoration Project (in Korean).

Seoul Museum of History (2006). The Excavation of Cheonggyecheon, p. 141 (in Korean).

UNESCO, Memory of the World Register (1997). The Annals of the Choson Dynasty. http://www.unesco.org

Yoo, C., Kim, D., & Kim, H. (2007). Evaluation of major storm events measured by Chukwooki and recorded in Annals of Chosun Dynasty: 1. Qualitative approach. *Journal of Korea Water Resources Association*, 40(7), 533–543 (in Korean).

Yoo, C., Park, M., Kim, H., Choi, J., Sin, J., & Jun, C. (2015). Classification and evaluation of the documentary-recorded storm events in the Annals of the Choson Dynasty (1392–1901), Korea, *Journal of Hydrology*, 520, 387–396.

WMO (2021). Weather-Related Disasters Increase Over Past 50 Years, Causing More Damage but Fewer Deaths. https://public.wmo.int/

Chapter 18

Technographical review of embankments for dams and levees in Joseon Dynasty (1392–1910), Korea

Un Ji
Korea Institute of Civil Engineering and Building Technology

Hyoseop Woo
Sejong University

CONTENTS

18.1 INTRODUCTION

The East Asian monsoon carries moist air from the Pacific Oceans to the East Asia, where seasonal rain is formed and traverses back and forth around the Korean Peninsula. The Korean Peninsula is affected by frequent typhoons and monsoons from June to September, which causes large floods in rivers (Ji, 2006). Therefore, historically, irrigation and flood prevention have been considered important in a rice-farming-centered society because of the climatic characteristics of the Korean Peninsula. Conventionally, innovative flood mitigation and irrigation projects have been applied to develop agricultural productivity based on rice paddy farming (Encyclopedia of Korean Culture, 2021).

Historically, the most representative flood defense measure has been building reservoirs and river embankments. By building a reservoir embankment, water resources required for farming can be secured and prepared to foresee drought season;

DOI: 10.1201/9781003222736-21

by installing a river embankment, farming can be expanded on high-quality soil, and land in waterfront areas, people, and property can be protected from flooding. Considering the positive aspects, embankment installation for reservoirs and rivers is one of the representative hydraulic engineering techniques actively used in modern society.

Although the construction technology of embankments has existed and improved from ancient times, farming and other settlements, such as building houses along the riverside, have been considered challenging from the ancient days onward. The traces of our ancestors' efforts to strengthen the river and reservoir embankments, such as planting trees on the banks of the river, are found in the literature.

There are three conventional ways to plant trees in Korea as disaster risk reduction measures. The first type is the anti-flood forest strip, which involves planting trees on earthen levees on the riverside to protect their villages and farmlands from floods. The second type is the anti-storm coastal forest strip, which involves planting trees along coasts and beaches to protect them from storm surges and windstorms incurred from the sea. The third type is the anti-wind forest strip, which involves planting trees on the northwestern side of villages to protect them from cold northwestern winds during the winter season. All of these methods are modern green infrastructures for disaster risk reduction. At present, green infrastructures are vegetated, natural, or engineered system that mimics the natural ecosystem used for treating the urban storms at their source and a patchwork of natural areas that provide habitat, flood protection, cleaner air, and water. The concept started in the 1980s, first in the USA, and has been expanded geographically to other regions and semantically to flood risk reduction. Currently, 23 ancient, anti-disaster forest strips are being protected as natural monuments by the Cultural Heritage Administration in Korea.

Perhaps, the flood defense mission that encounters the current climate crisis era provides similar intensity to the mission experienced by people from the older generation when the river embankment technology could not sufficiently handle the flood. Due to climate change, abnormal floods (extreme floods) exceeding the number of floods levees can control are increasing (National Resources Defense Council, 2021). Therefore, controlling river flow and defending floods by conventionally stacking levees is not appropriate for combating the climate crisis, and alternatives are actively considered to further expand the floodplain to the riparian area outside the river embankment to prevent extreme flooding (MCST, 2021). Due to the influence of climate change, the space alongside the river is no longer considered a safe space for human activity. The Room for the River project is the most well-known case—a government design plan in the Netherlands intended to address flood protection, master landscaping, and the improvement of environmental conditions in riparian areas (van Alphen, 2020).

The paradigm of river management in Korea is transforming into a multifunctional floodplain project that improves the ecological disconnection of rivers and waterfront areas, reduces contaminants and pollutants from watersheds, improves water quality, and expands nature-based solutions that can be used as extra detention areas during extreme floods. In the new paradigm for river management, the riverside area is created as a space for nature and aquatic ecosystems. It is considered a more sustainable, safe, clean water environment that benefits human life. In addition, in Korea, alternative ways to improve the river environment are being discussed to further secure carbon absorption sources by creating and managing vegetation sections in rivers and waterfront forests in riparian areas, which is a critical national policy.

This paper reviews the water management policy during the 500 years of Joseon Dynasty in Korea by investigating the history of the construction and management of embankments. This study also focuses on the traditional practice of levee planting for river flood risk reduction in Korea, which has been replaced since the 20th century with the civil engineering practice of using compacted earthworks and concrete materials. Two cases, Gwanbangjerim and Sangrim, were examined in detail to study their characteristics as flood risk reduction measures at those times. Any feasible benefits and demerits of planting trees on levees during those periods were identified at the technical level.

18.2 DEVELOPMENT OF EMBANKMENT TECHNOLOGY IN KOREAN HISTORY

The history of embankment construction in the Korean Peninsula, both for water use and flood control, goes back to the Three Kingdoms era (2nd to the 7th century). An example is a historic site of the embankment at Yaksa-dong, Ulsan, in the southeastern region of the Korean Peninsula (Figure 18.1). This embankment is a 155 m long and 4.5–8 m high bank built on the Yaksa river to build a reservoir during the era of the Silla Dynasty (Cultural Heritage Administration homepage, 2018).

Figure 18.1 Aerial view of the Yaksa-dong embankment. (Here, we can see the drainage-inducing technique using twigs and leaves.) (Image source: Yonhap News.)

Several embankments were constructed and maintained during the era of Joseon Dynasty, which lasted from 1392 to 1910. In the earlier 1700s, the number of embankments reached 3,695. However, the embankment number was reduced to 2,680 in the 1900s, because the embankments were not properly maintained and were associated with political instability.

The Joseon Dynasty installed and operated several administrative offices for water management, such as the Gwonnonggwan and Jeunsa. Gwonnonggwan was installed in 1395, immediately after the foundation of the Dynasty in 1392, to maintain the Jeun (water embankment) and, eventually, to manage agricultural water, and Jeunsa was solely responsible for the management of all reservoirs in the country. In 1662 (Hyunjong's third year), Jeunsamok was established as Korea's first embankment management guideline. Jeongjeolmok—an advanced version of Jeunsamok, was proclaimed in 1778 (Jeongjo, second year). From the abovementioned investigation details, the water management policy during the Joseon Dynasty seems to have focused mainly on agricultural water use.

In Korea, conventionally, settlements are not recommended in flood-prone areas, as is written in the book of Taekriji in the 18th century, commenting, "It is not worthwhile living by the large river" (Woo, 2019). Nevertheless, some dike constructions were carried out to control floods, such as Gwanbangjerim in Damyang (Woo & Kim, 2017). It is a 2 km long tree-planted levee constructed along the Youngsan-gang River (Figure 18.2). Approximately 420 trees are still alive, as depicted in Figure 18.2, planted twice, some hundred years ago in the 17th and 19th centuries, respectively, to protect the village and the embankment from floods and strong winds. This can be considered a good example of embankment construction using ecological technology.

Figure 18.2 Gwanbangjerim (tree-planted levee) of the Youngsan-gang River in Damyang.

18.3 HISTORICAL RECORDS OF EMBANKMENTS IN THE JOSEON DYNASTY

During the era of the Joseon Dynasty, written records of embankments increased rapidly. "The Joseon Wangjo Sillok," or simply "Sillok," is the official, veritable chronicle records of the Joseon Dynasty, including 28 sets of each ruler's reign. The total volume of the Sillok is 1,187 books registered as UNESCO's Memory of the World Program in 1997 (National Institute of Korean History, 2018). According to the Sillok, both Jebang and Jeun were equivalently used for water embankments. In the Sillok, 1,265 words from Jebang, 816 words from Jeun, and 55 cases using both these words simultaneously were found.

In the Sillok (15th Volume of King Taejo, the first king), the first appearance of the words is as follows: "in early winter a Jebang should be built to prevent fire". This makes it difficult to judge whether Jebang is a water-blocking or water-reserving embankment. However, in April of the 7th year (in 1407) of King Taejong (the second king), the following phrase, "Let each householder living by stream build Jebang on both sides of the stream and plant trees," could be observed. In this sentence, it is evident that Jebang acts as a levee against flood control. After that, in November of the 12th year (in 1412) of King Taejong, there is the phrase, "building up Jebang along the high and low terrain, and confining water so that in each a small boat can float". In this record, it is evident that Jebang was used as a dam for the irrigation reservoir. In August the 15th year (1415) of King Taejong, the Sillok goes as "The Jebang saves the water and makes it through irrigation, so it is a good method to prepare for the tribulation and relieve the drought." Here, Jebang indicates the embankment for reservoir construction.

As mentioned earlier, throughout the "Silloksilos," we can see that both Jebang and Jeun were used to build reservoirs to impound water, and prevent river floods and tidal surges. Jebang generally refers to the embankment itself, whereas Jeun refers to both the embankment and the reservoir as a whole.

18.4 PRACTICE OF RIPARIAN FOREST STRIPS IN THE JOSEON DYNASTY

18.4.1 Temporal and regional distribution of riparian forest strip

Traditional riparian forest strips have been managed for various purposes and functions, such as anti-flood, anti-wind, and rest and amenity spaces (Lee et al., 2008). Conventional cases of riparian forest strips along the levees are distributed across South Korea. Trees planted in levees have positive effects, such as flood prevention and bank stabilization. Nevertheless, most traditional riparian forests have been lost because of modern river management works, redevelopment projects of agricultural fields, and road development. It is difficult to find riparian forest strips near rivers and streams in Korea. The most detailed records and notes on this topic were the Forest of Joseon, published by the Japanese Government-General of Korea in 1938. This report provides information on the existence of riparian forests and their historical background. Recently, Lee et al. (2008) analyzed historical changes in riparian forest strips

Table 18.1 Regional distribution of 152 riparian forest strips that existed in 2000

Status	*City*	*Town*	*Small village*	*Others*
Han-gang River	0	2	9	20
Dong-gang River	3	1	0	8
Nakdong-gang River	7	9	14	17
Hyoungsang-gang River	7	0	3	0
Mangyung-gang River	1	0	1	2
Dongjin-gang River	0	1	1	0
Youngsan-gang River	1	1	4	4
Seomjin-gang River	1	3	0	1
Small streams flowing into the East Sea	5	3	11	6
Small streams flowing into the West Sea	1	0	1	2
Small streams flowing into the South Sea	2	0	0	0

Data source: Lee et al. (2008).

based on the Forest of Joseon and examined their distributions by specific region, location, and river.

According to Lee et al. (2008), the total number of riparian forest strips in 1938 presented in the Forest of Joseon was 102: 71 existing, 16 ruined, and 15 disappeared. The number of riparian forest strips in 1938 declined from 71 to 25, and the ruined strips also reduced from 16 to 12 in 2000. Therefore, the number of disappeared riparian forests decreased from 15 to 65 in 2000. The number of riparian forest strips recorded in the Forest of Joseon in 1938 was significantly reduced by 65% in 2000. Considering the additional riparian forests found from the field investigation in 2000, the total number of riparian forests in 2000 was 152. Table 18.1 lists the regional distribution of the 152 riparian forest strips in 2000. Riparian forests were distributed especially in regions with rough terrains with frequent disasters and where river management works and redevelopment projects of agricultural fields were seldom conducted.

On the other hand, no riparian forest strip in the urban areas of the Han-gang River basin was found in the data in Table 18.1. This implies that urbanization, including the Seoul metropolitan region, has expanded greatly. However, many riparian forest strips were found in the watershed of small streams flowing into the East Sea. Due to geographical features, cities, towns, and villages were located in narrow plains, where small streams flow in the center.

Among the 23 ancient village forest strips protected as natural monuments by the Cultural Heritage Administration in Korea, seven ancient village forest strips were for anti-disaster and anti-wind, as listed in Table 18.2. In particular, Gwanbangjerim of Damyang-gun (Figure 18.2) and Sangrim of Hamyang-gun (Figure 18.3) represent the ancient village forests protected for flood risk reduction in Korea.

18.4.2 Effects of levee planting on river flood-risk reduction

The traditional method of planting trees on levees is one of the current limitations of river maintenance activities and practices. Planting trees on the levees in the past seems to conflict with current river management practices. The riparian forest strips listed in Table 18.2 were mostly planted in earthen levees. Some of the potential negative effects

Table 18.2 Ancient anti-disaster village forest strips protected as natural monuments by the Cultural Heritage Administration in Korea

Location (name)	*Type*		*Major species of trees*	*Ownership*	*Others*
	Pattern	*Function*			
Mulgunlee Bangjoeoburim of the South Sea	Strip	Anti-wind/ indigenous religion	Hackberry, Muku Tree, Zelkova	Namhae-gun	Anti-wind from the coast and ocean
Hamyang Sangrim	Strip	Confucianism / anti-disaster	White Oak, Zelkova	Hamyang-gun	Anti-flood
Wando Maengsunlee Snagroksurim	Strip	Anti-disaster	Silver Magnolia, Red wood Evergreen Oak	MSF	Anti-wind from the coast and ocean
Damyang Gwangbang-jerim	Strip	Anti-disaster	Muku Tree, Hackberry, Carster Aralia	MLIT	Anti-flood
Euisung Sachonlee Garosup	Strip	Anti-disaster	Oak, Zelkova	The Kim's of Andong	Anti-wind
Hadong Songrim	Rectangle	Anti-disaster	Pine Tree (Pure Forest)	MLIT	Anti-wind from the river
Yecheon Ge-umdangsil-song-rim	Strip	Anti-disaster	Pine Tree (Pure Forest)	Yecheon-gun	Anti-flood and anti-wind

Data source: Hwang and Kim (2017).
MSF, Ministry of Strategy and Finance; MLIT, Ministry of Land, Infrastructure, and Transport.

Figure 18.3 Sangrim of Hamyang-gun along the bank of the Wi-cheon stream. (Image source: gnnews.co.kr and joungul.kr.)

of woody vegetation and trees on river floods are that flow resistance increased by trees in the floodplains and levees can generate high flood risk due to increased flood levels; local scours can occur around the stump during floods; root growth penetrating and extending underneath the levees can provide a preferential route for seepage flow; root growth and decay may create macropores and subsequent piping; and large trees may be more susceptible to mass failure (Woo & Kim, 2017; USACE, 2011; Coutts, 2004).

However, healthy root systems planted in levees may strengthen the levee planes and effectively increase the slope stability of the system. Riparian trees stabilize banks along watercourses, thereby preventing erosion during flood flows. Riparian forests absorb water, further decreasing the flood flow and relieving the pressure on drains (Woodland Trust, 2013). In addition, woody vegetation and trees planted in floodplains can block and deflect the debris during floods and reduce the secondary damage to downstream sections. Riparian forests and vegetation may reduce the energy of flood flow. However, it is worth noting that USACE (2011) indicated that a tree could increase or decrease the safety factor concerning slope stability, depending on the tree's location on the levee.

There have been many publications on the effects of woody vegetation on slopes and riverbanks. USACE (2011) conducted an extensive literature review and initial research on the effects of woody vegetation, including trees, on levees. Among the literature reviewed, Greenway (1987) suggested that both hydraulic and mechanical factors should be considered while evaluating the effects of woody vegetation on a slope stability analysis. The five mechanical factors to be considered are as follows: the effect of roots in reinforcing soil is to increase soil shear strength; large roots will act as an anchoring system, holding the weaker upper soil layer to the more stable lower soil layer; tree weight will provide an additional vertical load to slope stability calculation; dynamic-wind force on the tree crown will convey horizontal, vertical, and moment loads to the slope; and roots will hold soil grains at the ground surface and resist erosion.

18.4.3 Conventional practice of levee planting

Gwanbangjerim is a forest along the left bank of the upstream Youngsan-gang River (Figure 18.2). It was designated as Korean Natural Monument No. 366 by the Cultural Heritage Administration in 1991 (Korea Cultural Heritage Administration, 2017). The 2.4 km long forest stretches from Dungeons Village in Navsari, approximately 123,000 m^2. It comprises large trees, namely, muku, zelkova, nettle, and cherry, approximately 200–300 years old. The size of the trees differs from a circumference of 1.1–5.3 m.

Historically, in the 26th year of King Injo (in 1648), Seong Ihseong, the region's official, first built levees to prevent flooding and then planted trees on the levee. Subsequently, in the fifth year of King Cheoljong (in 1854), Hwang Jonglim—the region's official, retrofitted the levees and planted additional trees. Therefore, it could be considered that the current large trees were planted in the 26th year of King Injo (in 1648), and the relatively smaller trees were planted in the fifth year of King Cheoljong (in 1854).

Based on the first investigation, by considering the historical record, oral tradition, arrangement between the levee and trees, the relative location of villages, etc., it was identified that the earthen dike was constructed to protect villages and

Figure 18.4 Layouts of Gwangbangjerim (left) and Sangrim (right). (Image source: Lee et al., 2008.)

agricultural lands from seasonal floods in the upstream Youngsan-gang River. The trees were planted to prevent erosion during high flow. As depicted in Figure 18.4, the levees' height and the trees' location imply that the possibility of levee instability, such as mass failure by overtopped trees, is very low. Positive effects of woody vegetation and trees planted on banks and levees are expected.

One of the countermeasures against bank failure in those days was the arrangement of live branches on the slopes of levees. Currently, it is called willow cutting—a soil bioengineering technique. Trees were planted on levees like Gwanbangjerim instead of willow cuttings because they expected the additional effects of a windbreak, landscape, amenity, and flood prevention. It is a favorite place as a riverside park for recreation and amenities for villagers and tourists.

Choi Chiwon, the governor-general of Hamyang province in the Queen Jinsung era (887–897) of the United Silla, created a riparian forest of Sangrim to prevent flood damages to the province (Korea Cultural Heritage Administration, 2017). Sangrim was designated as the Korean Natural Monument No. 154 by the Cultural Heritage Administration in 1962. Sangrim is a riparian forest located west of Hamyang-gun along the bank of Wi-cheon stream (Figure 18.3). The total rectangular area of the Sangrim is 119,008 m^2 (11.9 ha). It comprises various broadleaf trees: zelkova, fringe, oak, snowbell, loose-flower hornbeam, and dogwood trees. The irrigation channel is located in the forest, as depicted in Figure 18.4 (right). According to historical records, agricultural lands and villages suffered significant damage every year from the floods of Wi stream, which passed through the middle of large fields of Hamyang province. Therefore, it could be speculated that the levees were built to bypass water flow, and a riparian forest of Sangrim was developed to protect villages and lands from floods.

Wi-cheon stream has a trapezoidal cross-section without floodplains, as depicted in the layout of Sangrim (Figure 18.4 right), which is different from that of Gwanbangjerim. The trees in Gwanbangjerim were planted on the levees, whereas the trees in Sangrim were not planted on the bank or levees. It is speculated that Sangrim was built to reduce flood damage and not directly protect the levee. Therefore, a rectangular riparian forest might have been selected to protect riparian lands and villages.

18.5 CONCLUSIONS

In this paper, through a review of the historical records of the Sillok (official veritable, chronicle records of Joseon Dynasty), it was confirmed that several administrative offices such as Gwonnonggwan and Jeunsa operated during the 500 years of the Joseon Dynasty (1392–1910) of Korea to manage agricultural water use for farming. During the era of the Joseon Dynasty, several embankments, namely, Jebang and Jeun, were systematically constructed and maintained by administrative offices, mainly focusing on flood control and agricultural water use.

The main policy of the Joseon Dynasty on river floods seems to be rather passive, "not to live by the rivers". Nevertheless, some flood-protection embankments, such as Gwanbangjerim in Damyang, have also been constructed to protect villages and agricultural lands against river flow erosion by planting trees on the levee. Currently, river management still restricts planting trees in levees or floodplains because of the negative effects of piping, mass failure, and increased flow resistance. However, trees in floodplains and levees could increase or decrease safety according to hydraulic and mechanical factors, depending on the tree's location on the levee. Therefore, during the era of the Joseon Dynasty, regional officials conducted maintenance work, such as retrofitting levees and planting additional trees. Ironically, efforts to strengthen embankments with eco-friendly materials were also attempted using modern river engineering technology. The current embankment construction technology, which has been maintained using rigid concrete blocks, has been applied less. Using eco-friendly materials such as vegetation mats and biopolymers, natural embankments have been attempted. In addition, measures were considered to create a riparian forest in floodplains by retreating or relocating the levee in the river to prevent floods of scale that the embankments cannot defend in the climate crisis era. Levee retreats can be combined with nature-based flood protection technology to link with measures to create carbon sink forests to improve river waterfront areas' water environment and reduce carbon emission. To derive new challenging alternatives for river management in this era, a technographical review of embankment technology during the 500 years of the Joseon Dynasty would provide an important starting point.

As a way of implementing the old wisdom of "Ongoijisin," which means "review the old and learn the new," this study suggests investigating possible ways of revitalizing planting trees on the levee, riparian area, or waterfront area, not only for a flood-risk reduction but also for providing habitats for flora and fauna and amenity for human.

ACKNOWLEDGMENT

This article is based on two articles presented at the 2018 IAHR-APD Congress in Jogjakarta, Indonesia, and the 2020 IAHR-APD online Congress in Sapporo, Japan.

FUNDING

This research was supported by the major internal project of the Korea Institute of Civil Engineering Building Technology (Project Number -20220178).

REFERENCES

Coutts, M. P. (2004). *Wind and trees*. Cambridge University Press.

Cultural Heritage Administration. (2018). http://www.cha.go.kr/main.html (in Korean).

Encyclopedia of Korean Culture. (2021). http://encykorea.aks.ac.kr/ (in Korean).

Greenway, D. R. (1987). Vegetation and slope stability. In: M. G. Anderson, and K. S. Richards (Ed.), *Slope stability: geotechnical engineering and geomorphology* (pp. 187–230). John Wiley & Sons Ltd.

Hwang, D. G. and Kim, D. Y. (2017). Project for the database of village forest strips protected as Natural Monuments by the Cultural Heritage Administration in Korea. *Proceeding of 2017 Spring Conference for Korean Institute of Traditional Landscape Architecture, 17*(1), 1–4 (in Korean).

The Japanese Government-General of Korea. (1938). *Forest of Joseon: translated by Forest for Life in 2007.* GeoBook publishing (in Korean).

Ji, U. (2006). *A numerical model for sediment flushing at the Nakdong River Estuary Barrage.* Ph.D. dissertation, Colorado State University, Fort Collins, CO, USA.

Korea Cultural Heritage Administration. (2017). http://www.cha.go.kr (In Korean).

Lee, D. W., Kang, D. J., Ko, J. H., Kwon, J. O., Kim, J. H., Kim, H. S., Ryu, Y. R., Park, C. Y., Sung, D. H., Shin, J. H., Oh, C. H., Lee, G. I., Lee, M. W., Jang, D. S., Choi, W. S., and Han, P. W. (2008). *Korean Traditional Ecology*, Science Books Publishing (in Korean).

MCST. (2021). *Korea Policy Briefing.* Ministry of Culture, Sports and Tourism. https://www.korea.kr/news/pressReleaseView.do?newsId=156451457 (In Korean).

National Institute of Korean History. (2018). http://www.history.go.kr/en/main/main.do

National Resources Defense Council. (2021) https://www.nrdc.org/stories/flooding-and-climate-change-everything-you-need-know

USACE. (2011). *Initial Research into the Effects of Woody Vegetation on Levees.* ERDC TR TO HQUSACE, US Army Corps of Engineers, USA.

Van Alphen, S. (2020). Room for the river: innovation, or tradition? The case of the Noordwaard. *Adaptive Strategies for Water Heritage* (pp. 309). Springer, Cham.

Woo, H. (2019). A historical review of levees. *The Korean Society of Civil Engineers Magazine, 67*, 56–59 (in Korean).

Woo, H. S. and Kim, C. (2017). Technographical reviews of levee planting for flood prevention-focused on Gwanbangjerim in Damyang County. *Proceeding of 2017 Conference for Korea Society of Civil Engineers, 10*, 1884–1885 (in Korean).

Woodland Trust. (2013). Severn Rivers Trust Wood: In *Wise, Trees & Woodland in Water Management, Wood Wise, Conservation News, Winter*, 14–17.

Chapter 19

The ancient instrumental hydrological measurement device, *Chugugi* and *Supyo*, in Joseon Dynasty, Korea

Hyeonjun Kim
Korea Institute of Civil Engineering and Building Technology

Cheolsang Yoo
Korea University

CONTENTS

19.1 INTRODUCTION

Meteorological observations are used for the real-time preparation of weather analysis, forecast and severe weather warning, climate study, local weather-dependent operations, hydrology and agricultural meteorology, and research in meteorology and climatology (WMO, 2018). The rainfall measurement was started in India and Palestine before the 2nd century BC. About 3000 BC, an ancient instrument (called a Nilometer) was first used to measure the water level of the Nile River (Selin, 2008). From the early historical times, the Ancient Egyptians regularly measured the maximum height of the yearly flood and recorded the level in their royal annals (Bell, 1970).

The history of rainfall observation in Korea goes back to 1441, in the early period of the Joseon Dynasty (Wada, 1911; Chun and Jeon, 2005; Kwon, 2006; Strangeways, 2010). "*Chugugi or Chuk-Woo-Kee*" was invented by King Sejong and distributed nationwide. According to the Veritable Records of the Joseon Dynasty, heavy rains and severe drought alternated in the early 15th century. The invented instruments were lost, and the rainfall observation network collapsed after the two wars (1592–1598, 1636–1637) against Japan and Qing China. In 1770, King Yongjo made bronze rain gauges following the specification established by King Sejong in 1441. He also reconstructed

DOI: 10.1201/9781003222736-22

the nationwide rainfall observation network. Reliable precipitation data for Seoul have been recorded since 1778, representing one of the world's longest instrumental measurements in daily precipitation. Also, during this time, *Supyo* (water level gauge to measure flood stage in the stream, *Cheonggyecheon*) was installed. Wada firstly introduced *Chugugi,* and compiled its monthly precipitation data set and flood water level records measured by *Supyo* (Wada, 1911, 1917).

This study will introduce a brief history of the instrumental rainfall gauge—Chugugi, water level gauge—Supyo, and recent research on those two data.

19.2 THE INVENTION OF THE RAINFALL GAUGE, CHUGUGI (測雨器)

In 1441, the Minister of the Hojo (戶曹, Ministry of Taxation) reported that, according to reports from the provincial governors on the amount of rainfall, the conventional method of measurement was unable to distinguish the difference in the depth of rainwater on the ground when it was parched and when it was soaked. Prince Munjong designed the instrument and the rain gauge—*Chugugi.* The rainfall depth was measured by a ruler called *Chuchok* (周尺) (Chun and Jeon, 2005). The instrument was placed at the observatory, and the officials of the observatory measured the rainfall depth each time it rained. Similar instruments were distributed to the provinces and cantons, and the results of the observations were reported to court (Wada, 1911, 1917).

King Yongjo (1724–1776) finally determined to set up the network of *Chugugi* throughout the country and issued an edict to this effect. On May 1, 1770, a bronze rain gauge was made following the specifications established during the reign of King Sejong. The only known surviving *Chugugi* was made in 1837 and operated until 1910 (Chun and Jeon, 2005).

The *Chugugi* was a cylinder with a diameter of 14.7 cm and a height of about 45.5 cm. It was installed at the Palace and under the supervision of each province office in the local area. Later, the size was reduced to 30 cm deep and 14 cm in diameter (Figure 19.1). The depth of rainfall collected in the device is measured with a standard ruler.

Figure 19.1 The *Chugugi* and *Chugudae* (stand) made in 1837 at Gongju (Cultural Heritage Administration, Republic of Korea).

On April 23, 1791, the standard guideline for rainfall measurement was established as three times a day at 4, 12, and 20 o'clock. Before this guideline, it was recommended two times at dawn and dusk in a day. The most meaningful record among the Veritable Records of the Joseon Dynasty related to measuring rain can be found on May 22, 1799. King Jeongjo said that

> I made sure to record the amount of rain since January 1 of every year, and when I looked at the statistics for the last eight years from 1791 to 1780, the annual rainfalls were as follows by each; 1789mm, 1498mm, 935mm, 1227mm, 879mm, 879mm, 950mm, 1158mm. The monthly rainfall amount is just 40mm, compared to the same month of last year was almost 200mm. We can see traces of trying to understand the plight of the people along with concerns about farming compared with past rainfall data. It is impossible to know in advance what the harvest in autumn will be like, but the current situation of the people is very pitiful.

19.3 RESTORATION AND ANALYSIS OF ANCIENT RAINFALL

The rainfall measuring instrument—*Chugugi*, and the document of rainfall data were found and introduced by a Japanese meteorological scientist, Wada Yuji (1859–1918). In 1910, he firstly reported "the Korean meteorology—old and new" in *Nature* (Wada, 1911). He also reported the ancient meteorological observation of Korea in 1917. In this report, he analyzed monthly rainfall amounts and suggested a correction factor for ancient rainfall to modern. Tada (1938) analyzed the periodicities of annual rainfall, and Arakawa (1956) reported the reliability and secular characteristics of the data. Jeon introduced the meteorological history of the Joseon Dynasty (1963), and secular variation analyses were carried out by Kim (1976) and Cho and Na (1979).

The ancient rainfall data set was reconstructed through the comprehensive and massive investigation of historical documents, *Seungjungwonilgee* (承政院日記, The Diaries of the Royal Secretariat) and *Ilsungrok* (日省錄, The Records of Daily Reflections). Lim and Jung (1992) re-compiled the annual rainfall data and analyzed interannual variability using modern statistical skills. Jhun and Moon (1997) reconstructed daily precipitation records. The fluctuation of monthly rainfall for the monsoon season was also suggested (Ha and Ha, 2006; Wang et al., 2006). Additionally, Boo et al. (2006) found 18 years of rainfall data observed by *Chugugi* in Gongju from the ancient official document *Gaksadeungnok* (各司謄錄, Local Government Docum). According to Arakawa (1956) and Cho and Na (1979), the rainfall data measured by Chugugi are reliable and demonstrate well-established cycles. Figure 19.2 shows the annual rainfall variation in Seoul from 1778 to 2019.

Recently, more expanded investigations to find ancient rainfall data sets were performed (Lim and Moon, 1992; Kim et al., 2012; Cho et al., 2015). The records of *Wootaek* (雨澤) were found in the *Gaksadeungnok*. *Wootaek* is an indirect method of measuring the soil's rainfall. During King Yongjo, the nationwide rainfall observation network was re-established with 14 *Chugugi* observation stations and the 352 local authorities such as *Bu*, *Gun*, and *Hyeon* conducted by *Wootaek*. Later on, six more *Chugugi* stations were established (Figure 19.3). This kind of restored rainfall data

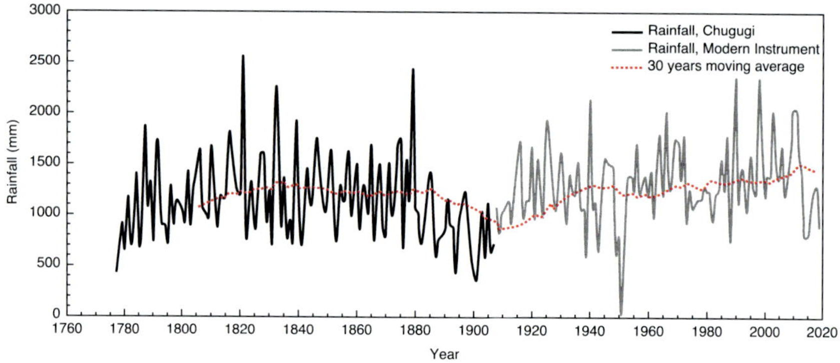

Figure 19.2 Variation of annual rainfall and 30 years moving average from 1778 to 2019 measured by ancient and modern rainfall gauge in Seoul, Korea. There was a big drought spell in the late 19th century, and one year (November 1950–November 1951) of rainfall was not measured during the Korean War (June 25, 1950–July 29, 1953).

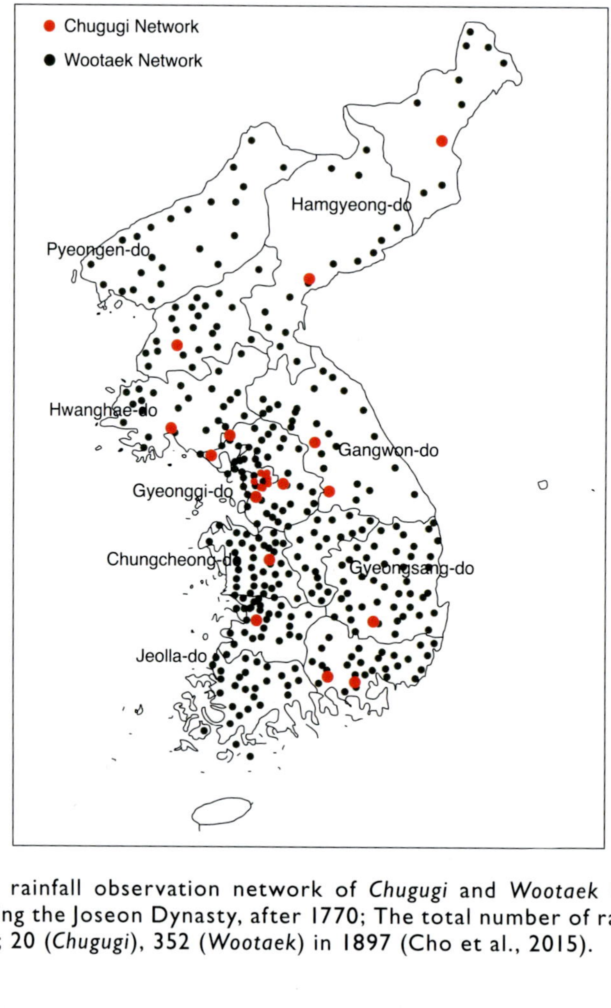

Figure 19.3 The rainfall observation network of *Chugugi* and *Wootaek* by the province during the Joseon Dynasty, after 1770; The total number of rainfall stations is 372; 20 (*Chugugi*), 352 (*Wootaek*) in 1897 (Cho et al., 2015).

Table 19.1 Research reports and articles on *Chugugi* and its rainfall data

Sources	*Year*	*Main findings*
King Sejong	1441	The invention of the rainfall gauge—*Chugugi*, a nationwide network
King Yongjo	1770	Rebuild *Chugugi*, a nationwide network
Wada	1911,1917	Introduce and re-compile monthly rainfall data set
Tada	1938	Annual precipitation
Jeon	1963	Introduce *Chugugi* to the Science History Society
Kim	1976	Precipitation
Cho and Na	1979	Secular variation of the rainfall
Lim and Jung	1992	Interannual variability with annual rainfall data
Jhun and Moon	1997	Reconstruction of daily rainfall data set
Chun and Jeon	2005	Introduce *Chugugi* to Meteorology History Society
Boo et al.	2006	Restoration of 18 years of rainfall in Gongju
Wang et al.	2006	Summer monsoon precipitation
Ha and Ha	2006	Interannual fluctuation with monthly rainfall data
Kim et al.	2010	Interdecadal variability of monsoon precipitation
Kim et al.	2012	Restoration 19c rainfall data in Wonju, Hamheung, Haeju
Cho et al.	2015	Review and introduce the rainfall observation network
Jang et al.	2017	Drought frequency analysis using annual rainfall data
Yoo et al.	2018	Comparison of annual maximum rainfall events
Kim and Jang	2019	Compare extreme floods with daily rainfall data

Table 19.2 Number of water level measurements by *Supyo*

Sources	*Station*	*1–2 Chok*	*2–4 Chok*	*4–6 Chok*	*6–8 Chok*	*8–10 Chok*	*10 Chok~*
Giujedeungrok (1633–1889) (KICT, 2005)	Jungbu Supyo	126	151	74	61	29	20
	Nambu Supyo	14	23	22	9	5	2
	Supyo	25	16	12	1	0	2
Research on ancient observation data during Joseon Dynasty (1554–1778) (Wada, 1911)	Junbu Supyo	0	6	27	40	19	14
	Nambu Supyo	0	3	7	9	4	0
	Supyo	0	0	1	2	2	0
Seungjeongwon-ilgi (1663–1774) (KICT, 2005)	Jungbu Supyo	0	2	1	1	0	0
	Nambu Supyo	0	2	3	1	0	0
	Supyo	8	9	9	1	2	4

can guide understanding of the climatic tendency of the 19th century in the Korean peninsula.

Kim and Jang (2019) reported the relationship between extreme rainfall and severe flood events. In addition, the research field using ancient rainfall was spread out to the hydrology and water resources and listed in Table 19.1 (Jang et al., 2017; Lee and Kim, 2018; Kim et al., 2018; Yoo et al., 2018; Kim and Jang, 2019). The largest daily rainfall observed in Seoul, 394 mm, was measured on July 16, 1885, and 520 houses were swept away. The flood was most severe on July 19, 1832, with 348 mm. Over 3,100 houses were washed out, and 64 persons were drowned (Table 19.2).

19.4 ANCIENT WATER LEVEL MEASUREMENT, *SUPYO* (水標)

The first people in history who measured the water level of rivers were the Egyptians, who measured the flood water level of the Nile to predict grain production around 3000 BC (Selin, 2008).

The water level of streams and rivers was measured from 1441 in the Joseon Dynasty. The watermark or water level gauge, Supyo, originally a wooden type column (2.5 m high), was invented in 1441 and located west of the Majon bridge in the Gaecheon stream. The Supyo was installed at the Gaecheon (now called Cheonggyecheon) stream in the capital city. Later in the era of King Songjong (1469–1494), the wooden watermark was improved using the granite-type column, which has a scale of 1 *chok* (about 20 cm) to 10 *chok* (about 2 m), and this is the model that remains today (Figure 19.4).

Wada Yuji investigated old historical documents of the Joseon Dynasty, including *Pungwoonki* (風雲記, Meteorological Observation Document), *Cheonbyeonchochuldeungrok* (天變抄出謄錄, Quarterly Meteorological Observation Report), the *Joseonwangjosillok* (朝鮮王朝實錄, the Veritable Records of the Joseon Dynasty), *Seungjeongwonilgi*, *Ilseongnok*, *Gukjobogam* (國朝寶鑑, the records of exemplary political accomplishments and actions of Joseon Kings), *Jeungbomunheonbigo* (增補文獻備考, the revised edition of the encyclopedia on Korea's civilization and institutions), and *Giujedeungrok* (祈雨祭謄錄, the Records of Rituals for Rain) to organize the records of the rainfall measured by *Chugugi* and the water level measured by *Supyo*. For 460 years since 1400, Seoul areas have flooded 172 times, with the highest occurrence in July at 75 times, followed by August (47 times) and September (19 times) (Wada, 1917).

The water level records of the *Supyo* were reconstructed (KICT, 2005) from the *Seungjeongwonilgi* and the *Giujedeungrok*. The *Giujedeungrok* contains records of the ritual for rain, the ritual for stopping rain, and the ritual for snow from 1633 to 1890, and consists of six volumes. As well as the ritual records, it includes water level records measured by *Supyos* (Figure 19.5).

There are two locations to measure the water level in the Gaecheon stream Jungbu Supyo and Nambu Supyo. In the Giujedeungrok, 690 water levels were recorded from

Figure 19.4 The *Supyo* (watermark) and *Supyogyo* (bridge) in *Gaecheon* stream in the early 1900s. (Photos, National Museum of Korea.)

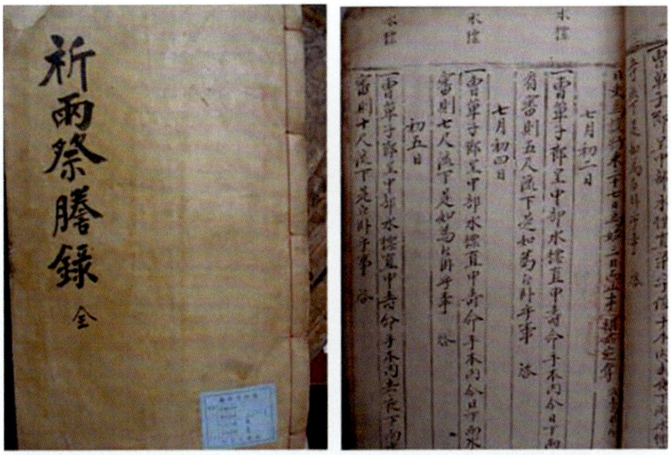

Figure 19.5 *Giujedeungrok* (祈雨祭謄錄, 1636–1889, Kyujanggak Institute for Korean Studies of Seoul National University); it contains not only the procedures and records of the ritual for rain, the ritual for stopping rain, and the ritual for snow performed from King Injo's reign to King Gojong's reign but also the water level measurement record of the *Gaecheon* stream and the *Hangang* River.

1636 until 1779. The flood was recorded every year from 1669 to 1719. However, from 1697, only the measurements of the *Jungbu Supyo* were recorded. Most records were concentrated in 70 years between 1680 and 1725 and between 1755 and 1780 (Table 19.2). Figure 19.6 shows the number of water level records of *Supyo* and *Chugugi* in the *Giujedeungrok*. After reconstructing the rainfall device, i.e., the *Chugugi* in 1770, the water level record decreased and finally ceased.

There are no remaining paintings of ancient floods. Still, now we can estimate the ancient flooding situation in the *Gaecheon* stream through modern information like a newspaper or photos from the 1900s. In July 1915, the Maeil Shinbo newspaper featured an article with a picture, reporting, "At 8:00 AM on 24th, two-thirds of the *Supyo* stone of the *Supyogyo* bridge was submerged." In addition, a Korean photographer took a photograph of the flooded stream. His photo illustrates that the waves were lapping right under the deck of the *Supyogyo* bridge, and even the upper part of *Supyo* was submerged in the Summer of 1953 (Figure 19.7).

19.5 CONCLUSIONS

The valuable and one of the world's longest ancient rainfall datasets in Seoul, Korea, has been investigated since the early 20th century. Successive research proved the reliability of the data; also, recent investigations revealed more nationwide rainfall datasets collected by direct and indirect rainfall measurement. The interannual periodicity and fluctuation of monthly rainfall using over 200 years of long-term rainfall records were compared with the recent meteorological characteristics.

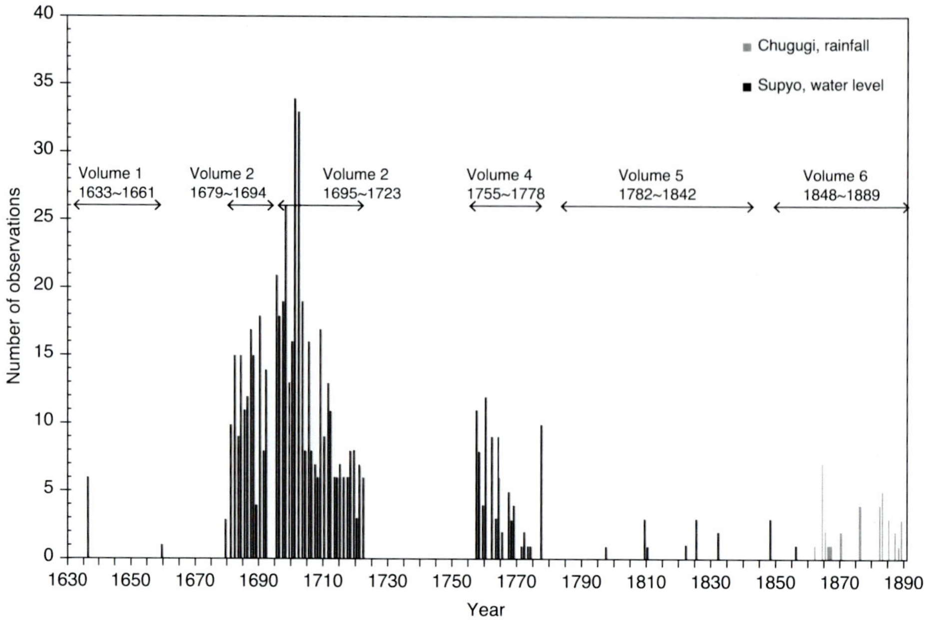

Figure 19.6 Number of water level and rainfall records in *Giujedeungrok* (祈雨祭謄錄, 1636–1889); after the reconstruction of the rainfall device—*Chugugi* in 1770, the record of water level decreased and ceased.

Figure 19.7 Flood at the *Supyogyo* bridge and *Supyo* on July 25, 1915, Maeil Sinbo (left), summer flood in 1953, Insik Im (1920–1998) (right).

The *Chugugi* and *Supyo*, invented in 1441 for hydrological observation, can be comparable to modern observation equipment. In particular, *Chugug*i is the world's first rainfall meter for scientifically measuring rainfall. It was invented 200 years before the modern rainfall measurement by Benedetto Castelli in 1639. The rainfall measured by *Chugugi* has been utilized as priceless data in relevant fields such as meteorology and hydrology. The water level was measured at the *Gaecheon* stream, and KICT (2005) restored the *Supyo* records in the *Seungjeongwonilgi* and the *Giujedeungrok*. The restored data includes the records Wada (1917) omitted and the list of observers. Most of the water level records were concentrated between 1633 and 1779. According to the records, the number of floods where the observed water level exceeded the maximum level of 10 *chok* (around 2.1 m) was reached 20 times.

As hydrology is a field studying the hydrological cycle of rainfall and runoff in nature, the records of droughts and floods are such precious data that an experiment cannot reproduce. Natural disasters have been around us and will be in the future. The historical records from our ancestors provide valuable information about the past droughts, floods, and typhoons, which is a heritage only Korea has. These records' help with understanding must be helpful to understand current natural disasters. This is something that hydrologists should not forget.

ACKNOWLEDGMENTS

This article is based mainly on the "A review of the recent restoration of the *Chugugi* rainfall data (1770–1910) in Korea," authored by the authors and presented in *22nd IAHR-APD Cong. 2020, Sapporo, Japan,* and has been rewritten fitting this monograph. Also, the history of water level observation by the *Supyo* was enhanced.

Ancient historical documents "The veritable records of the Joseon Dynasty" and the "Daily records of Royal Secretariat of the Joseon Dynasty" can be accessed at http://silok.history.go.kr and http://sjw.history.go.kr.

FUNDING

This research was initiated with the support of the 2005 SOC Project (05-GIBANGU-CHUK-D03-01) through the Design Criteria Research Center for Abnormal Weather-Disaster Prevention (DCRC-AWDP) in KICTTEP of MOCT. This research was also supported by the Basic Research Program (20160160-001, 20200041-001) of the Korea Institute of Civil Engineering and Building Technology.

REFERENCES

Arakawa, H. (1956). On the secular variation of annual total of rainfall at Seoul from 1770 to 1944. *Arch. Met. Geophy. Biol.*, 7(2), 205–211.

Bell, B. (1970). The oldest records of the Nile floods. *Geog. J.*, 136(4), 569–573.

Boo, K.-O., Kwon, W.-T., Kim, S.-W., & Lee, H.-J. (2006). Restoration of 18 years rainfall measured by Chugugi in Gongju, Korea during the 19th century (in Korean). *Atmos. Korean Meteor. Soc.*, 16(4), 343–350.

Cho, H.K. & Na, I.-S. (1979). Changes in climate in Korea during the 18th century – Centering on rainfall amounts (in Korean). *J. Korean Stud.*, 22, 83–103.

Cho, H.-M., Kim, S.-W., Chun, Y., Park, H.-Y., & Kang, W.-J. (2015). A historical review on the introduction of Chugugi and the rainfall observation network during the Joseon Dynasty (in Korean). *Atmos. Korean Meteor. Soc.*, 25(4), 719–734.

Chun, Y. & Jeon, S.-W. (2005). Chugugi, Supyo, and Punggi: Meteorological instrument of the 15th century in Korea. *Hist. Meteorol.*, 2, 25–36.

Ha, K.-J. & Ha, E. (2006). Climatic change and interannual fluctuations in the long-term records of monthly precipitation for Seoul. *Int. J. Climatol.*, 26, 607–618.

Jang, H.-W., Cho, H.-W., Kim, T.-W., & Lee, J.-H. (2017). Developing extreme drought scenarios for Seoul based on the long term precipitation including paleoclimatic data (in Korean). *J. Kor. Soc. of Civ. Eng.*, 37(4), 659–668.

Jeon, S.-W. (1963). On the method of measuring rainfall during the Choson Period. *Kagakushi kenkyu (Studies of the History of Science)* (in Japanese), pp. 49–56.

Jhun, J.G. & Moon, B.K. (1997). Restoration and analyses of rainfall amount observed by Chugugi (in Korean). *J. Korean Meteor. Soc.*, 33, 692–707.

KICT (2005). Improvement of design criteria of hydrologic structures for defense against abnormal floods, KISTTEP, MOCT (in Korean).

Kim, C.-J., Weihong, Q., Kang, H.-S., & Lee, D.-K. (2010). Interdecadal variability of east Asian summer monsoon precipitation over 220 years (1777–1997). *Adv. Atmos. Sci.*, 27(2), 253–264.

Kim, H. & Jang, C. (2019). A review on ancient urban stream management for flood mitigation on the capital of the Joseon Dynasty, Korea. *J. Hydro-Environ. Res.*, 22, 14–18.

Kim, H., Jang, C., & Hong, I.P. (2018). Analysis of ancient extreme floods in Chenggecheon urban river based of Chugugi rainfall data and historical documents during 19th century in Seoul, Korea, *Proceedings of the 21st IAHR-APD Congress 2018, Yogjakarta, Indonesia*, Special Session 1, 1–4.

Kim, S.-S. (1976). A study in summer precipitation and winter temperature at Seoul (in Korean). *J. Korean Meteor. Soc.*, 12, 1–6.

Kim, S.-W., Park, J.-S., Kim, J., & Hong, Y. (2012). Restoration of 19th-century Chugugi rainfall data for Wonju, Hamheung and Haeju, Korea. *Atmos. Korean Meteo. Soc.*, 22(1), 129–135.

Kwon, S. (2006). Rainfall observations in Korea by the world's first rain gauge. *Paddy Water Environ.*, 4, 67–69.

Lee, J.-H. & Kim, H. (2018). Long-term rainfall observation records through Korean history and its application for modern hydro-meteorological science. *Geophys. Res. Abstr.*, 20, EGU2018-5717-1.

Lim, G.H. & Jung, H.S. (1992). Interannual variation of the annual precipitation at Seoul, 1771–1990. *J. Korean Meteor. Soc.*, 28, 125–132.

Selin, H. (2008). *Encyclopaedia of the History of Science, Technology, and Medicine in Non-Western Cultures*. Springer, p. 1124.

Strangeways, I. (2010). A history of rain gauge. *Weather*, 65(5), 133–138.

Tada, F. (1938). Ueber Die Periodische Aenderrung Der Regenmenge in Chosen Seit Dem Jahre 1776, Compte Rendus, Congres Internationale de Geogr., Amsterdam, Tome II, Section C, pp. 305–308.

Wada, Y. (1911). Korean meteorology – old and new. *Nature*, 85(2150), 341–342.

Wada, Y. (1917). Reports on the Survey of the Ancient Records of Observation in Korea (in Japanese), 200 pp, Ilhan Print Cooperation, Seoul, Korea.

Wang, B., Ding, Q., & Jhun, J.G. (2006). Trends in Seoul (1778–2004) summer precipitation. *Geophys. Res. Lett.*, 33, 199–203.

WMO (2018). Guide to Instruments and Methods of Observation Volume I – Measurement of Meteorological Variables, p.573.

Yoo, C., Park, M., Kim, H., Choi, J., & Jun, C. (2018). Comparison of annual maximum rainfall events of modern rain gauge data (1961–2010) and Chukwooki data (1777–1910) in Seoul, Korea. *J. Water Climate Change*, 520, 387–396.

Part IV

Historical water projects and traditional water technologies in South Asia

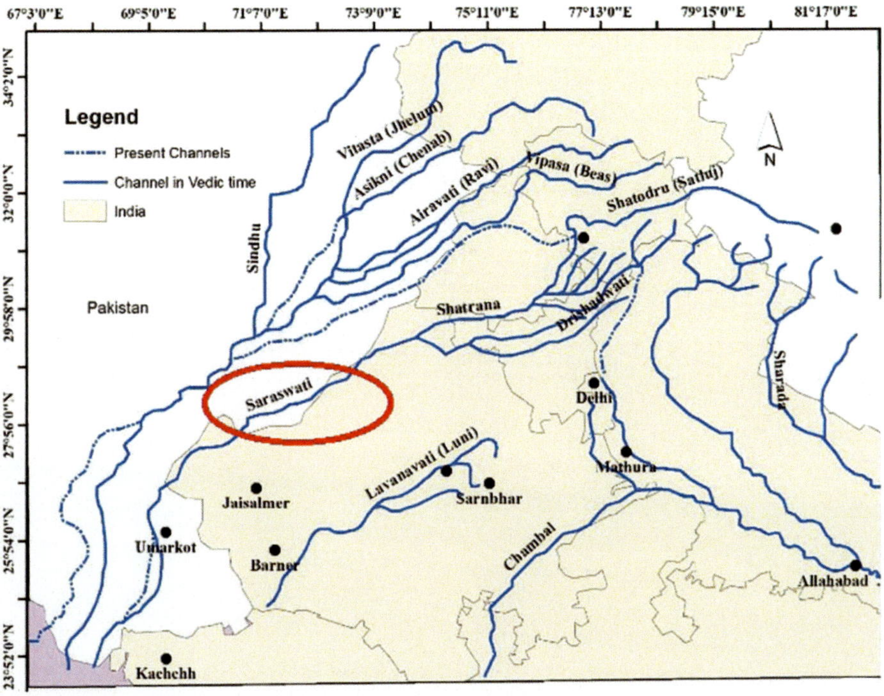

The Punjab land showing the rivers during the Vedic era (c. 1500–c. 600 BC) and the present: The river Sarawati vanished due to tectonic plate movement and high-intensity earthquakes, resulting in barren land and desert.

South Asian Region, even though small in world geographic area, consists of one-fifth of the world population, drains many large rivers, and has been one of the four cradles of civilization. The above factors give rise to not only a water-related culture but also to water-related technologies. The region is rich in agriculture, requiring adequate water supplies, which has driven the rise of water technologies. The articles in this part endeavor to bring out examples of these aspects from a historical perspective in this region. Punjab—a province of India and Pakistan, known as the land of the five rivers, consists of irrigation structures dating from 2600 BC. It had been a land of agriculture

DOI: 10.1201/9781003222736-23

since the ancient civilization. The irrigation practices in this region and the associated technologies for the betterment of society have been made clear in the article entitled "Overview of Irrigation Practices in Punjab." Sri Lanka, too, has many historic irrigation works and engineering marvels. The cascade system, the reservoir system tanks (weva), and the related irrigation works bear witness. Here, the Nuwara Weva reservoir and the related irrigation works have been analyzed, emphasizing a specific hydraulic feature used at the time, namely Bisokotuwa. The significance of the Bisokotuwa and its hydraulic benefit has been made clear in the article entitled "Investigation of Hydraulic Character for Ancient Inlet Sluice Barrel in Nuwara Weva Reservoir."

Chapter 20

An overview of irrigation practices in Punjab

Vivek L. Manekar and Ritica Thakur
Sardar Vallabhbhai National Institute of Technology (SVNIT)

CONTENTS

20.1 INTRODUCTION

The backbone of the economy of any country is agriculture, and irrigation is the backbone of agriculture. The development of the irrigation system has a long linkage with the evolution of humankind. There was much evidence readily available in Indian history regarding its existence and development. It is revealed from the literature that the irrigation systems not only existed but were well-designed and managed with proper infrastructure in place. One of the best examples to understand the historical aspect of irrigation development is the Punjab State of India. This monograph considers the historical evidence of the developed Indian Irrigation System and the Punjab Irrigation System. This study's consideration era is divided into pre-Colonial, 19th-century British Colonial, and post-Colonial.

Punjab is a state of India lying at the foothills of the Himalayas and blessed with the perennial rivers of the Himalayas, i.e., Sutlej, Ravi, and Beas. Rolling hills surrounding Punjab stretch along with northeastern Himachal Pradesh, Haryana, and Rajasthan in the south. The total area of Punjab is 50,362 km^2. It is blessed with a fertile alluvial plain and many other rivers as natural resources. It even receives water twice annually from the monsoon. It is a leading agricultural state of India, having 85% of its topographical area under agriculture with an average harvesting intensity of 189%. Punjab's agriculture is highly intensive and dependent on the heavy water requirement. The source of water is either groundwater or surface water from rivers. Natural water access is available to a limited area; hence, an intensive canal system

DOI: 10.1201/9781003222736-24

is made to irrigate the whole area. Punjab had faced a harrowing past despite being a land of rivers and blessed with fertile plains. In the 19th century, famine hit Punjab in 1883–84, 1896–97, and 1899–1900 due to a lack of monsoon. Punjab has a golden history of irrigation development.

20.2 IRRIGATION PRACTICES IN THE PRE-COLONIAL INDIAN ERA IN PUNJAB

India is an agricultural base country. Historians claim that the spread of knowledge of agriculture took place from here to worldwide. The first physical evidence of agriculture and its advancement in human history was found in the pre-Harappan civilization from c.7000 to c.5500 BC. This period is also spotted as the Neolithic Period, which refers to the last stage of the Stone Age (Indus Valley Civilization—World History Encyclopedia). Indus Valley Civilization, also known as the Harrapan civilization, was extended between 3300 BC and 1900 BC. This civilization was nearly extended in the 1.2 million km^2 area at present, most of which is a part of Pakistan (old Punjab and Sindh), Uzbekistan, and India (Rajasthan, west Gujarat, and Punjab) (Cotterell, 2011).

Harrapan civilization was settled across the perennial rivers. This area across rivers Sindhu (Indus), Jhelum, Chenab, Ravi, and Beas were later named Punjab—an area surrounded by five rivers. This area was highly fertile and suitable for agriculture. The primary source of irrigation was river water. Due to tectonic plate movement and high-intensity earthquakes, the river Saraswati vanished, and rivers Sutlej and Yamuna came into existence. This phenomenon results in large areas such as barren land and desert (Thar desert). Figure 20.1 depicts Punjab's land and shows the rivers during the Vedic era (c. 1500–c. 600 BC) and the present.

Many evidence of the irrigation structure (from circa 2600 BC) and water structures, such as drainage structures, reservoirs (artificial reservoirs at Girnar dated to 3000 BC), and canals, shows the innovative techniques and science used dated back to 2500 BC (Rodda & Ubertini, 2004). Initially, barley was the main crop for the region—a drought-resistance crop; with the improvement in irrigation, wheat and other crops replaced barley (Nene, 2007). This new revolution in the irrigation system and two seasons of yearly rainfall in the area leads to twice the cultivation of crops since its pre-Harrapan era. In the Neolithic era, animals were domesticated and used for plowing and extracting water from wells (Vishnu-Mittre, 1974). Agriculture was practiced as strip farming, and canal irrigation was practiced along the slopes.

After the declination of Harrapan culture in 1750 BC, Aryan came into existence in 1500 BC. This era is known as the Vedic era. In the Vedic era (c. 1500–c. 600 BC), many works of literature describe the technology and practices followed at that time. One well-known mythical literature, 'Rigveda' has described the agriculture practices and irrigation system. Majorly irrigation system is described as artificial wells [called 'Kūpa' with connecting channels (called 'Khanitrima')] to the fields. The water lifting was done with the help of a pulley and wheel (Chakra) arrangement using straps (Varatra) and a vessel (Kośa). These water channels were specifically used for agriculture practices in the Vedic era.

Further from the roughly 5th-century literature, evidence from Indian scholar Panini describes the tapping of many rivers for irrigation (Puri, 1968). Tapping of the

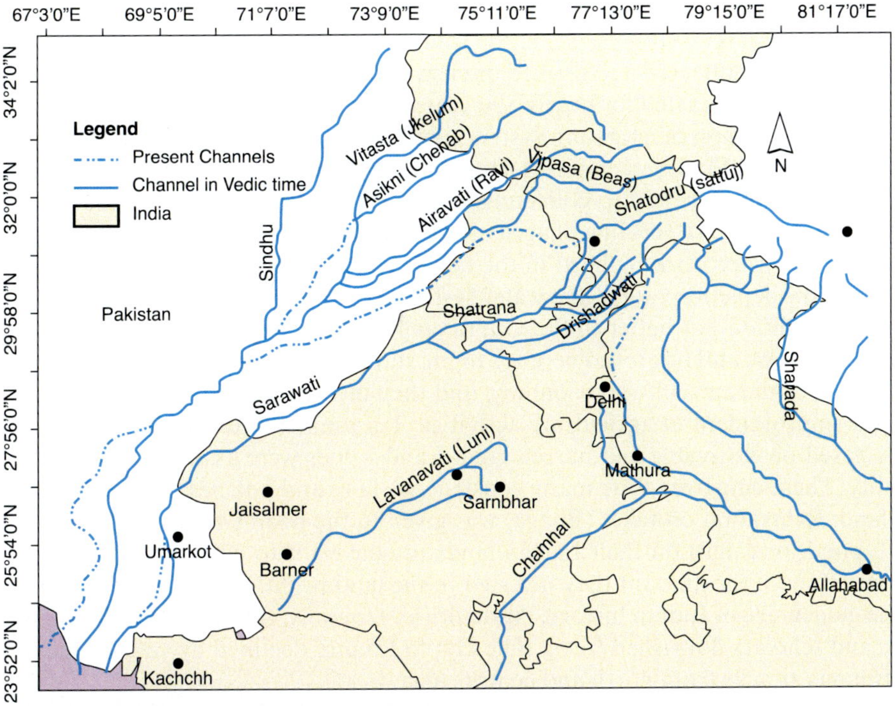

Figure 20.1 The land of Punjab: Presently in Pakistan (Sindh and Punjab province) and India (Punjab, Rajasthan, and Gujarat area), and rivers at the time of the Vedic era and present day.

river was the advancement in irrigation in that era. The great Indian Vedic scholar Patanjali mentioned periodic irrigation using canal systems for agriculture in his book 'Mahaabyasa.' The text 'Krishi-Parashara,' written by scholar Parashara, suggested maintaining the rainfall record from the beginning of the year by a trained person. It signifies their understanding and knowledge of climate and planning of water for irrigation (Sadhale, 1999). Many researchers believe that only these practices were spread to Central Asia from here.

In 600 BC, the Vedic era declined due to rigid social structure. Later this area came under many foreign invasions from the west. Persian rulers made few advancements in irrigation by constructing and maintaining the canal system in this region. Later Nanda–Mauryan period was a glorious period for agriculture. Many new canals were dug, and a state control irrigation system was maintained in this region. Mauryan empire introduced the coin system (punched marked), and revenue was generated from the farmers to maintain the irrigation canals. Small dams, levies, water tanks, and artificial ponds were constructed by the state and controlled by the superintendent of water houses (pãniyagharika). District-level offices were established to maintain the land measurement and irrigation records. The total taxes on the agricultural land were based on the type of irrigation system (state irrigation system or natural irrigation system, water drowning system) and the irrigation source (well, pond, water levy, and canal). Officials (Agriculture Superintendent) were appointed to cost and collect the yield according to the type of cereal (Haltmar and Ghoshal, 1930). Many laws regarding selling and buying water from privately owned sources and maintaining water sources in the presence or absence of the source owner were also made. The Mauryan era was highly engineered and worked intensively to construct and maintain hydraulic structures. This era lasted for nearly 800 years.

The construction of irrigation systems in the Mauryan era lasted until the 1st century BC. After the Mauryan empire declined, the land came under the Roman Empire, which was more interested in collecting and increasing their revenue to add to their treasure. This led to the decline of agriculture practices, as people did not have enough food left even after working hard in the fields. The famine and lack of water sources increased the depletion rate of the water system set by the Mauryan empire.

The condition got better in the Gupta empire in AD 318. The Gupta Empire was finely admirative and improved people's living standards in society. The Gupta Empire was aware of the agricultural economy, and they invested money in water structures and the improvement of agriculture activities. They made a record of land classifications based on cultivable and barren land. Land values were assessed based on their fertility. These emperors took many constructive laws and policies to reclaim the barren land. Relaxation on taxes (Bhaga) was given to the people to build water sources (well, pond, or canal) on their barren lands to convert it to cultivable land, till they generated twice the amount they invested in the land profit. The Gupta era is known as the golden age of Indian history. Agriculture, economy, trade, art, literature, scientists, and scholars flourished freely. The Gupta Empire declined by AD 500 due to the continuous attacks from north and central Asia.

Later the region came under the Arabs. Arabs kept land records for irrigated and non-irrigated land. So, revenue was generated. Arab agriculture revolution flourished throughout Central Asia from AD 700 to AD 1300. Many different types of crops, fruits, and agriculture techniques of the Islamic world were introduced in this region. Many canals and irrigation tanks were revived and newly constructed in this era. They introduced the vertical water-lifting technique called water wheel (sāqiyah) in India. The origin of this technique is apprehension either in Egypt, Iran, the Kush empire in southern Egypt, or India (Mokhtar, 1981). It is the vertical wheel with buckets arranged on the periphery, driven with the donkey's or cattle's help. Mughal administrators also constructed a few irrigation infrastructures and generated similar revenue. However, the investment made in the irrigation section was very little. Firoz Tuglok, in 1354, opened a new canal system at Dipalpur of a 48 kos (149 km) long canal from Satluj to Kaggar river for irrigation purposes only. Later, this canal system was extended to the river Shruti. Another canal was constructed in 1356 between Hansi and Hissar. Ali Mardan Khan, in 1639, also constructed a canal joining the river Ravi water to Lahore.

20.3 19TH-CENTURY IRRIGATION UNDER BRITISH COLONIAL

British explored the immense possibility of growth and revenue after assessing the natural resources of the Indian subcontinent. British was around Punjab since the beginning of the 19th century. Still, a large part of Punjab was under the influence of Maharaja Ranjeet Singh and the newly evolving revolutionary religion of Sikhism. Like the Hindu belief system, they also built water tanks in their religious places to conserve water. After the death of the King in 1839, the whole of Punjab came under the British colonial system.

In the 19th century, under the British rule until 1902–03, the total spending on irrigation was only Indian Rupee (INR) 430,000,000/- (₹43 crores), and, till 1905, the total spending on railways was INR 3590,000,000/- (₹359 crores) throughout India. After experiencing losses in generating revenue from the railway industry, irrigation becomes better than railways. Sir Author Cotton was one of that era's renowned scholars and engineers. He was famous as the Master of irrigation in British India. In his report to the British government, Sir Author Cotton (1854) mentioned Indian land as a "Land of immense possibilities." He persisted in putting more budget for the irrigation projects over the railway lines projects. The expected revenue ratio of agriculture concerning investment is far higher than the railways. British introduced a scientific and engineered approach to designing and planning the irrigation system across India.

A proper monitoring system for the rainfall was placed across the basins. The old canal systems and dams were rejuvenated. Many tanks were built to store water from floods or monsoons for later use in irrigation, especially for rice cultivation. Canals from large perennial rivers were constructed to get water in adverse climate months. They divided the canal system into the Inundation and Perennial canals. The Perennial canal had water storage throughout the year, and water was directly provided from the rivers in Inundation canals. Even the irrigation projects were classified under two heads according to their goals. Projects subject to reaching profits from revenue were classified as "Productive" and restrained famine situations as "Protective." According to British records of the Indian irrigation commission (1901–1903), from the 19th century, land under irrigation was 4,400,000 acres. Among them, 1,300,000 acres of land were irrigated by well irrigation, and 800,000 acres were under Tank irrigation; the rest of the 2,300,000 acres were under canal system. However, most of the work in the irrigation sector was done under the public work department (PWD) by the government only. Few works were done by the private sector also.

20.3.1 Irrigation system in Punjab in the 19th century

Punjab, surrounded by rivers, has flat land and is highly suitable for endorsing an agricultural economy. In the first years of capturing Punjab, around 1849, a British company found the area unsuitable for development due to its unsuitable social conditions (Piedmont & Lombardy, 1849). Only one canal named Bari Doab, starting from the river Ravi having a length of 450 miles (724.20 km), was constructed under the British rule on a budget of INR 1,35,85,502/-. Sir Robert Montgomery inaugurated it on 11 Apr 1859. The canal head was constructed at Madhupur and supplied the water to Gurdaspur, Amritsar, and Lahore. This area lies between the river Beas and river Ravi. Under the British rule, canal irrigation was intensively developed based on the type of crops planted in the area and the rainfall received. In Punjab, nearly 1,000,000 acres of the area were under privately constructed irrigation facilities (Piedmont & Lombardy, 1849).

Later in November 1882, the great Sirhind Canal was inaugurated, dedicated to the sole purpose of irrigation. It was a massive project of three stages: the central canal, the Abohar branch canal, and the Sutlej navigation channel. Later, the western

Jamna Canal was one of the pride projects. Few more canals were constructed. Swat River, Sidnai, and Chenab canals irrigated large catchment areas. Besides these Perennial canals, many inundation canals projects were also constructed named as Lower Sohag and Para Canal, the Sidhnai Canal, the Upper Sutlej series, the Lower Sutlej, and Chenab canal series, the Indus canal series, the Imperial canals across Shahpur District, and the Mozaffargarh canal series. Later in 1866–74, on the eastern Jamuna Canal, a simple approach of contractual water delivery was applied in which a contract for the delivery of the three-year water supply will be made between farmer and authority. However, it was an unsuccessful approach.

With due time, the canal network was expanded by adding branches, distributaries, sub-distributaries, minors, and watercourses. Still, the problem was persistent. In 1860, the water regulation work was given under the PWD department for a better distribution system. The canal system, especially in Punjab, has a problem with the siltation of canal head regulators and outlets. R G Kennedy, an Engineer in the Punjab PWD department, studied the erosion and deposition of the sediments in an unlined canal of the Upper Bari Doab Canal system (Bolding et al., 1995; Gilmartin, 2003). He came up with a theory for stable channel and critical flow velocity. Initially, a tremendous amount of labor was spent on maintaining the unlined channel. Later, R G Kennedy suggested that it was a desperate task and needed a better scientific approach. He describes a flow (velocity) state in which erosion and sedimentation balance over the depth in an unlined canal.

This theory became the basic principle of designing for the unlined canal. He continuously observed the Bari Doab Canal system and proposed the "Regime theory." This theory explains the natural construction process of a stable regime in the unlined canal through continuous silting and scouring over time. The stable canal's final regime was found to be rectangular. Though with the variation in the discharge, the regime seems to change slightly throughout its cross-section. This led to the conclusion that the well-designed regime of the canal does not need much maintenance work across the operation period, though it regulates itself and needs to be clear annually in case of heavy silting.

In 1873, Canal and Drainage Act was passed to limit the number of output and the accurate determination of the area to be watered from each channel, putting more focus on measurement of the discharge and channel efficiency maintenance. Hence R. G. Kennedy designed a gauge for this purpose; engineers were impeded for years by challenges with constructing modules that could withstand changes in canal levels and irrigator interference. Due to Kennedy's work on silt control, engineers were needed to reconstruct most irrigation canals regularly to preserve their efficiency in the face of silting concerns. Afterward, Mr. Kennedy was assigned to design and construct the Lower Gugera Branch of the Lower Chenab Canal under the Superintending Engineer, Mr. Sidney Preston (Gilmartin, 1994).

Later, the British government came up with the "Northern India Canal and Drainage Act, 1873". Under this amendment, the government can use and control the water of all rivers and streams flowing in natural channels and all water sources. In 1890, after the visit of Prince Elbert Victor to Punjab, many new irrigation projects got a grant from the government (Muhammad, 1891). Construction of the Sirsa Branch of the Western Jamna Canal Projects and subsidiary canals from the Ravi above the Sidhnai weir was started. R G Kennedy was then promoted as an Executive Engineer of Punjab PWD in 1895, considering his contribution to irrigation.

20.3.2 Irrigation system in Punjab in the early 20th century

In 1901–1928, the left canal of the Sindh Sagar Doab taking water of Indus to Dera Ismail Khan was constructed. Inundation canals from Chenab to Khadar (a northern part of Jhang District) and the Kabul River to Peshawar were constructed (Muhammad, 1891; Shahid et al., 1994).

Meanwhile, in 1906, RG Kennedy developed a gauge for outlets. Figure 20.2 shows the Kennedy gauge outlet. This gauge was unsuitable for installing on sites, as it consists of a vent pipe that can be tempered. Later, RG Kennedy was promoted to Chief Engineer in Punjab PWD Department. In the early 20th century, the concept of modularity appeared of keen interest among hydraulic scientists. Many studies took place to understand the sensitivity of the degree of change in outlet structure for a depth–discharge relationship of the modular canals. Using the power of control over water resources under "Northern India Canal and Drainage Act, 1873," the British government set a stepping stone for the irrigation departments named "Special Irrigation Division" by June 1916. In 1910, W G Bligh, Executive Engineer, PWD of India, introduced his Bligh's creep theory. This theory was adapted from Col. Clibborn's work, analyzing the field data of the seepage flow in Punjab barrages on permeable foundations. It describes the solution to reduce the seepage losses due to the head made by vertical obstructions in flowing water. He proposed that to reduce the piping failure, the length of the base along which the seepage travels should be 5–15 times the hydraulic structure's designed hydraulic gradient, depending on the type of the soil. This was the most popular theory used to design barrages worldwide. By 1928, CC Inglis developed standing wave pipe outlets and standing wave flumes. These revolutionizing structures control water discharge at the outlet of canals (Mahbub & Gulhati, 1951).

Similarly, many research and studies were conducted, and water control and measuring structures were improvised. However, significant development came from Gerald Lacey. He was a Professor of Civil Engineering (1915–17, 1928–32, 1945) and the last British Principal (1945–46) of the University of Roorkee near Delhi.

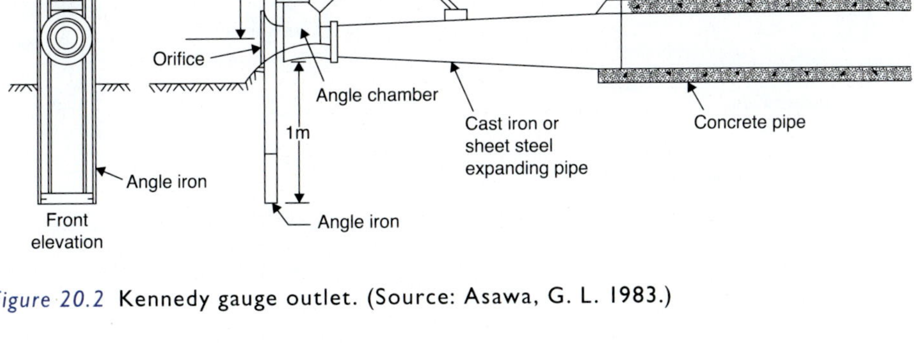

Figure 20.2 Kennedy gauge outlet. (Source: Asawa, G. L. 1983.)

The British government funded his research, which helped him contribute to regime theory. He made a significant amendment to the regime theory proposed by RG Kennedy. He gained international acclaim for his 'regime theory,' which was used to construct major irrigation canals. He also made significant contributions to Stable Channel Flow and gave technical counsel and insight that aided the development of irrigation systems in India. According to Lacey, the vertical component of eddies formed at any point of forces normal to the wetted perimeter keeps the silt suspended. He introduced the silt factor to the equation; this helped in giving a more practical approach to designing the canals.

Generally, water is distributed in the Punjab canal system by a continuous water supply with a share in rotations. Crump's open flume, Haighs, and Sharmas modified open flumes, Jamrao-type open flumes, Kennedy's Gauge Outlet, standing wave pipe, and standing wave flumes were installed upstream of the Gated Pipe Outlets and canals in Punjab (Mahbub & Gulhati, 1951).

From 1916 to 1940, practically the whole Punjab region went under a series of construction of many weirs and barrages across the streams for the controlled irrigation system. Figure 20.3 shows the positions and location of the barrages built by the British government between 1916 and 1940. In 1932–1939 Punjnad and Abbasia canal was commissioned from the river Chenab; later, the Sindh canal system was upgraded by constructing Trimmu Barrage Project, covering added Haveli and Rangpur canal system.

Meanwhile, many weirs and barrages came into the central irrigation board's limelight for their repairs, maintenance, and structure failure. The most highlighted case was of failure of the Khanki Weir in 1895, which was constructed on the Chenab

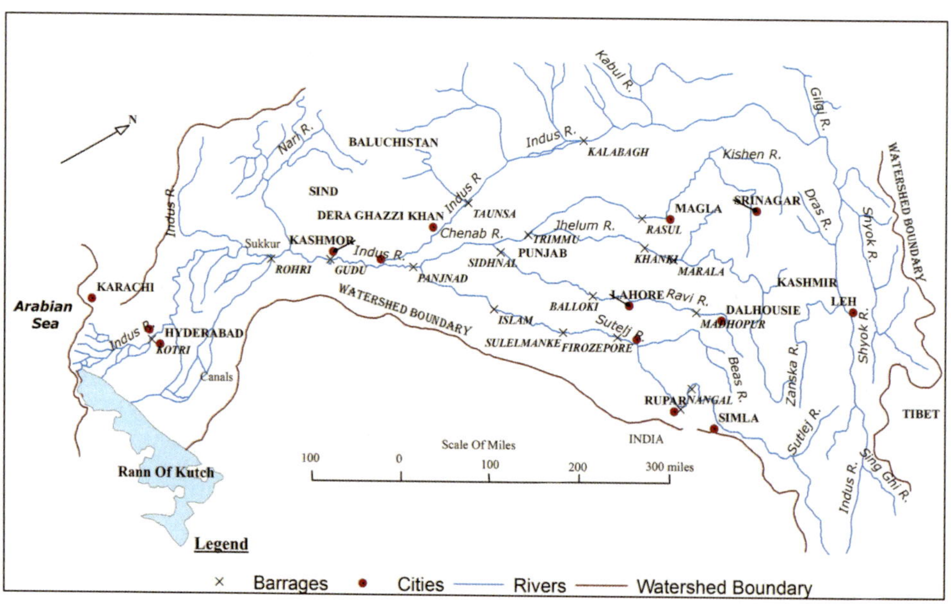

Figure 20.3 The Indus River Basin and the various barrages built between 1916 and 1940 to irrigate the command area.

River in 1892 only. Hence in 1934, the central irrigation board conducted its first meeting, which took the initiative of discussing the failure of the weirs and barrages. Dr. Ajudhia Nath Khosla was responsible for conducting the analysis and finding the cause of the failure. He contributed significantly to highlighting the flaws of Bligh's theory and modified the basic principle of Bligh's theory of water movement along the hydraulic gradient to movement of the water along with the set of streamlines. His theory is termed Khosla's exit gradient theory. He further evolved the "Method of Independent Variable" to design hydraulic structures across the permeable foundation. His theories are still valid and popular for designing hydraulic structures on permeable foundations. Dr. Ajudhia Nath Khosla was an engineer from India and served as Chairman of India's Central Waterways Irrigation and Navigation Commission. Later from 1954 until 1959, he also served as Vice-Chancellor of the University of Roorkee. He was also responsible for conducting a survey and investigation for the Bhakra Dam project. The Bhakra Dam was erected on the same axis line that he identified in 1917 (Khosla et al., 1954). Later, the amendment in the construction designs was made concerning the uplift pressure forces generated on the foundation of the weirs and barrages.

In 1947–1948, the canal of Kalabagh took water from river Indus to supply a supported irrigation system for the added requirements of Sindh Sagar Doab (area between river Indus and Jhelum). Though canals are a significant investment for the British colonials, their biggest challenge is running and managing them. In the first days, the farmers were directly responsible for operating water from the canals by cutting the canal and operating them. Revenue was generated according to the area under irrigation and the type of crop grown under the rotational water supply warbands system (Bolding et al., 1995). This led to the problems of waterlogging and poor management of water. Later this system was replaced by the Bhandara system. Under this system, the catchment area was divided into four parts, and four significant crops were supposed to grow. All the farmers were allotted a share of the land in each part. This was purposed to control water waste and better water management. However, it failed due to local inference and other problems (Bolding et al., 1995). By 1994, nearly 40,000 outlets were there to irrigate 14 million acres (56,656 km^2) of Punjab. The Chenab canal pipe outlet made of cast iron was adopted for the first time. In the British era, with increased investment in the irrigation sector, the people's socioeconomic condition also improved. A large amount of the area came under an irrigation system that led to a better yield.

In the late 19th century, there was an increment in the rate of change in agriculture product prices. Between 1870 and 1913, the rate was positive at 1.10, which later lowered to 0.44 between 1913 and 1938 (Federico, 2010). Federico believes this happens as, in the early era, the demands in the market were high due to consecutive famines and wars experienced by the world. Later, with better productivity and stable conditions, these rates got reduced. Hence with the better irrigation facilities and reasonable pricing of the agricultural products, the condition of the Punjab farmers improved significantly. Even Punjab's semi-arid, thinly populated areas got new flourishing and emerging cities with better roads, communication, and schools. This improvement's ripple effect results in a better-governed society with fewer crime rates, better law and order conditions, better infrastructures, factories, industries, and higher living standards for Punjab people.

Table 20.1 Details of canal infrastructure after independence

Sr. no.	*Project name*	*Year*	*River*	*Location*
1	Nangal Dam	1948	Satluj	Downstream (Bhakra Dam)
2	Bhakra Main Line Canal	1950–54	Satluj	Extension of Nangal Hydel Channel
3	Old Sirhind Canal System	1952–54	Satluj	Ropar Headworks
4	Ferozepur Feeder	1952–53	Ravi–Beas	Harike Headwork
5	Nangal Hydel Channel	1954	Satluj	Nangal Dam
6	Harike Headwork	1954–55	Satluj–Beas	Harike
7	Madhopur Beas link	1955–57	Satluj–Beas	Madhopur
8	Rajasthan canal	1958–61	Satluj–Beas	Harike Headworks
9	Bhakra Dam	1963	Satluj	Bhakra (H.P.)
10	Pong Dam	1974	Beas	Pong
11	Beas Sutlej Link	1977	Beas–Satluj	Pandoh (H.P.)
12	Mukerian Hydel Channel	1982	Beas	Shanehar Headwork
13	Shanehar Headwork	1983	Beas	Downstream of Pong Dam.
14	Ranjit Sagar Dam	2000	Ravi	Upstream of Madhopur Headworks
15	Shahpur Kandi dam	2006–07	Ravi	Downstream of Ranjit Sagar Dam

Source: Punjab Irrigation Department, Govt. of Punjab

20.4 POST-COLONIAL DEVELOPMENT

Indus basin forms rivers Sind, Chenab, Jhelum, Ravi, Beas, and Sutlej, with a total basin area of 1,165,500 km^2. In 1947, at the time of independence, India was divided into India and Pakistan; Punjab got a painful partition experience. Under the Indus water treaty of 1960, the Sind, Chenab, and Jhelum were granted to Pakistan, and Sutlej, Ravi, and Beas were granted to India. It was a 56% basin area settlement to Pakistan and 31% of the basin area to India. The Indian Indus basin includes 321,289 km^2 (Singh & Bhangoo, 2013).

The river of Punjab is a perennial river with a source Himalayas. The canal system feeds the water to Himachal Pradesh, Punjab, Haryana, and Rajasthan through the canal system. The primary source of water for irrigation is still groundwater. The pump systems known as Tubewell upgraded from the pre-Independence Persian well system to draw water. Post-Independence, many new large multipurpose dams were proposed and constructed. Table 20.1 gives the details of multipurpose dams. Many maintenance works, re-modeling of existing canals, and added irrigation infrastructure were conducted. Many other small important canal works were also constructed in this era under the state to generate hydropower and irrigation facilities. Table 20.2 explains the details of the capacity of major canal works. Figure 20.4 shows the location of all the dams and canals across the states of India.

Apart from the various multipurpose dams, a few important dam works were also taken under construction. Table 20.3 details the additional dam constructed in the Punjab region.

20.5 IRRIGATION WATER DELIVERY SYSTEM

Like pre-Independence, the warabandi system is only used for post-Independence, but with a better approach. The warabandi approach is made at three levels, namely, Khuli-wari (open turn), Panchayati-wari, and weekly-wari. Farmers only act as field staff

Table 20.2 Capacity (discharge and length) of main canals of Punjab

Sr. no.	*Canal*	*Discharge (cusecs)*	*Length (km)*
1.	Sirhind Canal	12,622	59.44
2.	Nangal Hydel Channels	14,500	20.12
3.	Combined Branch	7,635	3.22
4.	Sidhwan Branch	1,751	88.01
5.	Abohar Branch	3,027	109.75
6.	Bathinda Branch	2,890	152.40
7.	Ferozpur Feeder	11,192	51.42
8.	Sirhind Feeder	5,264	136.53
9.	Rajasthan Feeder	18,500	149.53
10.	Abohar Branch Lower	1,693	46.37
11.	Bikaner Canal	2,720/3,027	112.01
12.	Eastern Canal	3,929	8.02

Source: www.pbirrigation.gov.in
Note: Total C.C.A. 30.88 Lacs Hectare.

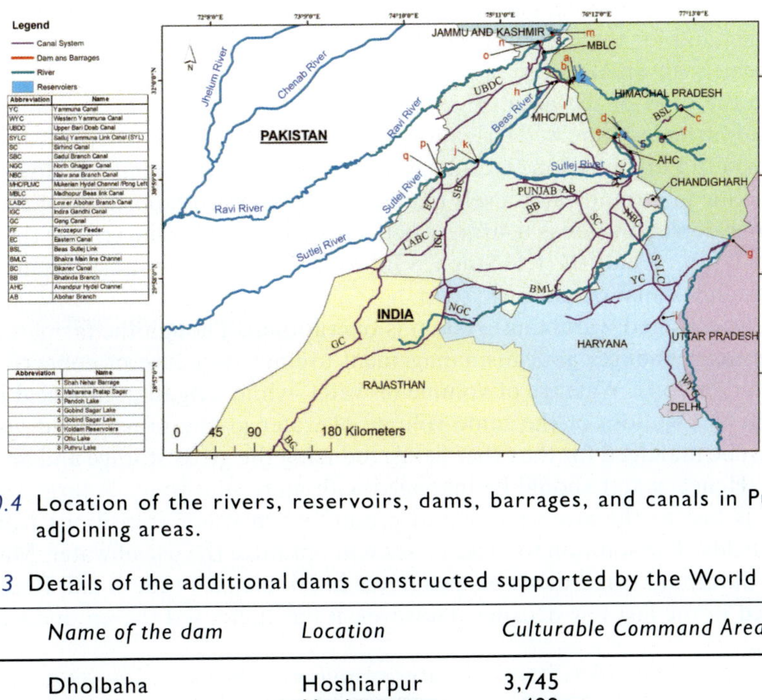

Figure 20.4 Location of the rivers, reservoirs, dams, barrages, and canals in Punjab and adjoining areas.

Table 20.3 Details of the additional dams constructed supported by the World Bank

Sr. no.	*Name of the dam*	*Location*	*Culturable Command Area CCA (ha)*
1.	Dholbaha	Hoshiarpur	3,745
2.	Janauri	Hoshiarpur	492
3.	Maili	Hoshiarpur	914
4.	Damsal	Hoshiarpur	1,920
5.	Chohal	Hoshiarpur	900
6.	Saleran	Hoshiarpur	365
7.	Patiari	Hoshiarpur	730
8.	Thana	Hoshiarpur	1,160
9.	Perch	SAS Nagar	400
10.	Mirzapur	SAS Nagar	970
11.	Siswan	SAS Nagar	950
12.	Jainti	SAS Nagar	500

Source: Singh and Bhangoo (2013).

and fix their turns, duration, and water quality among themselves. Their decisions are basses on the landholding of the farmer, location of land from the canals, and type of crops.

20.6 PRESENT AND FUTURE NEEDS OF IRRIGATION SYSTEM

The need for yielding cash crops is also increasing with the increasing population. Fields of Punjab are highly potential for producing more grains. Punjab has become the country's paddy–wheat producer to meet the food demand. It led to an increase in the water demand for agriculture. The area under cultivation of paddy rose from 9% to 72%, between 1980 and 2015 (Hima Bindhu 9 Jul 2021). At the same time, growing industrialization also increased water demand. It increases the water stress conditions in Punjab, regardless of having many dams. Table 20.4 shows the data on Punjab's water requirement and deficit in 2007.

Presently, a 150-year-old planned canal system is still working for the fertile plains of Punjab. However, the government is spending a significant amount on regular canal maintenance. Still, the canals cannot carry the discharge for which they were designed. Even the demand for the present needs is relatively high, and the performance of the canal system is abysmal. Many operational measures and automation of the canal system are the need of the hour. Due to the underperformance of the canals, many farmers had turned to tube well irrigation to overcome the water deficit. Nearly 72% of the irrigation in Punjab is done by tube well to meet the water demand. However, this led to a decline in groundwater levels in the area. It has been reported that the annual groundwater depletion rate is 0.40 m, which is a considerable concern for the government and environmentalists (KG Singh, 8 Jul 2019). Hence, more stress should be given to improving the canal irrigation system.

Still, the old-aged warabandi system is operational. Though the farmers are well managed, miss conducts and mismanagement are always cause of concerns due to human interventions. Wastage of volume of water, while irrigation of a field is equivalent to the double loss of the same volume of water as the same added volume of water, must be managed for the other fields, reducing the gross storage and area under irrigation. Hence, water should be managed with more sensitivity. A large volume of the water is lost to the atmosphere and ground when water transitions from canal system to fields. The solution to these losses will optimize the use of water. Many multipurpose dams and small dams on each river in Himachal Pradesh and Punjab were constructed in the last few decades. However, it has increased the groundwater level

Table 20.4 Status of water resources in Punjab

Detail	*M ham*
Annual canal water at headworks	14.54
Annual canal water at outlets	1.45
Annual groundwater available	1.68
Total annual available water resources	3.13
Annual water demand	4.40
Annual water deficit	1.27

Source: Jain and Kumar (2007).

of the surrounding area. Nevertheless, more groundwater recharging systems are still the need of the hour.

Another problem with the Punjab irrigation systems is the siltation of the canals. Many technical measures and multistage soil conservation programs must be considered in the upper catchment area of the canal system and dams. The government takes even many works and measures to date. However, more technical approaches and solutions are needed for this problem. These steps will increase the canal system's life span and reduce the canals' maintenance.

20.7 CONCLUSIONS

Punjab has been the land of agriculture since the beginning of human civilization. It is a land blessed with high fertility plains and water resources. It is evidenced from the literature that advanced irrigation practices were well developed and adopted in India, since the last stage of the Stone Age (Indus Valley Civilization). The Punjab irrigation practices of India during the pre-Colonial and colonial eras were the source of knowledge spread all over Asia. The irrigation practices designed, evolved, and adopted in Punjab are still alive. They set the example of excellent ancient engineering knowledge and the golden glory of the Punjab irrigation system. The research and theories developed under British India's reign in Punjab were an outstanding contribution to the Society of Hydraulic Engineers worldwide. Many interlinking canal and command area development projects were executed across the Punjab region with a scientific approach. It increased the culturable command area in Punjab from 1869 to 1899 from 8.1 to 9.4 million ha (Siddiqi, 1986).

In the spirit of meeting the needs of food security under changing climate scenarios, and the present growing demands of food and industrialization, Punjab farmers face a massive water crisis challenge. The change in the food habits and increase in paddy crop farming from 9% to 72% leads to the high water demands for irrigation. The present developed canal system is high maintenance and low performing and does not meet irrigation needs. This resulted in farmers turning toward tube well irrigation systems (nearly 72% of agriculture area). Groundwater is a natural water source and gets accumulated at a very low rate over a significant period. Hence, it is necessary only to provide surface water irrigation using a canal system. Despite the intense canal system present in Punjab, farmers face water shortage caused by mismanagement of the water resources, polluted canal water, highly silted canals, the high maintenance cost of the canal system, and underperforming canal irrigation system (Jairath, 1985). Many old canal systems need reconstruction as per the future needs of agriculture.

It is evident that if a society has flourished in the past century, it has been through better agriculture practices and producing better agricultural products. It only happens with the enormous support and interest of governing bodies to support the scientific community to develop solutions to problems. Solutions become effective only if it preaches the root of the problems, thereby need enormous efforts to impart training to the farmers, and promote new advanced techniques for farming and irrigation, which will reduce water losses. Like the saying "a penny saved is a penny earned," every drop of water saved to the losses is the addition of water to irrigate. Hence, more research must be done to improve the canal system and recharge the groundwater to attain a sustainable water resource for us and the next generation ahead.

REFERENCES

Asawa, G. L. (1983). *Irrigation engineering.* Wiley Eastern Ltd, 568p.

Bolding, A., Mollinga, P. P., and Van Straaten, K. (1995). Modules for modernization: colonial irrigation in India and the technological dimension of agrarian change. *The Journal of Development Studies, 31*(6), 805–844.

Cotterell, A. (2011). *Asia: a concise history.* John Wiley & Sons.

Cotton, A. (1854). *Public works in India: their importance. with suggestions for their extension and improvement.* WH Allen & Company.

Federico, G. (2010). *Feeding the world.* Princeton University Press.

Gilmartin, D. (1994). Scientific empire and imperial science: colonialism and irrigation technology in the Indus basin. *The Journal of Asian Studies, 53*(4), 1127–1149.

Gilmartin, D. (2003). Water and waste: nature, productivity, and colonialism in the Indus Basin. *Economic and Political Weekly*, 5057–5065.

Haltmar, K. and Ghoshal, U. N. (1930). *Contributions to the history of the Hindu revenue system* (Book Review). *Archív Orientální, 2*(2), 381.

Hima Bindhu. *Punjab's Water Crisis Due to Paddy Cultivation.* Grain Mart India. https://www.grainmart.in/news/punjabs-water-crisis-due-to-paddy-cultivation/. Accessed 9 Jul 2021.

Indian Irrigation Commission. (1903). *Report of the Indian Irrigation Commission 1901–1903.* Part II: Provincial. Calcutta, Superintendent of Government Printing.

Jain, A. K. and Kumar, R. (2007). Water management issues – Punjab, North-West India. In Proc paper In: *Indo-US Workshop on Innovative E-technologies for Distance Education and Extension/Outreach for Efficient Water Management.* ICRISAT, Hyderabad.

Jairath, J. (1985). Technical and institutional factors in utilization of irrigation: a case study of public canals in Punjab. *Economic and Political Weekly*, A2–A10.

Khosla, A. N., Bose, N. K., and Taylor, E. M. (1954). *Design of weirs on permeable foundations.* Central Board of Irrigation, New Delhi.

Muhammad, L. S. (1891). *History of The Panjab.* Calcutta Central Press Company, Limited.

Mokhtar, G. (Ed.). (1981). *Ancient civilizations of Africa.* Unesco.

Nene, Y. L. (Ed.). (2007). *Glimpses of the agricultural heritage of India.* Asian Agri-History Foundation.

Puri, B. N. (1968). Irrigation and agricultural economy in ancient India. *Annals of the Bhandarkar Oriental Research Institute, 48*, 383–390.

Rodda & Ubertini (2004). The basis of civilisation-water science? *International Association of Hydrological Science.* ISBN 1-901502-57-0.

Sadhale, N. (1999). Krishi-Parashara (Agriculture by Parashara). Asian Agri-History Foundation.

Siddiqi, A. H. (1986). Agricultural changes in Punjab in the nineteenth century: 1850–1900. *GeoJournal, 12*(1), 43–56.

Singh, K. G. (2019). Strategies needed to conserve groundwater in Punjab. *The Times of India.* https://timesofindia.indiatimes.com/city/ludhiana/new-strategies-needed-to-conserve-groundwater-in-punjab/articleshow/70120612.cms. Accessed 8th July 2019.

Singh, I. and Bhangoo, K. S. (2013). Irrigation system in Indian Punjab. *Munich Personal RePEc Archive (MPRA)* Paper No. 50270, posted 30 September 2013.

Vishnu-Mittre, A. (1974). Palaeobotanical evidence in India. In J. Hutchinson (Ed.) *Evolutionary Studies in World Crops*, 3–30. Cambridge University Press.

Chapter 21

Water heritage of ancient Sri Lanka

Gregory S. De Costa
Open Polytechnic of New Zealand

Ruwan Rajapakse
Consultant in NYC

CONTENTS

21.1 INTRODUCTION

Due to the spatial and temporal variation in rainfall distribution, the main source of this essential resource for human civilization, the optimal use of the available resource was and is of paramount interest. This gave rise to water management strategies to cope with floods, droughts, and water purification methodologies.

As communities increased in numbers and became residents away from water resources, it gave rise to water channeling. Further, excess water needed to be removed, leading to sewer and sewerage systems. Traditional technologies in all these aspects are evident worldwide, and they inspire modern hydro-environmental engineering right through the ages until today.

We note the ancient irrigation schemes, which consisted of earth banks and basins, regulated sluices, and a canal network used to manage the waters of the Nile for irrigation purposes. It is also said that the ancient Egyptians built dams around 2750 BC. Water sourcing technology dating back to 1500 BC has been discovered in the deserts of Israel. We also have Qanats constructed in Persia and the Iraqi region, dating back to 1000 BC, which were an underground network of tunnel channels transporting water from an upstream aquifer interconnected by a good system that delivers water to the surface under gravity. Sri Lankan and Indian Engineers were also at the forefront of Water Engineering.

DOI: 10.1201/9781003222736-25

Water not only has been and is a necessity of life, but it also has played a major part in the cultures and religions of the world. From time immemorial water has been considered a protector, cleaner, and so on in religious and cultural activities worldwide. And Sri Lanka and India are no exceptions.

While there is much evidence for traditional water technologies worldwide, we concentrate on historical and traditional water technology in Sri Lanka in this chapter.

21.2 OVERVIEW OF TRADITIONAL WATER TECHNOLOGIES IN SRI LANKA

Ancient chronicles such as the Mahavamsa date back to Sri Lanka's traditional water technologies to 500 BC. Sri Lanka—an island with a tropical climate, consists of one-third of its landmass receiving 2,500 mm of rainfall per year, while two-thirds receive 1,750 mm per year, theoretically making it rich in water. While this is correct, the spatial and temporal distribution of the rainfall brings with it times of abundance and shortage. The climate conducive to farming and agriculture throughout the year exacerbated the need, benefit, and drive for water management and technologies.

Irrigated agriculture and paddy cultivation thrived during this period by constructing reservoir (Tank/Vewa)-based irrigation system—a complex irrigation system of the ancient world. King Parakrama Bahu 1, who ruled over Sri Lanka during the period AD 1153–1186, stressed the importance of this resource by saying to his people, "Let not even a drop of rainwater go to the sea without benefiting man." Figure 21.1 called Parakrama Samudra—a tank constructed during the reign of this king.

Water is released and conveyed from the tank through spillways and channels to connected agricultural land, mainly paddy fields. Each plot of land has its drain delivering water. It was a tank cascade system (Figure 21.10).

This guarantees a continuous flow of water throughout the year. And it was possible to manage to supply water in the right quantities, avoiding oversupply and

Figure 21.1 Parakrama Samudra.

Figure 21.2 A tank constructed during the Anuradhapura era.

Figure 21.4 Buffalos grazing among birds (left) and elephants among birds beside a tank (right).

inundation. Even today, some of these tanks are used, and there are over 10,000 tanks being the main source of the irrigation scheme.

Even before was the Anuradhapura era of Sri Lanka. 500 BC to AD 1000. During this period, too, many tanks were built. Figure 21.2 indicates one such tank.

Built around the irrigation systems were thriving communities with places of religious worship, schools, etc. And this, needless to say, not only gave rise to social and administrative governance systems but also developed an ecosystem of its own. Figures 21.3–21.5 elaborate on the ecosystem.

Figure 21.3 An ecosystem within a tank.

Figure 21.5 Kantale bund. (An immense amount of soil that was brought in and compacted.)

These irrigation systems consisted of bunds constructed by compacting soils at the right compositions to avoid seepage while ensuring strength. Sluice Gates are basically of two types: one of the simple cylindrical logs for small tanks and very complex valve pit tower construction for large tanks. The sluice gate for large tanks consisted of two components: the inlet at the bottom of the tank and the other at the exit. The reasons are to expel silt/mud/debris and reduce the pressure. They also include surplus weirs to remove excess water and stone liners to reduce erosion.

The noteworthy point is the larger reservoirs (Tanks/wewa). The sluice consisted of a gate mechanism and a large chamber called the Bisokotuwa, where the water first enters from the tank. This chamber aims to reduce the head by causing additional frictional forces, thereby reducing the pressure and negatively impacting the exit and downstream. This chamber is performed as an energy-dissipating method during the release of water under pressure. This chapter here details this feature.

Sri Lankan irrigation schemes include stupendous reservoirs such as Kala wewa and Parakrama Samudra, reservoir cascade systems, cuts to divert water from perennial rivers, long and meandering canals past rivers and many other obstacles, large spillways, and technically advanced structures such as Bisokotuwa.

Some of the major works of ancient Sri Lanka are Jayavapi—a pond deepened by King Pandukabhaya (310 BC), reservoirs such as Abhaya Wewa (Tank)—King Pandukabhaya (310 BC), Tissa Wewa (Tank)—King Devanam Piya Tissa (250 BC), Nuwara wewa (Tank)—King Wattagamini or Walagambahu, Kutakannatissa—Balaluwewa (AD 40), King Wasabha (AD 65), six major reservoirs.

Mahasen (AD 300) speaks of 16 major reservoirs [i.e., Minneriwewa (Manihira in Mahavamsa), Hurulu wea (Challuru in Mahavamsa), Mahakandarawa (Kanu vapi in Mahavamsa), Maminiyawewa (Mahamani in Mahavamsa), Niramullawewa (Kumbalaka in Mahavamsa), Padawiyawewa (Rattamalakandavapi), Kaudulla tank (Tissavadhamanakavapi), Nachchaduwa Reservoir (Mahagallakavapi), Karampankulam (Kalapasanavapi)].

King Dathusena (AD 450) Kala wewa (Kala vapi) incorporated Balaluwewa, Yodha ala (Jaya Ganga of Mahavamsa), an 80 plus km long canal with 15 cm per 1.6 km gradient.

The evolution of irrigation systems in Sri Lanka began with ponds built by excavating the ground. Next, a bund (Wall) was built around the pond. Kaudulla reservoir (Tissavaddhamanaka of Mahavamsa) has a bund approximately 40 m wide at the base. The height of the bund of this reservoir is measured to be 15 m. In AD 1680, Padawiya Reservoir (Ratmalakandaka of Mahavamsa) was another large reservoir built by King Mahasen. The bund of this reservoir is much larger than the bund of the Minneriya reservoir, measuring 23 m in height, 60 m at the base, and 10 m at the top, with a length of 18 km approximately.

Kaudulla reservoir (Tissavaddhamanaka of Mahavamsa) has a bund of 40 m wide at the base, extending to a greater height than the bund of Minneriya. The height of the bund of this reservoir is measured to be 16 m. In AD 1680, an Englishman named Mr. Pybus had written that this reservoir was in working order.

Padawiya Reservoir (Ratmalakandaka of Mahavamsa) is another huge reservoir built by King Mahasen. The bund of this reservoir is much larger than the bund of the Minneriya reservoir, measuring 23 m in height, 60 m at the base, and 10 m at the top, with a length of 18 km.

> Padawiya reservoir, most gigantic works of all the reservoirs – 10 m broad at the summit, 60 m at the base, and 23 m high. Its construction must have occupied a million people for 10 to 15 years.
>
> Governor Henry Ward (Ref: R.L Brohier, Ancient Irrigation Works of Ceylon)

A sluice was then designed to acquire water from pond/reservoirs. A sluice is a tunnel built through the bund for the water to travel and come out. As the sluice gates got displaced by water pressure for tanks with tall bunds, a chamber to release water pressure (Bisokotuwa) was designed.

Sir Emerson Tennent, Governor of Sri Lanka in 1847, wrote,

> Padawiya sluice is a remarkable work, not merely from its dimensions but the ingenuity and the excellence of its workmanship. It is built of layers of hewn stones varying from 2 to 4m in length and still exhibits a sharp edge and every mark of the chisel. These rise to a ponderous wall immediately above the vents that regulate the water's escape. Each layer of the work is kept in its place by the frequent insertion, endwise of long plinths of stone, whose extremities project beyond the surface.
>
> Sir Emerson Tennent (Ref: R.L Brohier, Ancient Irrigation Works of Ceylon)

If the spill is too low, it will decrease the total volume of the reservoir. On the other hand, if the spill is too high, water may go over the bund. Hence, we need a spill that would not let the water go over the bund and would not reduce the reservoir capacity. This can be achieved by providing a longer and higher spill.

Kalawewa ancient spill (pitawana) is 72 m in width and 57 m in length. This was built using rock. For comparison, the width of the spill is more than the three-quarters length of a soccer field. The spill was built using hammered granite—a solid structure one could imagine. Each block of granite is shaped precisely to fit its neighbor. The whole structure eventually acts like one huge rock. The embankment of Kalawewa is 20–27 m high, with a base of more than 67 m and a length of 5 km. When George Turnour (The first to translate Mahavamsa to English) visited the site, he could not understand why such a huge spill was built. Turnour wrote, "One of the most stupendous monuments of misapplied labor in the country." Emmerson Tennent expressed the same sentiment. Later it was found that such a large spill and an embankment are necessary to withstand floodwaters during the heavy monsoon season.

21.3 BISOKOTUWA

Bisokotuwa (Figure 21.6) is the crowning glory of ancient Sri Lankan reservoir building. Water would enter the tunnel at one corner, come through the tunnel, and then rise in the Bisokotuwa.

Bisokotuwa served two functions.

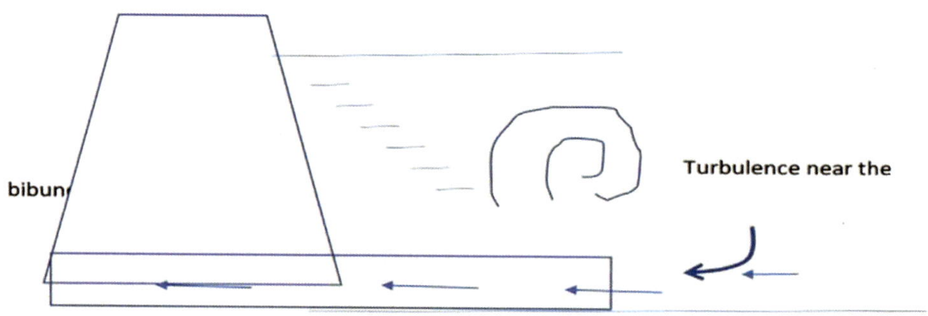

Figure 21.6 A sketch of turbulence near a bund.

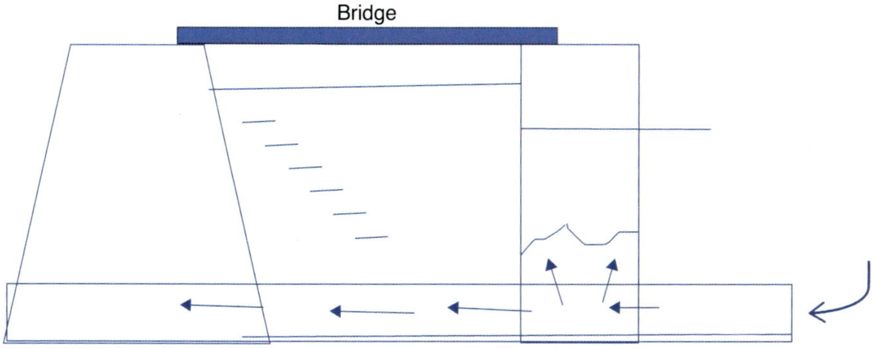

Figure 21.7 Function of the Bisokotuwa.

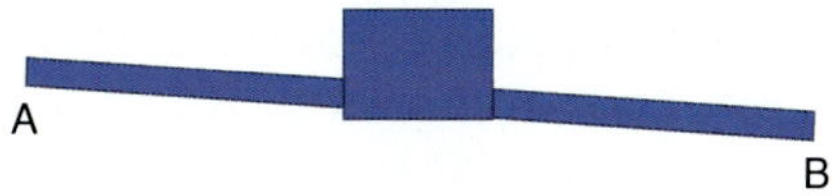

Figure 21.8 Sketch of water flow through Bisokotuwa.

A. Water pressure is reduced in the sluice.
B. Water going out of Bisokotuwa was controlled using gates placed inside the Bisokotuwa.

There would be turbulence (Figure 21.6) near the bund when water flows out of the sluice due to high pressure. Turbulence could damage the bund and ultimately cause failure.

Bisokotuwa solved the turbulence problem. Inside the Bisokotuwa, water pressure is reduced; hence, water can flow at a reduced and controlled rate (Figures 21.7 and 21.8).

In some reservoirs, Bisokotuwa is placed closer to the bund, and it is placed far away from the bund in some other reservoirs.

21.3.1 Hydraulics of the Bisokotuwa

When water flows from point A to B
The energy at point A = EA
The energy at point B = EB
Pb = Pressure at point B
VB = Velocity at point B
Energy due to gravity (height) = h
Energy due to velocity = $V2/2.g$
Energy due to pressure = p/γ
Energy loss = EL

$$E_A = E_L + E_B$$
$$hA + VA^2/2g + pA/\gamma = hB + VB^2/2g + p_B/\gamma + E_L \text{ in tunnel} + E_L \text{ in Bisokotuwa}$$
$$E_L \text{ in Tunnel} = C/D^{4.87}.\ L.\ (V.A)^{1.852}$$

A = Area of tunnel

$$EL\ (Bisokotuwa) = f\,(V_i, V_o, A_{IT}, A_{BK}, A_{OT})$$

A_{IT} = Area of the inlet tunnel
A_{BK} = Bisokotuwa
A_{OT} = Area of outlet tunnel

$$hA + VA^2/2g + pA/\gamma = hB + VB^2/2g + pB/\gamma + C/D^{4.87.}\ L.\ (V.A)^{1.852} + E_L \text{ in tunnel} + f\,(V_i, V_o, A_{IT}, A_{BK}, A_{OT})$$

$$hA + VA^2/2g + pA/\gamma = hB + VB^2/2g + pB/\gamma + C/D^{4.87.}\ L.\ (V.A)^{1.852} + E_L \text{ in tunnel} + f\,(V_i, V_o, A_{IT}, A_{BK}, A_{OT})$$

Canals have to cross rivers on their pathway. Long and meandering canals were built to transport water. Spills were built to protect the canal bund. Sluices were built to obtain water. Below is the route of the Elahara canal.

21.3.2 Path of the Elahara Canal

Milepost 0.0: Amban river was dammed, and water was diverted to the Elahara canal (Figure 21.9).

Milepost 4.0: Canal crosses the Kongeta Oya (river).

Milepost 5.0: Canal crosses the Kirandagalle Ela (river).

Milepost 6.2: Canal is provided with an overflow weir (Galwana—16m wide with wing walls). Galwana would let flood water escape without damaging the canal bund.

Milepost 6.5: Canal crosses the Heerati Oya (24m wide breach in the canal bund).

Milepost 7.5: Second overflow weir (or spill).

Milepost 9.0: Sluice to feed tributary canals.

Milepost 10.0: Canal crosses the Kottapitiya Oya (17m wide breach in the canal bund).

Milepost 12.0: The second sluice to obtain water.

60 Milepost 14.5: Overflow weir (Galwana).

Milepost 16.0: Canal crosses the Athanakadawela Oya (23m wide breach in the canal bund).

Milepost 16.5: Overflow weir (Galwana).

Milepost 16.8: Meegolla Ela (river) enters the canal.

Milepost 17.0: Meegolla Ela (river) departs the canal.

Milepost 18.0: Canal crosses the Radawige Oya.

Milepost 19.0: Canal feeds the Konduruwewa reservoir.

Milepost 20.0: Overflow weir (Galwana).

Milepost 20.7: Main canal branches off to three subcanals.

Milepost 21.0: Canal feeds the Minneriya reservoir.

Milepost 22.5: Canal feeds the Rotawewa reservoir.

Milepost 25.0: Canal feeds the Matalewewa reservoir.

Milepost 26.0: Crosses the Gal Oya–Polonnaruwa road (Canal bund is 3m high.).

At the 54th milepost, the canal enters the Kantalai reservoir.

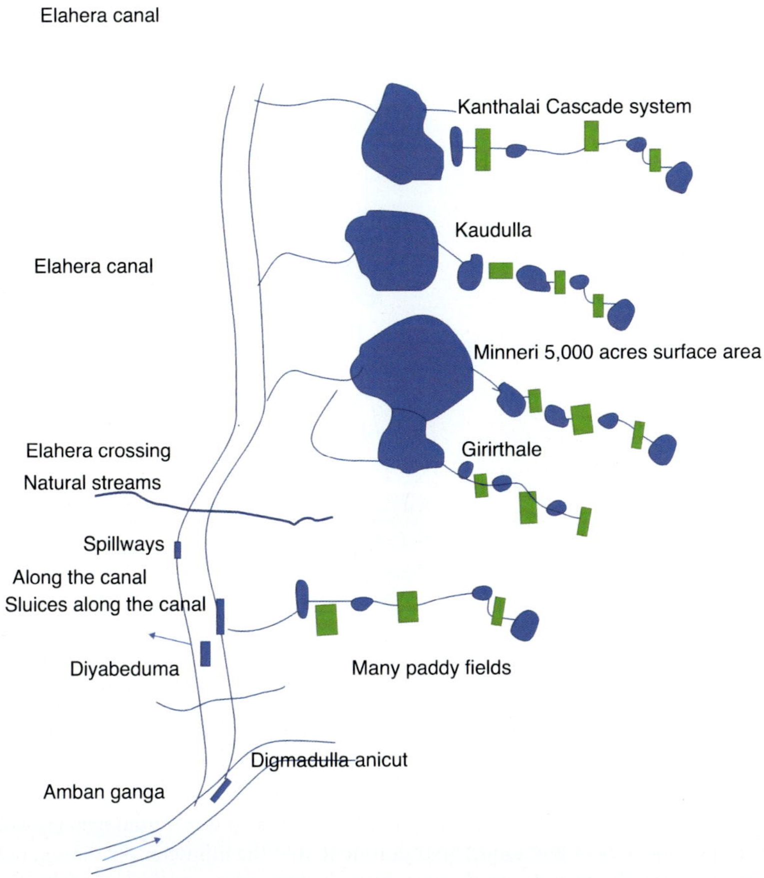

Figure 21.9 Irrigation systems around the Elahara canal.

After that, reservoirs were interconnected using canals to create a chain of reservoirs. This way, the reservoirs were captured, spilling water downstream from upstream reservoirs, basically a cascade system. Next, they built anicuts at ephemeral streams to divert water to reservoirs. As ephemeral streams dry out during the dry period, it is possible to construct anicuts to divert water. Figure 21.10 gives a schematic diagram of the cascade system.

21.4 CONCLUSIONS

Traditional water technologies have been prevalent around the world. This edition details some specific water technologies in Sri Lanka.

Traditional knowledge and wisdom have been at the core of the evolution and development of water technology. Given issues related to climate change and adaptation, it is more evident now than before that we give thought to traditional wisdom and water technologies as humans forge our future and modern water technology.

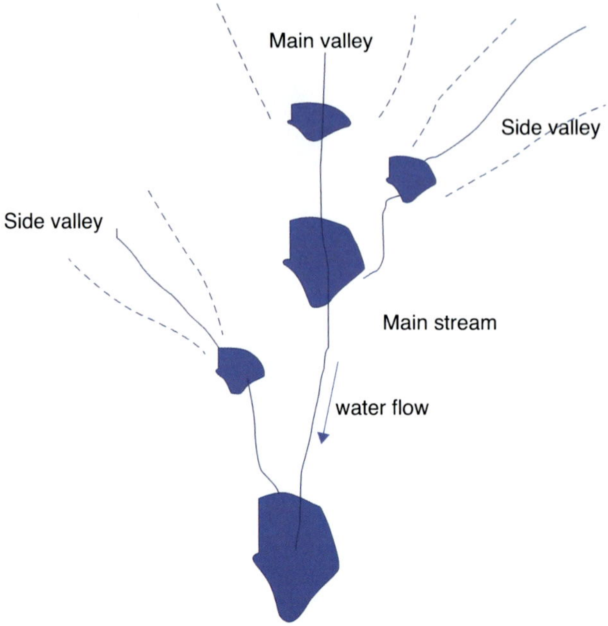

Figure 21.10 A schematic diagram of the Cascade system.

BIBLIOGRAPHY

Abeywardena, N., Bebermeier, W., and Chutt, B. (2018). Ancient water management and governance in the dry zone of Sri Lanka until abandonment, and the influence of colonial politics during reclamation. *MDPI Open Access Journal.* https://www.mdpi.com/2073-4441/10/12/1746/htm

Ancient Egypt water engineering. https://ancientengrtech.wisc.edu/ancient-egypt-water-engineering/Quanat. Wikipedia. https://en.wikipedia.org/wiki/Qanat. Accessed November 8, 2021.

Brohier, R. L. (1935). *Ancient irrigation works of Ceylon.* Smithsonian Libraries and Archives.

Dilmah Conservation. Sri Lankan flora and agricultural heritage, irrigation system in Sri Lanka | Traditional tank irrigation of Sri Lanka. www.dilmahconservation.org. Accessed September 8, 2021.

Geekiyanage, N. and Pushpakumara, D.K.N.G. (2013). Ecology of ancient tank cascade systems in island Sri Lanka. *Journal of Marine and Island Culture*, 2(2), 93–101. https://www.sciencedirect.com/science/article/pii/S2212682113000322

Peiris, K. and Wijesinghe, S. (2008). Introduction to the function of Bisokotuwa in ancient vewa, Engineer. *The Institute of Engineers Sri Lanka*, XXXXI (3), 24–28.

Smart Water Magazine. A journey through time. https://smartwatermagazine.com/news/smart-water-magazine/a-journey-through-time-how-ancient-water-systems-inspired-todays-water. Accessed November 21, 2021.

Vithanage, S., Perera, S., and Kallesoe, M. (2005). The value of traditional water schemes, *IUCN Water, Nature, and Economics Tech.* Paper No. 6, https://core.ac.uk/download/pdf/48023167.pdf

Chapter 22

Physical modeling of flow in the ancient inlet sluice barrel of Nuwara wewa reservoir, Sri Lanka

G. N. Paranavithana, R. S. Ranasinghe, J. M. Jayasundara, H. W. Harindra, and W. D. Ranasinghe
The Open University of Sri Lanka

CONTENTS

22.1 INTRODUCTION

Sri Lanka comprises a south-central mountain range of 2,100–2,400 m, influencing the island's climate. The southwest wind prevails from May to September, and, as it meets the high ground, monsoon precipitation occurs, reaching 508 cm on the mountain slopes. From October to January, the wind blows from the northeast over the Bay of Bengal and results in the monsoon in the plains of the north with an average rainfall of 127–513 cm. Thus, there is a wet zone to the south and central part of the island, and a dry zone in the north and south-east where the people were initially settled. They had to overcome the twin challenges of drought and monsoon to survive and thrive (Turpin, 2006). The ancient Sri Lankans had to manage and utilize available water with great care to overcome the above challenges.

The growth of the Sri Lankan hydraulic civilization developed new techniques in water management systems, such as reservoirs and irrigation canals. Reservoirs were built to keep the water table up by storing water by clay embankments across valleys. This water has been conveyed as irrigation water, creating a life cycle that has existed

DOI: 10.1201/9781003222736-26

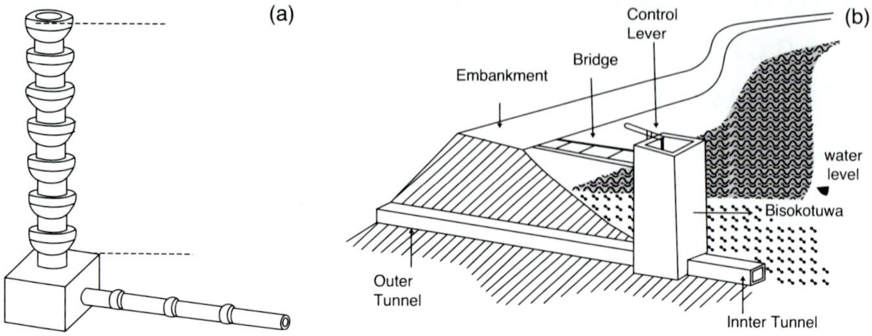

Figure 22.1 Keta sorowwa (a) components of an ancient sluice system (b).

for many centuries (Arumugam, 1969). This is considered one of the most advanced irrigation systems that have survived throughout history (Parker, 1909).

The sluice and sluice gates are important components in a reservoir, and these components are the control devices of a reservoir. Control devices save the reservoir bund without damaging it by the velocity and pressure created by water outflow. The outgoing water from small village tanks was controlled using temporary cuts through earthen bunds. In medium-size tanks, this was performed by using a technique called "Keta Sorowwa" (Figure 22.1a). A number of "Ketas" controlled the flow rates of the outgoing water from the reservoir, keeping one on top of the other (Ausadhahamy, 1999). This method was used to control the headwater entering the sluice. Keta Sorowwa cannot be used to control the outgoing water in much bigger reservoirs because of the excessive head. To counter this problem, ancient Sri Lankans invented Bisokotuwa, which was attached to the reservoir bund. It could help control water effectively without existing high velocity and pressure on the reservoir bund (Ausadhahamy, 1999).

Parker described that there are three components of ancient sluices of large reservoirs, as shown in Figure 22.1b.

1. A rectangular Bisokotuwa chamber was built near the point where the water level met the inner slope of the embankment.
2. An inlet tunnel (conduit) through which the water passed into this Bisokotuwa chamber.
3. A discharging tunnel (conduit) from the Bisokotuwa chamber to the foot of the outer slope of the bank.

As stated by Parker in "Ancient Ceylon (1909)", there is a special arrangement before Bisokotuwa, which is called the inlet barrel extended toward the middle of the irrigation tank, and some of these barrels are extended more than 30.5 m and spectacularly "Nuwara wewa" ancient sluice barrel is extended up to 44.8 m. Therefore, there must be an unrevealed hydraulic mystery about this kind of sluice barrel in some reservoirs in the ancient Anuradhapura and Polonnaruwa era, since the ancient Sri Lankan engineers may not have misspent their time and labor for such kind of construction. Therefore, it is easily realized that every reservoir component was built with a special need, and some of those systems are still functioning. What were the concepts, theories, and knowledge that they possessed? How did they determine the dimensions of "Bisokotuwa" according to the reservoir's capacity? Why did they build such a long inlet sluice barrel? What are the theories behind such designs? We must find answers to these questions to "determine the purpose of construction such inlet sluice barrel." However, there is no valuable research on the hydraulic performance of the inlet sluice barrel, built at great cost and effort during the 1st century BC. Hence, the prime intention of this research is to uncover the mystery of constructing an ancient inlet sluice barrel with the motive of regenerating purposeful engineering knowledge. The ancient reservoir "Nuwara wewa," which was built in 1st century BC, is selected for the prime analysis, since Henry Parker cataloged all the details in 1909 before the sorrowful fate of the structure in the face of modern engineering. Therefore, this study investigates the performance of the ancient inlet sluice barrel of the Nuwara wewa reservoir.

22.2 METHODOLOGY

22.2.1 Field measurements

Parker recorded all the dimensions of the ancient sluice at Nuwara Wewa in 1909 before the demolition of these structures by the British to construct new structures. An extract from the actual drawing from Parker (1909) is shown in Figure 22.2.

Ancient sluice is a granite structure made out of stone slabs. The most significant property of the hydraulic conveyance of this material is the surface roughness, which governs the resistance to the flow. Measuring surface roughness values of ancient rock surfaces, which were used to construct inlet sluice barrels, was essential to obtain the scaled-down value for the physical laboratory model. Several currently available archaeological concurrent sluice sites were investigated, and the surface roughness values were measured during the study since the Nuwara wewa sluice was shattered for various reasons. The map shown in Figure 22.3 indicates the locations of the available ancient sluices.

The surface roughness gauge (BOYN, SRT-6223) was used to measure the roughness of all these locations (see Figure 22.4). When measuring the roughness of a surface using SRT 6223, the sensor is placed on the surface and then uniformly slides along the surface by driving the mechanism inside the tester. The sensor gets the surface

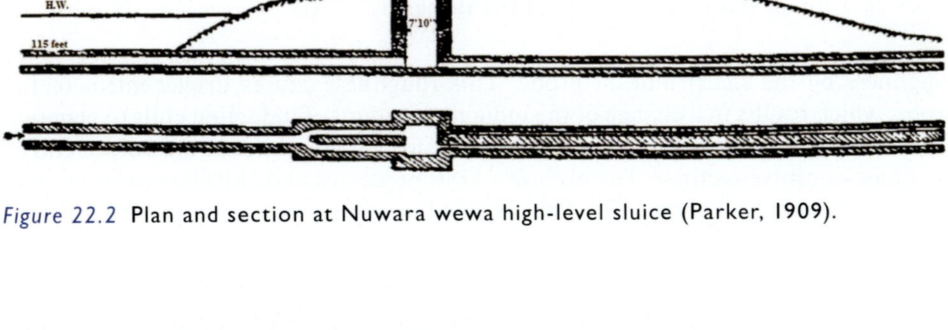

Figure 22.2 Plan and section at Nuwara wewa high-level sluice (Parker, 1909).

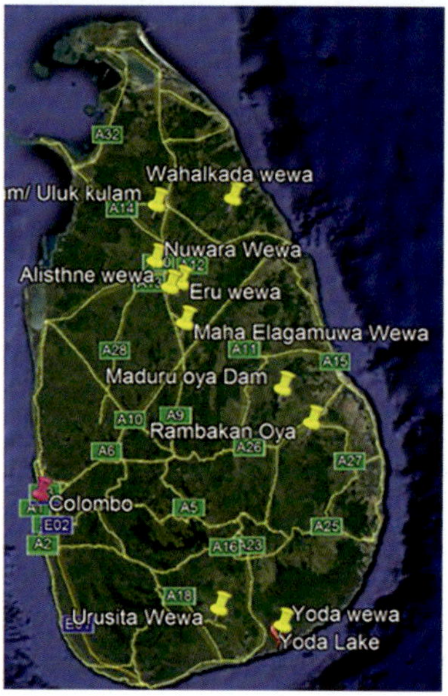

Name of the Reservoir	Coordinates
Urusitana Wewa	6°19'39.00"N,80°56'18.00"E
Yodha Wewa	6°15'36.34"N,81°18'35.54"E
Maha Elagamuwa Wewa	7°59'28"N,80°37'6"E
Alistane Wewa	8°11'35.69"N,80°31'25.52"E
Wahalkade Wewa	8°44'20.53"N,80°51'10.23"E
Eru wewa	8°13'21.00"N,80°34'57.00"E
Rambakan oya	7°27'22.12"N,81°23'55.88"E
Maduru Oya	7°38'42.50"N,81°13'1.80"E

Figure 22.3 Locations of the studied ancient sluices in Sri Lanka.

Figure 22.4 Roughness measurements of the sluices.

roughness by the sharp built-in probe. This roughness causes displacement of the probe, which results in a change of the inductive amount of induction coils to generate an analog signal, which is in proportion to the surface roughness at the output end of the phase-sensitive rectifier. The exclusive DSP processes and calculates and outputs the measurement results on LCD. The obtained results are presented under results.

22.2.2 Dimensional analysis

A three-dimensional representation of a person, a thing, an existing structure, or a proposed structure, typically on a smaller scale than the original, is called a model. A model is used to find out the behaviors of the original structure and carry out different studies related to it within the context of space and user needs. Models can be similar to their prototypes in three different ways, namely, geometric similarity, kinematic similarity, and dynamic similarity. For the similarity between the model and the prototype, we have to consider all three similarities. For this study, a model of the Nuwara wewa high-level inlet sluice barrel is built. As the flow along the barrel is similar to pipe flow, the flow behavior governing factor is the Reynolds number. Buckingham's pi theorem is used to substantiate the relationship between the Reynolds number of actual higher-level sluices and the model by dimensional analysis (Yunus, 2016).

22.2.3 Laboratory experiment setup

Based on the scaling factors obtained through the dimensional analysis, which is 1:20, the model's dimensions and surface roughness values are derived. The roughened Perspex sheets are used to develop the model, since there is a need to perform flow visualization to understand the hydraulic functionality.

The dimensions of the model are shown in Figure 22.5. The challenge was to establish the scaled-down surface roughness of 31.8 μm. To overcome this issue, a uniform application of roughing was performed on the surfaces of Perspex sheets (Figure 22.6). The developed laboratory model was then placed in a constant head water stilling basin to conduct the experiments. The experiments were performed by stabilizing the water flow along the channel. A schematic diagram of the experimental setup is sown in Figure 22.7. The water is fed to the system under gravity from an overhead tank into a glassed wall channel where a constant water head is maintained. A series of hydraulic baffles were utilized at the point of entry to the channel to dampen the pool level fluctuations.

The structure's purpose has been a mystery since the first inspection. Flow visualization was performed to study the behavior of the flow line, thus trying to conclude the functionality. Hydraulic analysis for the laboratory model was conducted in two main sections,

- Flow velocity variation of the sluice structure
- Study of the flow patterns at the inlet barrel

Flow velocity variation of the model sluice structure was conducted using two determinant methods: flow discharge analysis and particle tracking velocimetry (PTV). In flow discharge analysis, the discharged water volumes were measured for each water head for a known time duration, and computations were conducted to evaluate the velocity profile. Accordingly, the Reynolds number was obtained to determine whether the flow is turbulent ($Re > 4{,}000$), transitional ($2{,}100 < Re < 4{,}000$), or laminar ($Re < 2{,}100$) (Trinh, 2017). In PTV analysis, tracer particles were allowed to enter the sluice structure, and cameras were used to film the particle travel in the tunnel. Flow velocities

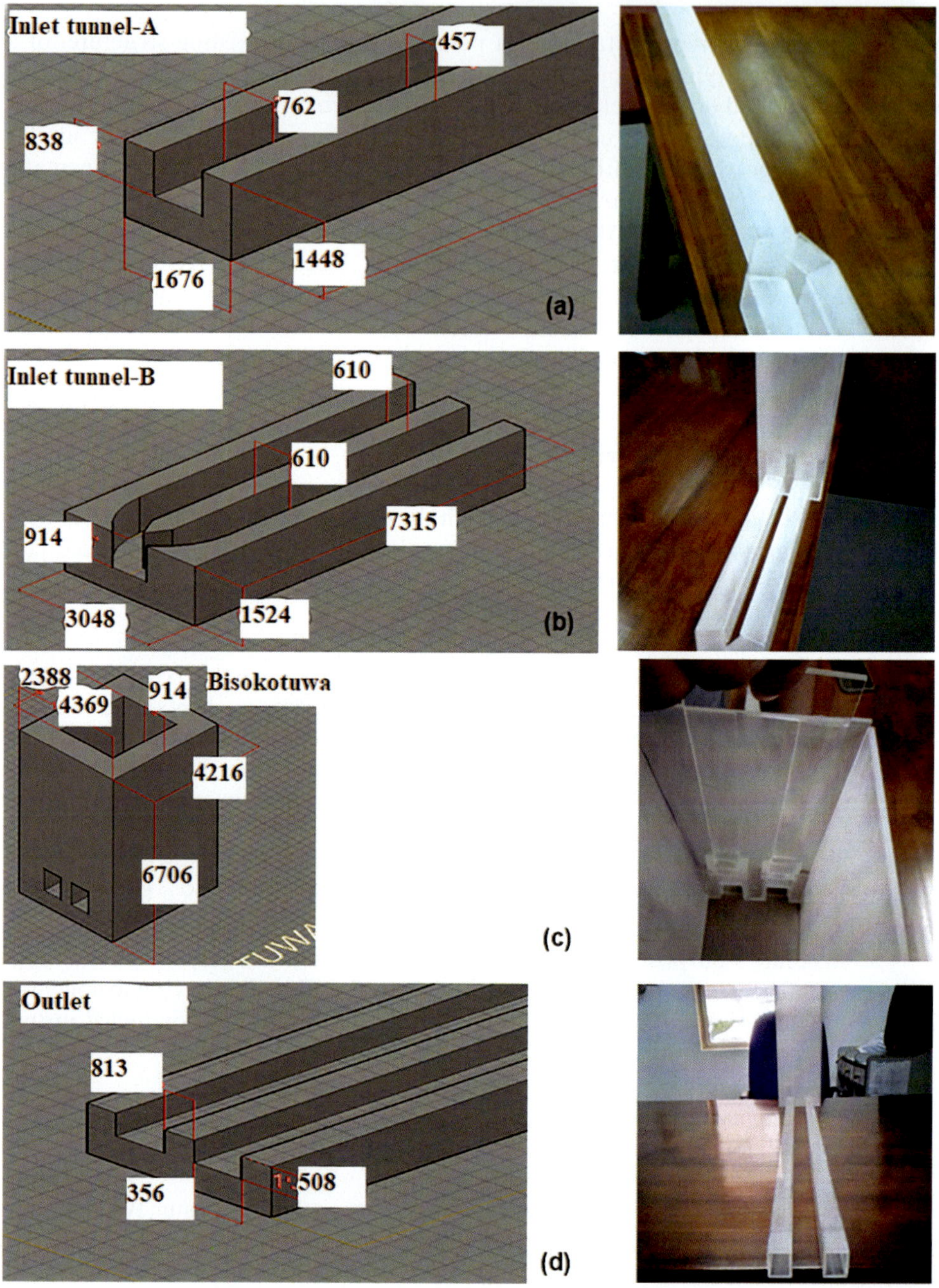

Figure 22.5 Dimensions of the scaled model at the single-barrel inlet (a), double-barrel inlet (b), Bisokotuwa (c), and outlet barrel (d).

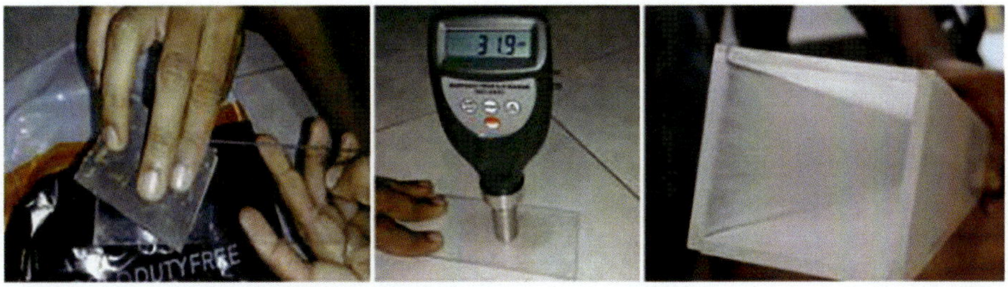

Figure 22.6 Roughening process of Perspex sheets.

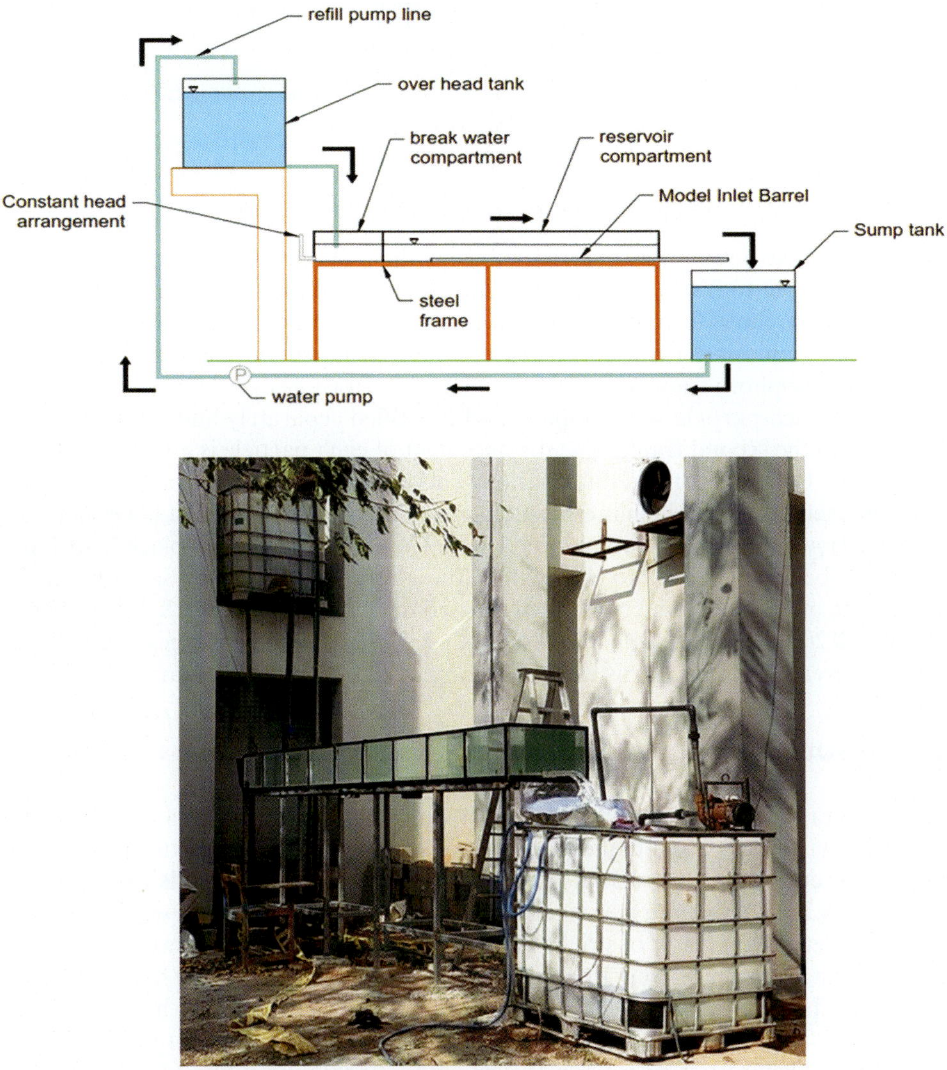

Figure 22.7 Experimental setup.

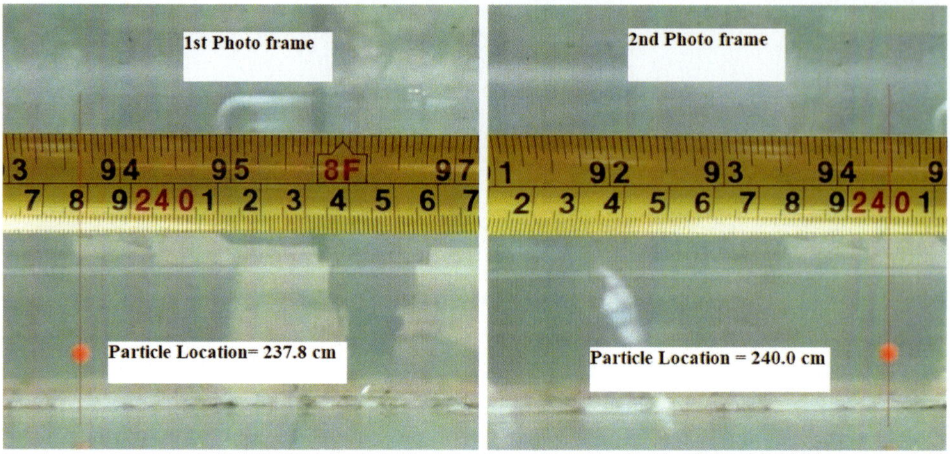

Figure 22.8 Particle tracking velocimetry technique for flow velocity measurements.

along the sluice structure were determined using the PTV method. Video images were converted into photo frames, and the traveling distance of the tracer particle between two consecutive photo frames was measured, as shown in Figure 22.8. Then the flow velocities were obtained by dividing the measured distance by time lap between the two consecutive photo frames.

In PTV, each particle is recognized and identified separately and tried to be rediscovered in the second frame. The displacement of each particle is effectively determined using the neighborhood information. This means PTV provides a displacement vector for each particle, which is efficient in terms of processing time at lower particle numbers. However, the vectors obtained by PTV are not regularly placed, so false vectors can be more difficult to eliminate. However, the resulting vector fields enable the tracking of particles, which can then be used to calculate (Socoray & Jozsa, 2008). "Mathlab R2018a", Simulation software was used to analyze the images, while "free studio" converting software was used to convert 1/30 seconds photo frames.

22.3 RESULTS AND DISCUSSION

The "Nuwara wewa" reservoir was provided with two sluices: one at a low level and the other having a sill 94 cm higher. At the low-level sluice, the Bisokotuwa measured 3.35 m in the culvert line and 4.58 m in a transverse direction; it had walls 107 cm thick, which rose 4.27 m above the sill. It was lined with stone slabs. There were two inlet and two outlet culverts built of stone. The former was only 5.34 m long and were separated by a masonry wall 198 cm thick; they were 61 cm wide and 127 cm high. An open paved inlet channel, 21.8 m long and 30cm wide, led up to them; this had side walls 107 cm thick. The outlet culverts were about 47.6 m long and were separated by a wall 2.14 m thick. They rested on a floor 45.7 cm thick. Each culvert was 61 cm wide and 83.9 cm high; their outer walls were 45.7 cm thick and covered with large stone slabs.

Table 22.1 Average surface roughness values of the ancient sluices

Name of reservoir	*Period of construction*	*Average roughness value (μm)*
Urusitana Wewa	3rd Century BC	612.2
Yodha Wewa	1st Century BC	603.6
Maha Elagamuwa Wewa	–	635.0
Alistane Wewa	–	632.0
Wahalkade Wewa	–	657.6
Eru wewa	–	655.6
Rambakan oya	3rd Century BC	636.6
Maduru Oya	1st Century BC	649.3
Average roughness value of the granite		**635.2**

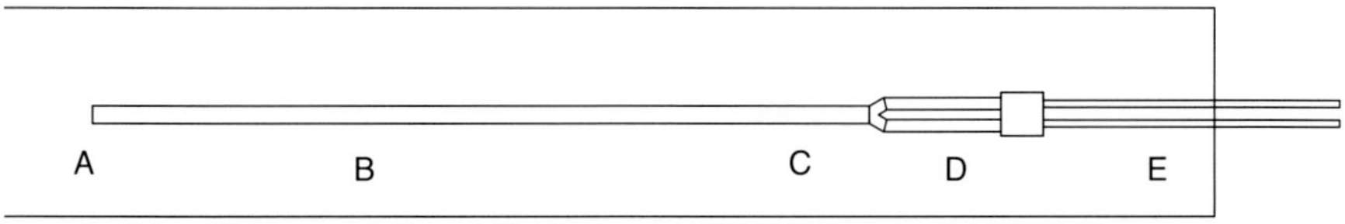

Figure 22.9 Particle tracking velocimetry (PTV) observation points for internal flow velocity.

The Bisokotuwa of the high-level sluice was built of brick and not lined with stone; it measured 2.54 m transversely and 2.39 m in the line of the culverts. It was 6.71 m high and had walls 91.5 cm thick (Parker, 1909). In order to estimate the roughness of the model, an accurate measurement of the roughness of stone slabs is required. To fulfill this need, roughness values of existing stone culverts were measured, and the measured values are shown in Table 22.1.

The inlet channel of the ancient sluice (see Figure 22.4) acts like a conduit to convey water. Based on this feature, the selected effective parameters are the friction force of pipes (F), the density of water (ρ), the velocity of flow (V), the height of inlet/outlet (d), and the dynamic viscosity of water (μ) (Yunus, 2016). The dimensional analysis reveals that $F/\rho\, d^2V^2 = \varnothing\, (\rho Vd/\mu)$, and since the subjected fluid for the study is water, the parameters ρ and μ are the same for both model and prototype. The Reynolds number is similar for the model and the prototype to maintain the dynamic and kinematic similarities. The linear dimension, d, is determined by geometric similarity after considering a reasonable scale factor for the model. For this purpose, the dimensions of the original inlet tunnel were considered, and the scale for the model was decided according to the convenience of constructing and the ability to study flow behaviors in the model. Thus, it leads to a scale factor of 1:20 for the model to study the flow behaviors of the proposed model.

Two aspects were studied to determine the purpose of constructing the inlet sluice barrel. The first aspect considered is the internal flow velocity variations of the sluice structure (measured at the points shown in Figure 22.9). Per the flow variation patterns shown in Figure 22.10, velocity up to the Bisokotuwa reduces through its run and increases after the Bisokotuwa. Since the ancient gates were constructed in timber, it is extremely important to possess amicable velocities at the gates. It indicates that the construction of an elongated sluice barrel was introduced to attain this purpose.

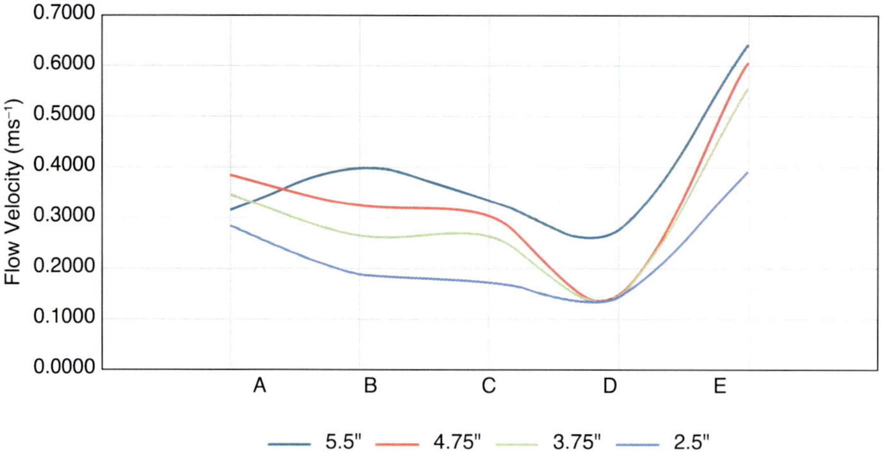

Figure 22.10 Averaged cross-sectional flow velocity variations along the sluice barrels.

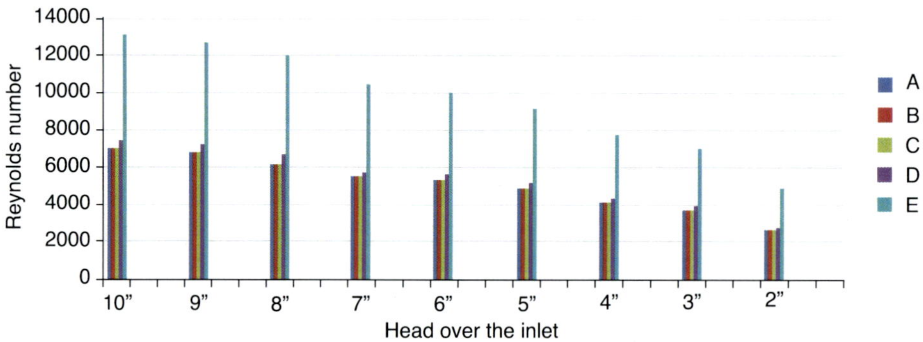

Figure 22.11 Variation of Reynolds number at different locations concerning the height of the water column.

According to Figure 22.11, for low water heads such as 1.52 m and 1.02 cm in the reservoir, which were obtained from equivalent model heights of 7.6 and 5.1 cm, Flow Reynolds number behavior for both inlet and outlet tunnel is in transition flow condition, while the double sluice barrel is at laminar flow. Therefore, it can be concluded that the double sluice barrel is directly imposed in the velocity reduction prior to Bisokotuwa. This double-barrel arrangement is the main protecting component of gates and the Bisokotuwa.

The second aspect studied to determine the purpose of the inlet sluice barrel is the study of flow patterns at the entry to the inlet sluice barrel. It concludes a very important discovery of the functionality of the inlet barrel. Figure 22.12 indicates that there is a formation of a vortex, starting from the water surface leading into the inlet. The figure shows results from the PTV study, which indicates the formation of a free-surface vortex and the mixing of all the layers of water in the reservoir.

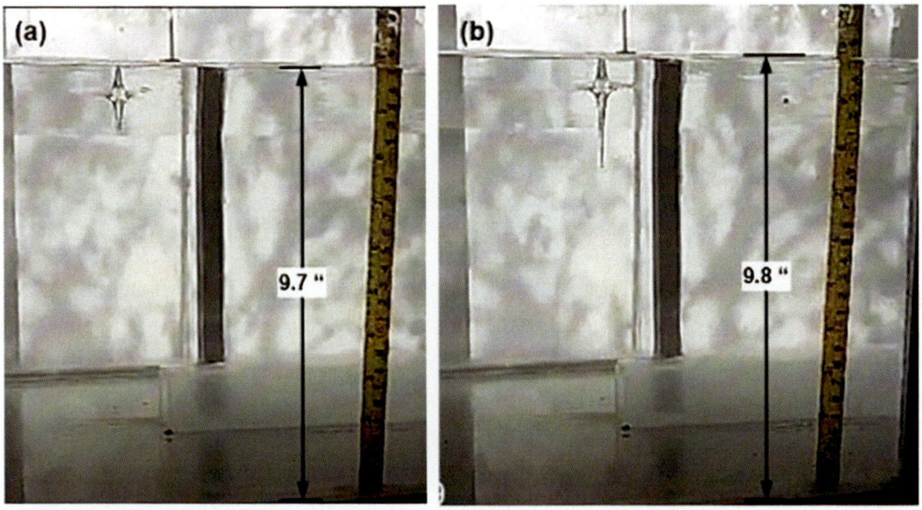

Figure 22.12 Formation of the vortex at the inlet barrel of the sluice when the water head is 24.9 cm (a) and 24.6 cm (b).

The formation depth of water in the model was 24.9 cm, which represents a water height of 4.97 m. According to Parker's records, this is almost equal to the full supply level of the ancient Nuwara Wewa reservoir. This is an unexpected discovery in this research.

According to Figures 22.13 and 22.14 vector fields from PTV, layer mixing in the water column clearly indicates that the ancient inlet sluice barrel has facilitated the breakdown of lake stratification. Further, it revealed that the vortex mixing is stopped with the reduction of the height of the water column up to 127 cm (6.4 cm in the model), which leads to a laminar flow. Due to this laminar flow, sediment escape might have been reduced. To conclude the behavior of the sediment concerning the inlet barrel, further studies with PTV is required, which we will conduct in the next phase of our research.

22.4 CONCLUSIONS

This research investigated the functionality of an ancient granite sluice barrel widely used by Sinhalese for large reservoirs. It is found that the utilized sluice barrel constructed into the reservoirs was highly beneficial in mixing all the strata of the reservoir by creating a vortex at the inlet to the barrel. Further, it can be concluded that this vortex formation is present in most of the reservoir's water levels but subsequently disappear when water levels reach below 1.22 m.

ACKNOWLEDGMENTS

We take this opportunity to thank Archeologists Mr. A.A. Wijerathne and Mr. G. Samaranayake for their guidance in finding existing sluice locations in Sri Lanka.

Further, we sincerely thank Mr. Udula Bandara Ausadhahamy for his conceptual contributions, which encouraged us to conduct this research.

NOTATION

g = Gravity acceleration (m/s²)
Re = Reynolds number (–)
t = Time (s)
Averaged surface roughness (μm)
F = Forces in pipes (N)
ρ = Density of water (kg/m³)
V = Velocity of flow (m/s)
d = Height of inlet/outlet (m)
μ = Dynamic viscosity of water (kg/m/s)

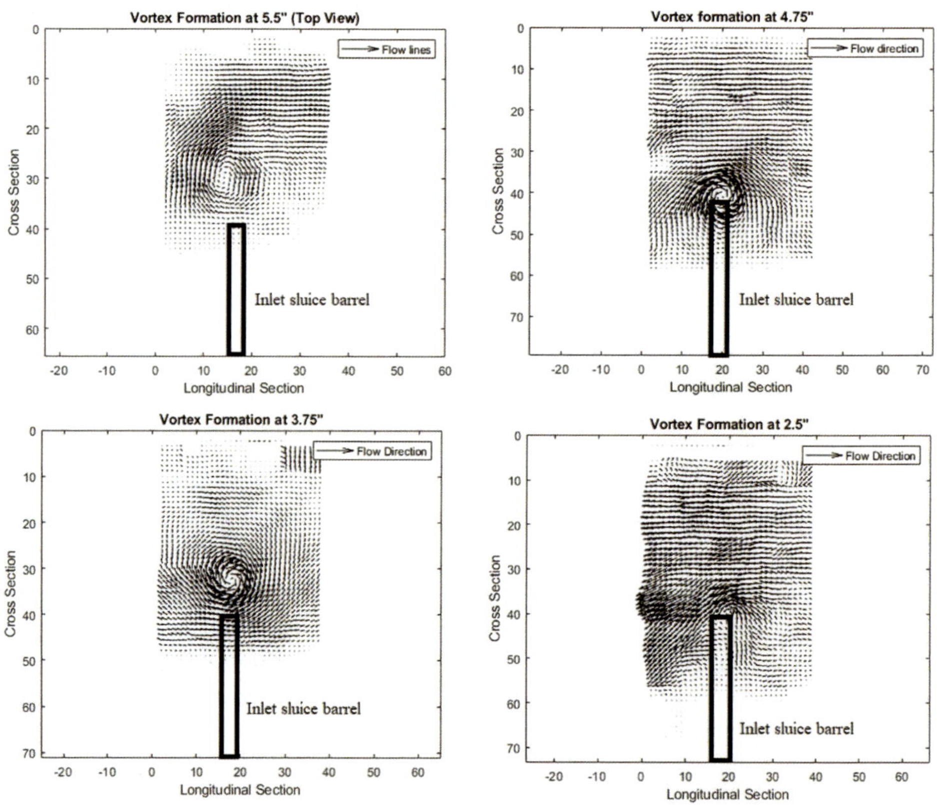

Figure 22.13 Plan view of the velocity vector fields obtained from particle tracking velocimetry (PTV) for different heights of water columns.

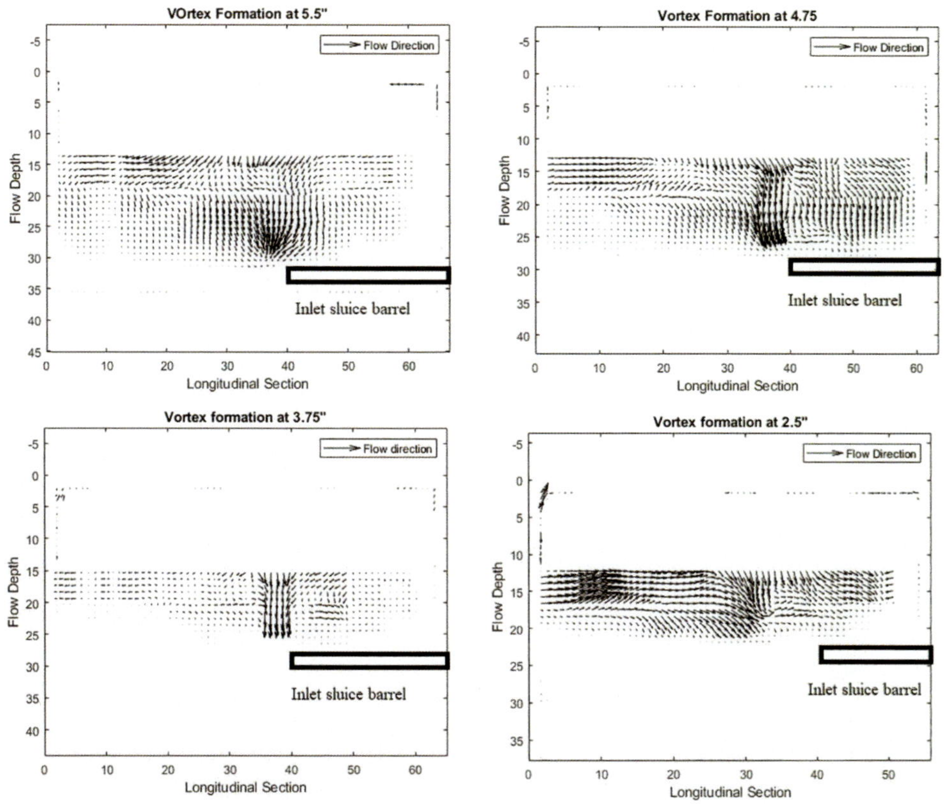

Figure 22.14 Sectional view of the velocity vector fields obtained from particle tracking velocimetry (PTV) for different heights of water columns.

REFERENCES

Arumugam, S. (1969). *Water resources of Ceylon*. Colombo, Sri Lanka: Water Resources Board.

Ausadhahamy, U. B. (1999). *Wewa*. Sri Lanka: Sri Printers.

Parker, H. (1909). *Ancient Ceylon*. London: London Luzac & Co. Publishers to the India office.

Socoray, B. J., Jozsa, J. (2008). Particle tracking velocimetry (PTV) and its application to analyse free surface flows in laboratory scale models. *Periondica Polytechnica*, *52*(2), pp. 63–71. doi:doi: 10.3311/pp.ci.2008-2.02

Trinh, K. T. (2017). *On the critical Reynolds number for transition from laminar to turbulent flow*. Retrieved June 30, 2021, from ResearchGate: https://www.researchgate.net/publication/45927628_On_The_Critical_Reynolds_Number_For_ transition_from_laminar_to_turbulent_flow

Turpin, T. (2006). Traditional water management in Sri Lanka & its relevance to the UK for climate change, *Winston Churchill Travelling Fellowship*.

Yunus, A. (2016). *Fluid mechanics in SI Units* (3rd ed.). Chennai, Tamil Nadu, India: McGraw Hill Education (India) Private Limited.

Part V

Historical water projects and traditional water technologies in Southeast Asia

The colmatage system in Cambodia, developed more than 1,000 years ago, conveys inundated water to the floodplains by cutting natural levees and providing soil dressing.

Southeast Asia, composed of a large landmass and many islands, has been influenced by the lower part of the Asian continent to the north and the Indian and Pacific Oceans to the south and the east, respectively. The former comprises Thailand, Cambodia, and Vietnam, while the latter includes Malaysia and Indonesia. This part includes various articles introducing historical waterworks activities and technologies embedded in those activities in the region, but some are only from the early 20th century.

In Cambodia, the colmatage—a series of comb-shaped narrow canals to the main river course excavated from the natural levee in the 19th century was very useful for irrigation, local navigation, and flood passage. In Thailand, the Rangsit Canal—the first main canal for the first irrigation project at the dawn of the 20th century covered about 2,000 km^2. It has been proven for multipurpose uses in the past and present, and it will undoubtedly be used as a resilience structure for flood disaster reduction in the future. For Malaysia, this part introduces the Wan Man Saman Canal, completed in the late 19th century, which transformed useless swamp lands of the coastal plain into another vast area for rice cultivation in Kedah, earning the nickname "Malaysia's rice bowl." Last but not least, this part introduces Subak—a traditional irrigation system well preserved on Bali island, Indonesia, a social–religious communal organization of farmers and its irrigation system.

DOI: 10.1201/9781003222736-27

Chapter 23

Water use of flood in Cambodia

So Kazama
Tohoku University

CONTENTS

23.1 INTRODUCTION

In the capital Phnom Penh in Cambodia, the Tonle Sap River joins the Mekong River at a junction referred to as the Four Arms and then splits into the Bassac River. Flooding is widespread during the rainy season in this country of tropical monsoons, but since ancient times people have lived their lives well-adapted to this regular flooding. The annual rainfall in Phnom Penh is 1,400 mm, and the average annual temperature is 28°C. According to the Köppen-Geiger climate classification, it is considered to be Aw. In recent years, Phnom Penh has developed and urban expansion is proceeding rapidly.

Hydrologically, Cambodia has two unique characteristics. The first is Lake Tonle Sap, which is the Mekong's lungs. Lake Tonle Sap is upstream of the tributary Tonle Sap River. Water flows back from the Mekong River during the flood season, where it stores this water and then releases it during the dry season. Freshwater organisms have a life cycle that responds to this flooding, and many aquatic products can be harvested from the floodplains. The other is a water conveyance system known as the colmatage canals. Inundation water is provided to the flood plains from the river by cutting a natural levee and providing soil dressing (Figure 23.1). This colmatage system was developed more than 1,000 years ago. The Cambodian flood plain has expanded over time with the colmatage canals and seasonal flooding, which provides many benefits for people. To quote Herodotus, Cambodia is a gift of the Mekong.

DOI: 10.1201/9781003222736-28

Figure 23.1 Colmatage. Left: Colmatage canals from a main river. Right: A fishing net on the inlet of a colmatage canal.

In this chapter, the hydrological and hydraulic characteristics of the Cambodian floodplain are explained in the context of past research. In Section 23.2, the history and mechanism of the colmatage system are explained. In Section 23.3, details of the study area are provided, along with the models used to understand the floods and the functions of the colmatage system. In Section 23.4, the benefits of flooding are explained, and in Sections 23.5 and 23.6, the effects of fertilization and the effects of flooding on the groundwater are explained, respectively. In Section 23.7, waterborne infectious diseases are discussed, and the future of the colmatage system is explained in Section 23.8.

23.2 HISTORICAL BACKGROUND OF COLMATAGE SYSTEM

The details of the colmatage formation are unknown well. Agriculture using flooded water has been practiced for over 1,000 years, and cultivated land expanded during the Khmer dynasty. The colmatage excavated the natural levee and used flood water actively, and has occurred before and after the French rule in 1863, according to the common view. This is because it became necessary to expand the agricultural land for the increase in population due to the release from the external pressure on Siam and Vietnam (Delvert, 1998). The purpose of colmatage is to obtain water and sediment. The deposition rate of suspended solids in a flood plain by colmatage was 20 mm/year, and this soil dressing expanded the cultivated land to the back marsh and wetlands (Hori, 2000). The waterway created by excavating a natural levee is called prek; it is also used to mean a village, because a village is formed along the waterway. Modern heavy machinery accelerates the number of colmatage, and 57 colmatages were produced from 1952 to 1958. The agricultural land created in 1967 is 180 km^2, and the beneficiary area of colmatage in 1994 is 230 km^2. Only Kandal Province has 153 km^2 with 247 preks for colmatage farmland (Kakudo et al., 1995). Most waterways (preks) have a depth of 2–6 m, a width of about 10 m, and a length of 1–3 km. In some areas of the Basak River, there are preks at close intervals of 1–2 km (Figure 23.1). Since 1997, water gates have been installed at some preks, and the flow rate is also controlled. In Prey Veng City and Takeo City, the water level difference between the rainy and dry seasons is close to 10 m, and some reservoirs are developed using the water level difference, which is

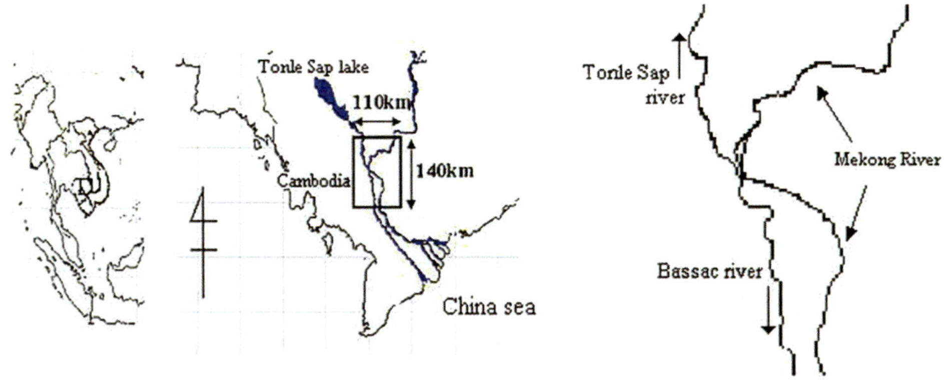

Figure 23.2 Study area for the model simulations.

used for agricultural water in the dry season. The reservoir is called Tum Nub, which means embankment. Even in the Mekong Delta area in Vietnam, many reservoirs are developed on the flat ground called a unit dike—a combination of a ring levee and a watergate. The colmatage system is used for acquiring agricultural water and fishery and boat transportation and is an important production place.

23.3 STUDY AREA AND MODELS FOR THE EVALUATION

The rainy season in Southeast Asia brings frequent rainfall from July to December, and the Mekong River discharges an average of 25,000 m^3/s during this season. From January to June, the dry season has little rainfall, with only 6,000 m^3/s on average during this season. The hydrological and hydraulic phenomena in this chapter are explained in Figure 23.2. It is a unique area characterized by a junction of rivers and flooding by the colmatage system. Hydrological and hydraulic models are based on the 2D non-uniform flow, Darcy's law, and potential evaporation. As the boundary condition, the water level was obtained at 4 points, located upstream in the Mekong River and the Tonle Sap River, and downstream in the Mekong and Bassac Rivers. These models express the material transport of sedimentation, coliform, and phosphorus, for example. For more detailed information on each application, please refer to the articles at the end of this chapter.

23.4 FLOOD BENEFITS IN THE MEKONG RIVER IN CAMBODIA

One of the design functions of the colmatage is to spread water effectively to the flood plains. The flooding events are well-utilized by locals for the maximum benefit, explained by a study conducted after the flooding of 2000. A fixed netting method referred to as "Dai," and a casting net method are used by locals to catch catfish and carp species at the intake of colmatage canals from the river during the seasonal migrations of the fish. These fish spawn on the flood plains during flood periods, and the hatched fish then migrate to rivers during the flood retention season. Kazama et al. (2003)

Table 23.1 Benefits per area by the 2000 flood magnitude at 2020 prices

	Benefits (USD/km^2)	*Damage (USD/km^2)*
Paddy field	41,400	4,000
Fishery (inundated area)	22,500	
Fishery (paddy field)	15,000	
Industry	220	

surveyed market prices in aquatic products and rice production, and found that the locals had benefited significantly from the flooding. Since the prices of the aquatic product vary spatially and temporally, an average consumer price was used in the 2000 study. The price, at 1 USD/kg, was taken from the official report from the Ministry of Agriculture, Forestry, and Fisheries. In earlier literature, investigations on the fixed netting method in the Tonle Sap River reported a correlation between fish catches and flood peak levels. The prices and the correlation were used to calculate the benefit of flooding to the locals regarding aquatic products.

Another benefit of flooding is rice production. Rice production is correlated to the extent of the flooding, because a high-water level in the river equates to an expanded land area for cultivation. Local people obtain the irrigation water through the colmatage naturally. The market price of rice and the correlation between rice production per area and flood level are indicators of how the people benefit from agriculture. The groundwater used by beverage companies is recharged during the flood season in the industrial sector. The benefit for the industry can be estimated from the records and market prices.

These data were collected around 2000. Prices rose for fish, rice, and the industry from 2000 to 2020 by factors of 1.5, 1.0, and 1.3, respectively (Table 23.1). The inflation rate was applied to the industry sector value. These results clearly show the significant financial benefit of flooding in Cambodia, and that this benefit outweighs the cost of flood damage. However, changes in this benefit are expected with national growth in Cambodia (Kazama et al., 2009). Other benefits such as those to ecosystems and navigation were not considered, and the disruption to transportation and education was also not considered. A more detailed analysis that takes all of these factors into consideration would provide a more thorough understanding of the financial impact of flooding in Cambodia.

23.5 FERTILIZATION BY FLOOD

Generally, the colmatages provide nutrient-rich agricultural water that serves as a natural irrigation system for paddy fields in flood plains. In the rainy season, the water from the main rivers includes a significant amount of suspended nutrient-rich solids. Some areas in the flood plain store the water in the rainy season for use in the dry season. The storage reservoirs have a water gate that changes the water level to around 5 m. The reservoirs are called "Tum Nub"—an embankment made of earth.

Figure 23.3 shows the total phosphorous (TP) concentrations in the water reservoir and in the paddy field located downstream from the reservoirs in November 2010 and 2011, which is the beginning of the dry season (Amano and Kazama, 2014).

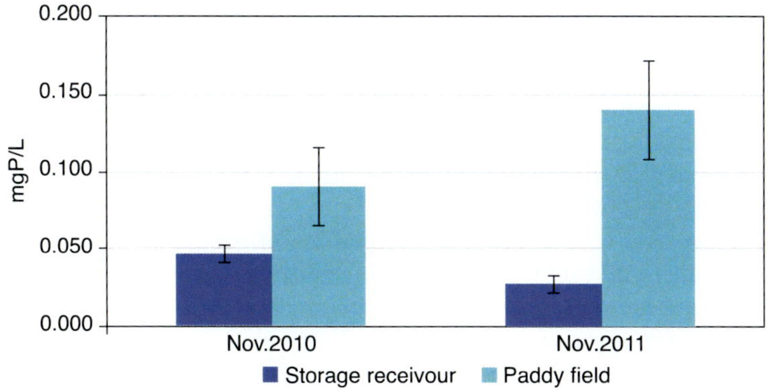

Figure 23.3 Total phosphorous (TP) concentration in the water reservoir and the paddy field in November 2011.

The observation area is near Takeo, which is connected to the Bassac River. The TP concentration of surface water in the storage reservoir is lower than that of the paddy field. The TN concentration expresses similar results. In November, the water level in the paddy field decreases, but the reservoir discharges inundation water to the paddy field as irrigated water through the sluice gate. At the bottom of the reservoir, the nutrient concentration of the water is higher than at the top due to the deposition of nutrients. This nutrient-rich water is provided to the reservoir through the sluice gate from the paddy field. The sluice gate is efficient as both a water and a nutrient supply system. While rubber gates are recently being used in some areas for the water storage reservoirs, in terms of nutrient deposition, the sluice gates function better than these new rubber gates.

Measurements of the nutrient level were taken in the soil, water, and harvested rice plants in Takeo and Kandal in Cambodia (Amano and Kazama, 2014). Table 23.2 lists the calculated nutrient balance. The input and output components are expressed with + and −, respectively. The contribution ratios are defined as the total nutrients of the inundation water divided by the harvested nutrients. The contribution ratio of inundation to rice production can be used to evaluate the fertilization effects of floods and inundation. Since the sum of the available nutrients in the soil and inundation

Table 23.2 Nutrient balance estimated at observation points

		N ton/km²		*P Ton/km²*	
		Kandal	*Takeo*	*Kandal*	*Takeo*
Soil	Available	9.9	12.0	1.6	2.0
	Total	210	140	29	15
Inundation water	Available	+ 0.64	+ 0.89	+ 0.10	+ 0.06
	Total	+ 1.5	+ 1.1	+ 0.14	+ 0.07
Harvested rice plants		− 3.2	− 3.2	− 0.60	− 0.60
Sum of available nutrients		11.0	13.0	1.7	2.1
Contribution ratio		48%	33%	24%	11%

water was larger than the nutrients absorbed by the harvested rice plants, it can be concluded that the amount of available nutrients is sufficient for rice production. The amount of nutrients in the soil is 25 times greater than the amount of harvested nutrients. This indicates that the soil contains many nutrients compared with the harvested nutrients amounts at a soil depth of only 15 cm from the ground surface. The contribution ratios are 48%, 24%, 33%, and 11% for nitrogen in Kandal, phosphorus in Kandal, nitrogen in Takeo, and phosphorus in Takeo, respectively. Because the contribution ratio can be considered an index to evaluate fertilization effects quantitatively, it is possible to evaluate the fertilization effects of floods and inundation. Continuing the current system of no or little fertilizer use will threaten agricultural sustainability, because the contribution ratios are not 100%, which means that the nutrient balances are negative.

23.6 GROUNDWATER RESOURCES

Remote areas from channels rely on groundwater as municipal and industrial water in the dry season. Huge inundations recharge the water widely into the subsurface zone during the rainy season. Colmatages contribute to both the inundation and the recharge.

Three different magnitudes of flood in 1993, 1997, and 1998 were selected to develop an understanding of groundwater storage using model simulations (Kazama et al., 2007). The model simulations expressed the change in groundwater storage as influenced by the large, middle, and small inundations of 1997, 1993, and 1998, respectively (Table 23.3).

In the case of 1997, the mean groundwater storage in the study area was calculated as 52 km^3 (52 billion m^3). When this groundwater volume is divided by the average land area in the study areas, it is found that the average groundwater resource height in the area was about 3.4 m. This is almost double the average annual rainfall of the area. The groundwater resources are sufficient to provide water resources for 440 million people. This is well in excess of the amount required by the Cambodian people.

The potential of groundwater resources in the Lower Mekong region has a great potential as a water source. Tonle Sap Lake, the upstream end of the Tonle Sap River, has a large reservoir function that accepts water from the Mekong River during the rainy season. Morishita et al. (2004) reported that the lake's storage was almost 42.0 km^3 in 2002 and 28.0 km^3 in 2003. These amounts are similar to groundwater storage. These two storage functions contribute to the base flow in the downstream region of the Mekong River during the dry season.

Due to rapid development and urbanization, a reduction in inundation is required. This requires removing the colmatages and building dams in the upstream areas of

Table 23.3 Effect of flood control on groundwater storage

Year	*Inundation area (km^2)*	*Reduction ratio (%)*	*Groundwater storage (km^3)*	*Reduction ratio (%)*	*Population served (thousands)*
1997	5,200	0	52	0	440
1993	4,200	19	36	31	300
1998	2,900	44	30	42	250

the Mekong River. These measures will decrease the inundation area and groundwater resources, and reduce the baseflow downstream in the dry season. However, because some of the flood plains do not get surface water, people in these areas have a higher dependence on groundwater. This is at odds with the adoption of strategies to reduce groundwater resources. Also, the immoderate use of groundwater involves the risk of arsenic contamination. It is important to carefully consider a more sustainable approach to the long-term use of groundwater over a wide scale.

23.7 RISK OF WATERBORNE INFECTIOUS DISEASES

Besides the benefits, flooding in Cambodia is also associated with significant risks. Substances can spread widely with water in inundation areas where people touch the water directly for bathing, washing dishes, and even drinking. Waterborne infectious disease is one of the highest risks for local people in the flood plains. The distribution of coliform bacteria was expressed by a numerical model consisting of the hydraulic model, and a coliform transport model expressing the concentration of coliform bacteria was developed. The numerical and dose–response models evaluate the risk according to different flood magnitudes for different water sources (Kazama et al., 2012).

Figure 23.4 shows the correlation between the annual integrated coliform group count (CGC) simulated for a year and the mortality rate for children under five years of age in three Cambodian provinces within inundation areas. There is a correlation between coliform concentration and infant mortality. This is evident that the concentration of coliform bacteria can be a risk index for waterborne infectious diseases.

The dose–response and hydraulic models show the spatial and temporal distribution of the probability of infection of a host as the risk of waterborne infectious disease (Kazama et al., 2012). The annual spatial risk in the cases of a small, medium, and large floods is shown in Figure 23.5.

A middle-scale flood shows a 0.21 on the annual average risk in the study area. A small-scale food shows 0.94 on the annual average risk in residential areas with a high risk of infection, because the small flood event makes a higher coliform concentration around residential areas by storage and enrichment due to the shallow water. Restricting surface water use around residential areas effectively reduces the risk, except in low water seasons, when the water level is below 0.3 m. This water level increases public health risks in floodplains. This macroscale evaluation will be useful for policymakers

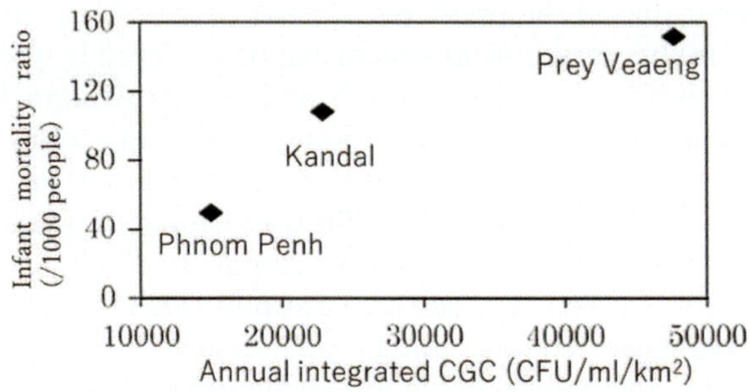

Figure 23.4 The relationship between the annual integrated CGC and the annual infant mortality rate in each province.

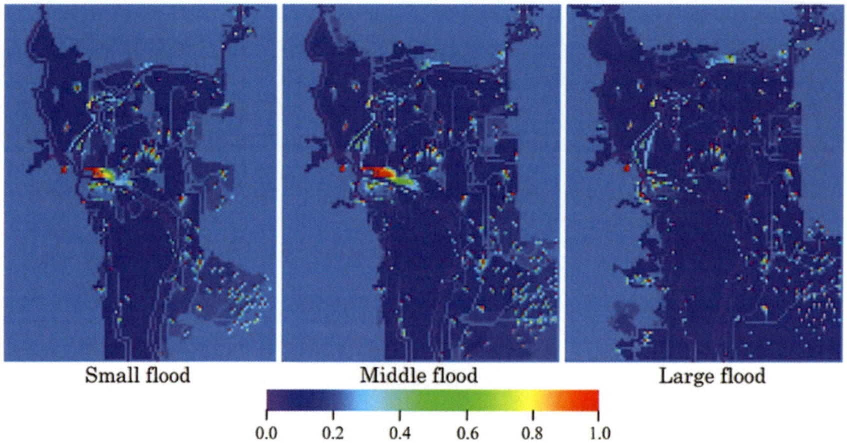

Figure 23.5 The annual infection risk, depending on flood magnitude.

to consider countermeasures at a regional level to reduce the risk of waterborne infectious diseases. Also, by combining microscale evaluations with macroscale evaluations, practical countermeasures can be considered for protecting public health in inundation areas (Amano et al., 2013).

23.8 THE FUTURE OF THE COLMATAGE

The expansion of urbanization equates to an increase in landfills in the floodplains, the removal of the colmatage system, and, therefore, the loss of its functionality. According to Hiraga et al. (2018), landfill increases 1 m on the water depth around Phnom Penh and 3.8 m downstream during floods, and increases the risk over a wide area. This has resulted in higher flood risk in the downstream regions due to reduced recharge in inundation areas and decreased base flow (Kazama et al., 2007). Also, water quality is deteriorating in the expanded urban area (Hiraga et al., 2020).

Rural people still rely heavily on the colmatage system for water resources, aquatic products, and navigation. Despite this, a number of colmatages near cities are expected to disappear due to the pressures of urbanization and modernization. It is possible that a sustainable agricultural system can be achieved by providing irrigation water in the dry season from reservoirs (Tum Nub) in the flood plain, based on changes in the water level due to the seasonal inundation drawn by the colmatages.

Each Southeast Asia country has a similar traditional system to the colmatage system in Cambodia. However, in the Chao Phraya Delta in Thailand, it is difficult to see evidence of this traditional system after development. Thi et al. (2012) have compared the Chao Phraya, the Mekong, and the Irrawaddy Deltas with different development stages, and have suggested flood mitigation measures based on a natural system. The colmatage remains an excellent green infrastructure for flood mitigation. It is necessary to conduct further studies to ensure that the value of the colmatage on economics, ecology, and cultures is fully understood.

ACKNOWLEDGMENTS

We thank the Department of Hydrology, Ministry of Water Resources and Meteorology, Cambodia, and Mekong River Commission for providing data.

FUNDING

This work was supported by Grant-in-Aid for Scientific Research of the Ministry of Education, Science, Sports and Culture (15760372, 18404008, 22404008, 15H05218) and a Grant from the Sumitomo Foundation.

REFERENCES

Amano, A. and Kazama, S. (2014). Evaluation of nutrient condition and agricultural production in the inundation areas of the Mekong River, *Environmental Hydraulics*, 7, 375–378.

Amano, A., Sakuma, T., Kazama, S., and Gunawardhana, L. N. (2013). Evaluation of diarrhea disease risk attributed to inundation water use on a local scale in Cambodia using hydrological model simulations, *River Systems*, 20(3–4), 185–196.

Delvert, J. (1998). *Le Cambodge*, Presses universitaires de France (in French).

Hori, H. (2000). *The Mekong: environment and development*, UN University Press.

Hiraga, Y., Kazama, S., Ekkawatpanit, C., and Touge, T. (2018). Impact of reclamation on the environment of the Lower Mekong River Basin, *Journal of Hydrology: Regional Studies*, 18, 143–155.

Hiraga, Y., Kazama, S., Ekkawatpanit, C., and Yamada, K. (2020). Evaluation of the impact of landfill on floodplain water quality in a tropical monsoon region, *Hydrological Research Letters*, 14(1), 48–55.

Kakudo, H., Kawai, T., Goto, A., and Mase, T. (1995). Colmatage system as an appropriate technology in Cambodia, *Journal of JSIDRE*, 63(4), 357–362.

Kazama, S., Aizawa, T., Watanabe, T., Ranjan, P., Gunawardhana, L., and Amano, A. (2012). A quantitative risk assessment of waterborne infectious disease in the inundation area of a tropical monsoon region, *Sustainability Science*, 7(1), 45–54.

Kazama, S., Hagiwara, T., Ranjan, P., and Sawamoto, M. (2007). Evaluation of groundwater resources in wide inundation areas of the Mekong River basin, *Journal of Hydrology*, 340, 233–243.

Kazama, S., Kono, T., Kakiuchi, K., and Sawamoto, M. (2009). Evaluation of flood control and inundation conservation in Cambodia using flood and economic growth models, *Hydrological Processes*, 23(4), 623–632.

Kazama, S., Morisugi, H., and Sawamoto, M. (2003). Integrated evaluation of the Mekong River flood using benefit calculation, *Journal of Hydroscience and Hydraulic Engineering*, 21(2), 85–92.

Morishita, K., Garsdal, H., and Manusthiparom, C. (2004). *Hydrological function of the great lake and Cambodian floodplains*. Proceedings of International Conference on Advances in Integrated Mekong River Management, 40–46.

Thi, M. M., Gunawardhana, L. N., and Kazama, S. (2012). A comparison of historical land-use change patterns and recommendations for flood plain developments in three delta regions in Southeast Asia, *Water International*, 37(3), 218–235.

Chapter 24

From irrigation perspective to disaster risk reduction using nature-based solution

The Rangsit Canal, Chao Phraya River basin, Thailand

Sutat Weesakul
Hydroinformatics Institute

Uruya Weesakul
Thammasat University

Phruetthiphong Thatanchuleekun
Royal Irrigation Department

Sirapee Ditthabumrung
Panya Consultant

CONTENTS

24.1 INTRODUCTION

Thailand is an agricultural country. Irrigation is a backbone for supplying water to farmers, especially for rice cultivation—the country's primary income in the past. The irrigation work was gradually developed more than 100 years ago. The starting point is the Rangsit area—a flood plain, swap, and abandoned area. The Rangsit Canal was the first irrigation canal to be dredged and still functions today. The study aims to recognize the historical Rangsit Canal by demonstrating its multifunction and benefits from the past to the present. The methodology is based on a literature review of

DOI: 10.1201/9781003222736-29

historical records and technical publishing. A mathematical model is applied to certify the nature-based solution.

24.2 HISTORY OF THE RANGSIT CANAL

In the past, boat transport was a primary and effective activity for communication and trade. Many canals have been dredged since Ayuthayaera. King Rama V had a royal thought to improve river transport in Thailand. The government assigned a company entitled "Siam Canals, Lands and Irrigation Company," established in 1890, to dredge the Rangsit Canal. The cost was paid by providing the property of land rights, including dredging material along the canal, to the company. This is the first time Thailand has a compensation cost for dredging, and equipment was used instead of labor. One of the objectives is to initiate an irrigation area in Rangsit wasteland, covering about 2,000 km^2. Rangsit Canal was then initially dredged in 1887 and completed in 1904. The full name is Rangsit Phrayunsak Canal. Canal with 56 km, 3 m depth, and 16 m width linked Chao Phraya River and Nakorn Nayok River. It is initially used for navigation, river transport, and especially rice cultivation for export. Figure 24.1 shows the location of the Rangsit Canal and its irrigation projects in the north and south of the canal in 1906. There are two main regulators: the Chulalongkorn regulator on the west and Sawapapongsri Regulator on the east. The regulator's name honors King Rama the V and Queen, respectively. Eighteen irrigation subcanals in the south of Rangsit Canals and 14 canals in the north were completed dredging in 1903. The irrigation area in the vicinity of the Rangsit Canal was designed as an inundation irrigation type. Additional water can be pumped from the Chao Phraya and Nakorn Nayok rivers on both the sides of the canal. The water quality in the paddy field in the Rangsit area was found to be acidity contributed by local soil, so the rice productivity was low. People moved out of the area so that the intention of the government to promote Rangsit as a big irrigation area could not be achieved.

The Thai government had asked an expert from the Netherlands, India Waterstate Company, to technically study concepts and guidelines for irrigated water development and drainage in Lower Chao Phraya and related rivers. Homan Van Der Heide (1903) produced a general report on irrigation and drainage in the Menam Valley. It presented the principles of irrigation engineering in the low area, which is fundamental for the irrigation development of the country. The system of irrigation canals can be developed in this area, where the slope is around 1/4,000. Several irrigation projects were introduced, including the first gravity irrigation type by planned diversion dam, the Rama VI barrage, completed in 1924. This is the improvement of the Rangsit area where only rainfall and inundation water were used. The newly diverted water flowed from the Pasak River, passing a limestone mountain.

Consequently, alkali water mixing with local acid water finally turned to neutral water. This concept worked very well to assist the Rangsit area with adequate water in the drought year and alleviate acid soil. Figure 24.2 shows the Rama VI barrage and diversion channels from the Pasak River to the Rangsit area, divided into North and South Rangsit irrigation projects. Raphiphat Canal is the main diversion canal controlled by the Phra Narai regulator, with the maximum discharge of about 200 m^3/s. There are two regulators; the first is Phrasrisin (with the maximum discharge of about

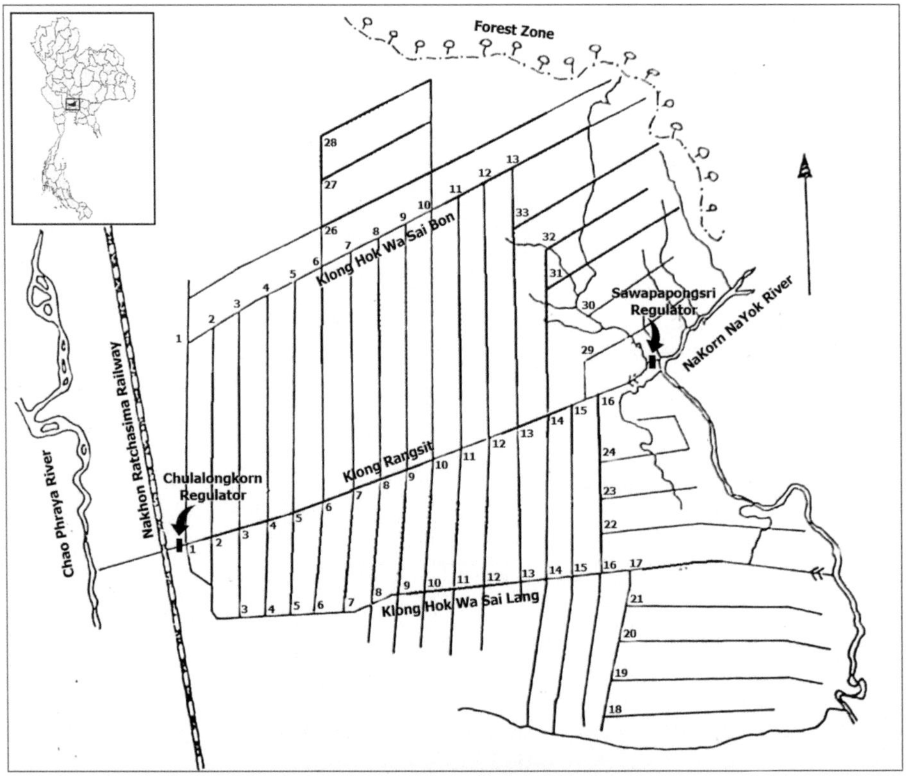

Figure 24.1 Flood plain and Rangsit Canals.

85 m^3/s) to control irrigated water to Raheiphat-yaek-tok for the North Rangsit irrigation Phrasrisoawapak regulator (with the maximum discharge of about 95 m^3/s) to Klong 13 to South Rangsit Irrigation project. There are 14 subdistribution canals in the North and 16 in the South. Klong Rangsit is a boundary of those projects, and functions as the main canal to balance the volume of water and return flow between them. The Greater Chao Phraya project was implemented in 1948 with 35 irrigation projects in the central part of Thailand. The main hydraulic structure is the Chao Phraya Diversion dam, as shown in Figure 24.3. The diversion canal "Chainat-Pask," as a part of the project, was constructed to divert water in front of the diversion dam to Pasak River with a length of more than 100 km. It joined the Pasak River upstream of the Rama VI barrage to operate to divert water to the Rangsit area, as shown in Figure 24.2. In 1964 and 1977, the main storage dams in Thailand were wholly constructed, i.e., the Bhumibhol dam and Sirikit dam, with 13,462 and 9,510 million m^3 (MCM). Those two dams have been named after King Rama IX and Queen Sirikit. They regularly supplied water to the Chao Phraya dam to obtain the completed system of storage and diversion type. Pasak dam was constructed and operated in 1999. Rama VI barrage can then be operated to regularly supply stored water from both Pasak and Chao Phraya Rivers to the Rangsit area. Multisources of water allocation provided stability for water management.

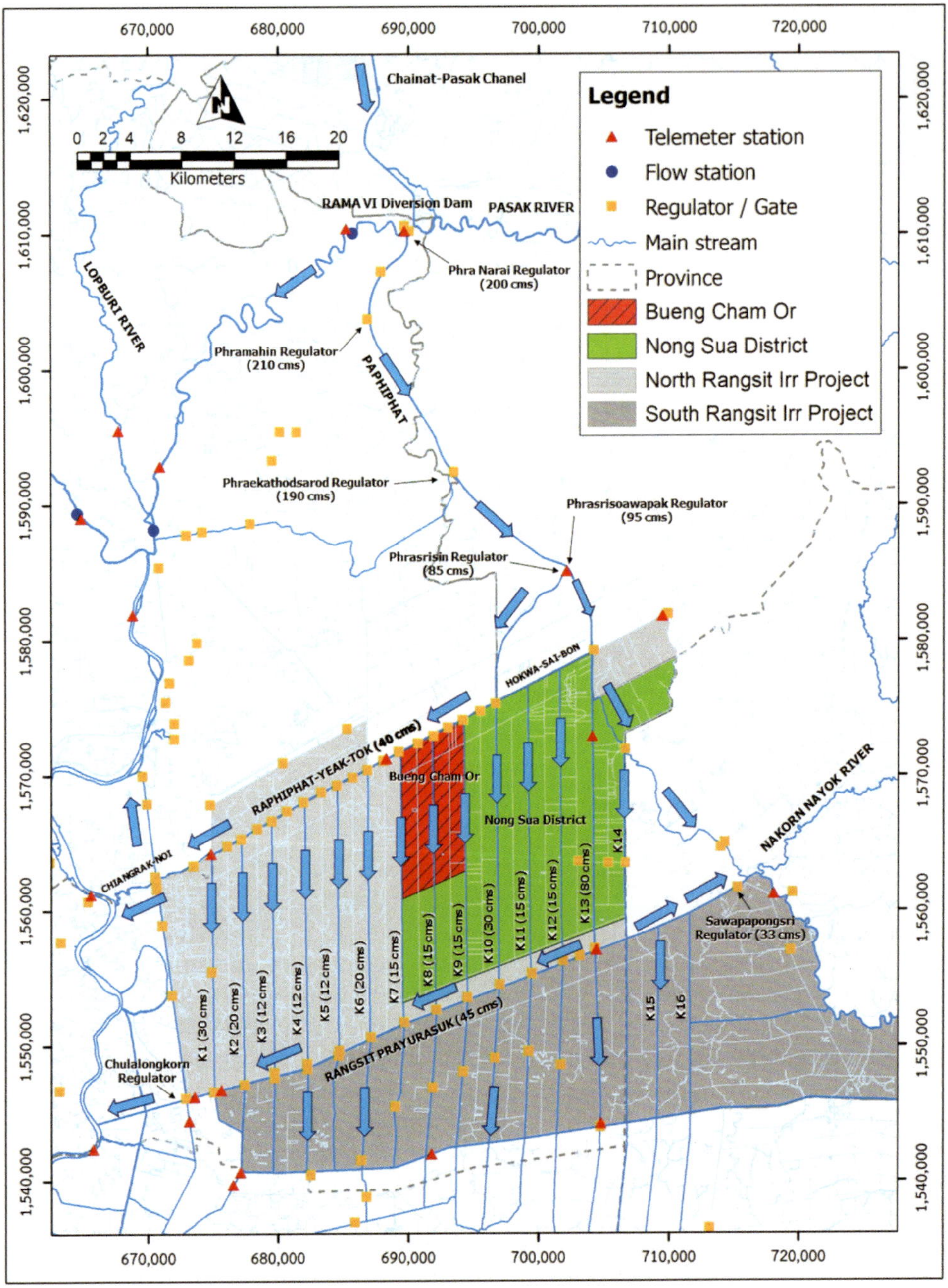

Figure 24.2 Location of North and South Rangsit irrigation projects.

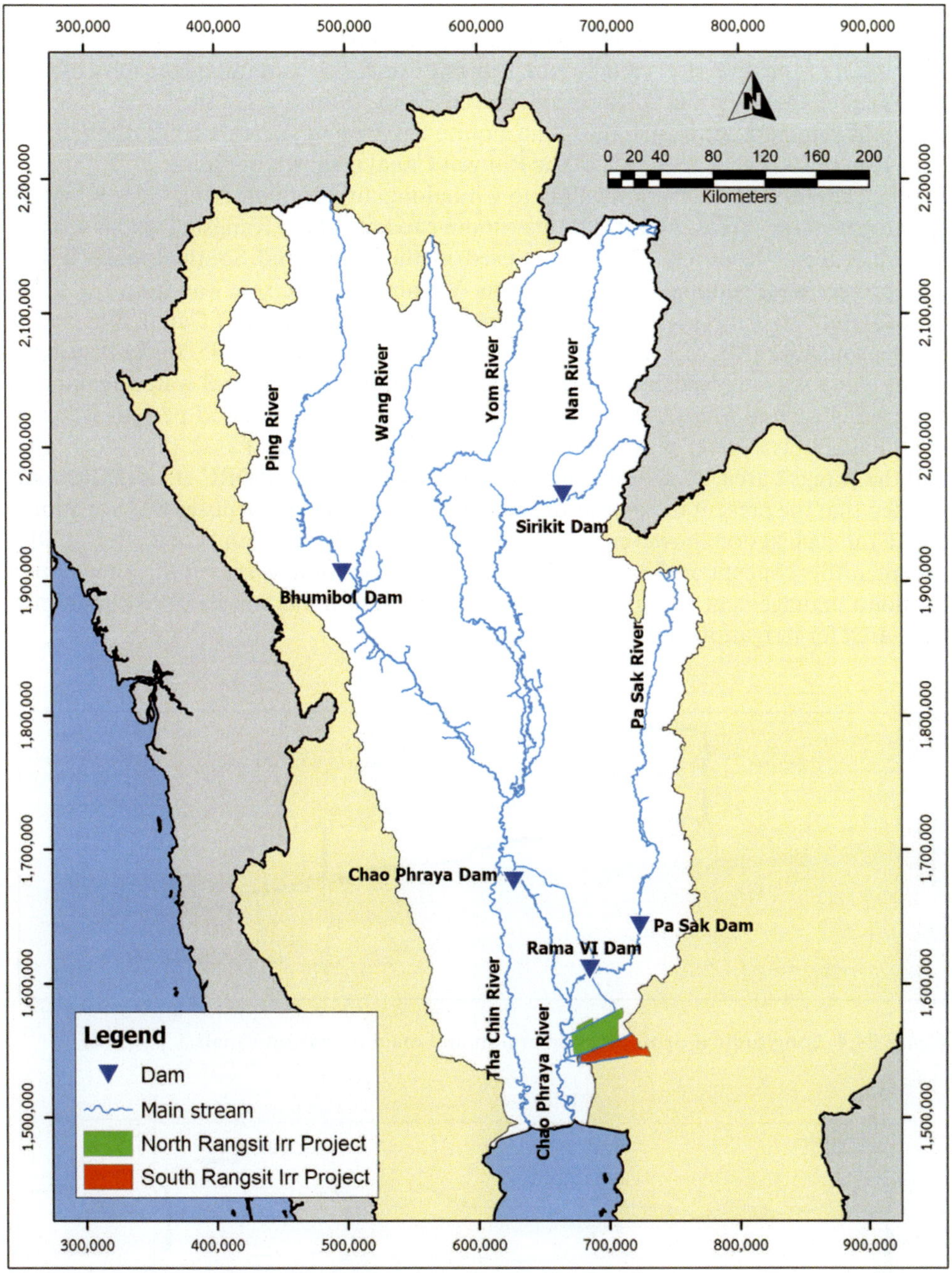

Figure 24.3 Dams in the Chao Phraya River basin.

24.3 CANAL PHYSICAL CONDITION AND WATER MANAGEMENT

The gravity flow is the dominant flow for the Rangsit irrigation project. The longitudinal profile of the diversion canal to the Rangsit Canal (with an initial capacity of Chinat-Pasak Canal of about $210\,m^3/s$ and increased to $320\,m^3/s$ after the Great Flood in 2011) is shown in Figure 24.4, revealing a quite flat ground slope with elevation difference 16 m in 200 km. The profile of the Rangsit Canal is shown in Figure 24.5, illustrating the canal bed slope from Klong 3 to Chulalongkorn regulator and from Klong 14 to Saowapaphonhsri regulator. The maximum capacity of the Rangsit Canal is $45\,m^3/s$.

Most water resources (70%–80%) used in the North and South Rangsit irrigation project were supplied from the Rama VI barrage. The rest was from the Chao Phraya and Nakorn Nayok rivers, pumping directly to the Rangsit Canal. During the dry season, 1,020 MCM is allocated to the project area, while 1,235 MCM is utilized in the rainy season (especially during erratic rainfall). Agricultural water demand in the area is around 91% of the total consumption. During the flood period, water in the Chao Phraya River was usually diverted to the Chainat–Pasak diversion canal and the Rangsit area, such as during the Great Flood in 2011. IDRC et al. (2016) commented that the present infrastructure designed for irrigation purposes was not appropriate for such action; however, it can work satisfactorily up to an extent. JICA (2013) accepted this concept and proposed expansion of diversion as one of the alternatives for flood management in this area. Therefore, Rangsit irrigation became a multipurpose area for irrigation and flood management.

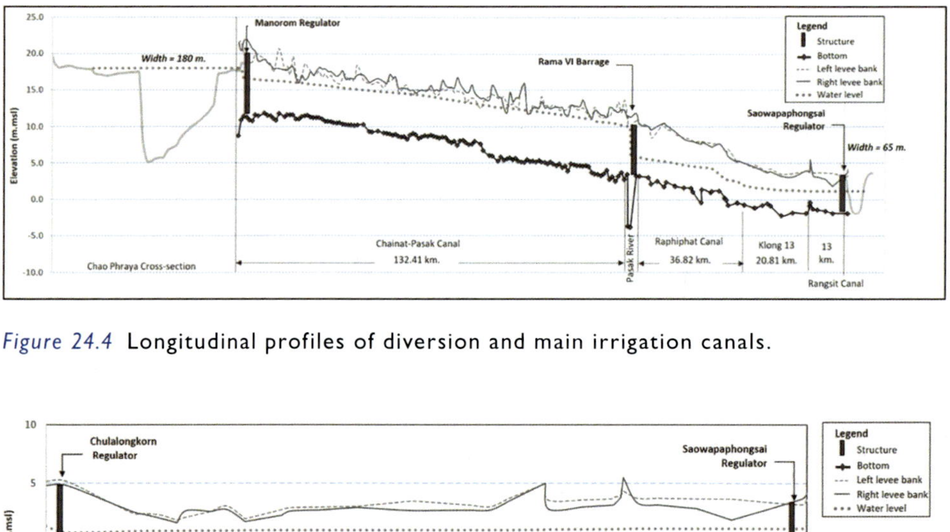

Figure 24.4 Longitudinal profiles of diversion and main irrigation canals.

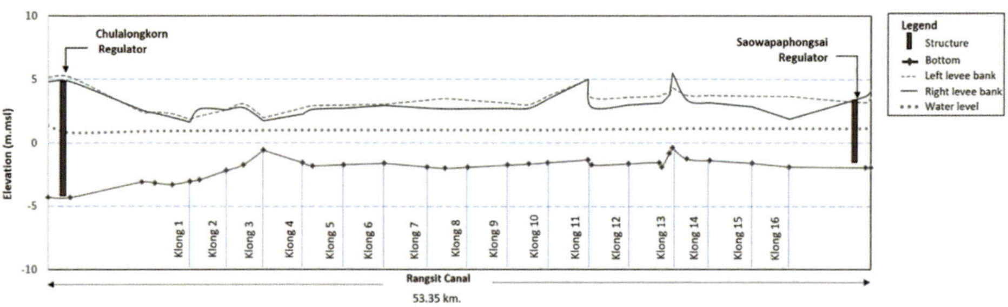

Figure 24.5 Longitudinal profile of Rangsit Canal.

24.4 RICE PRODUCTIVITY

Asawai and Limpanart (1997) indicated that the average cultivated rice area increased from 304 km^2 in 1901 to 534 km^2 in 1906, when the Rangsit Canal was completed in 1904. The rice productivity did not maintain its yield due to a water shortage in 1906, and droughts occurred in 1911 and 1912. In addition, the concession of the Rangsit Canal was expired, and the existing physical conditions have deteriorated. The canal could not be used for navigation, and water storage was reduced, so that only 40% of cultivated rice could be kept. Suksri and Prasongsom (1997) stated that an improvement of the irrigation system in the Chao Phraya River basin area was attempted by requesting Sir Thomas Ward, a water expert from the United Kingdom, to develop the irrigation projects in 1913. The government chose the South Pasak project, despite the high investment construction cost of the Rama VI barrage in the Pasak River. One reason is that the irrigated water could be supplied by barrage to the Rangsit area, which has already been a paddy field. The achievement of water security in the area was then shown by reducing the percentage of the damaged area, which were 5.5% and 1.92% in extreme drought years 1925 and 1927, respectively, while the other irrigation area had damage areas of 22.2% and 16.2% in years 1925 and 1927, respectively. The cultivated area increased to 1,111 km^2 from 1937 to 1978 because of the high rice demand during World War II. Farmers changed traditional paddy-sown fields to transplanting rice cultivation to increase productivity. However, it has been found that, since 1978, agricultural land use has been gradually changed to other crops. National Statistical Office of Thailand (1989) showed that the paddy field area was remarkably reduced by 303 km^2, which changed to horticulture, vegetable, and perennial 160 km^2 and orange farm 110 km^2.

All-season gross productivity of rice cultivation from 2005/2006 to 2019/2020 by Office of Agricultural Economics (2007) and (2021) is depicted in Figure 24.6. The average five years show that productivity was reduced from 428,636 tons to 377,845 tons or 12%. In-season productivity is almost the same, which declined from 223,585 to 219,821 tons. Therefore, off-season productivity is the cause for the reduction from

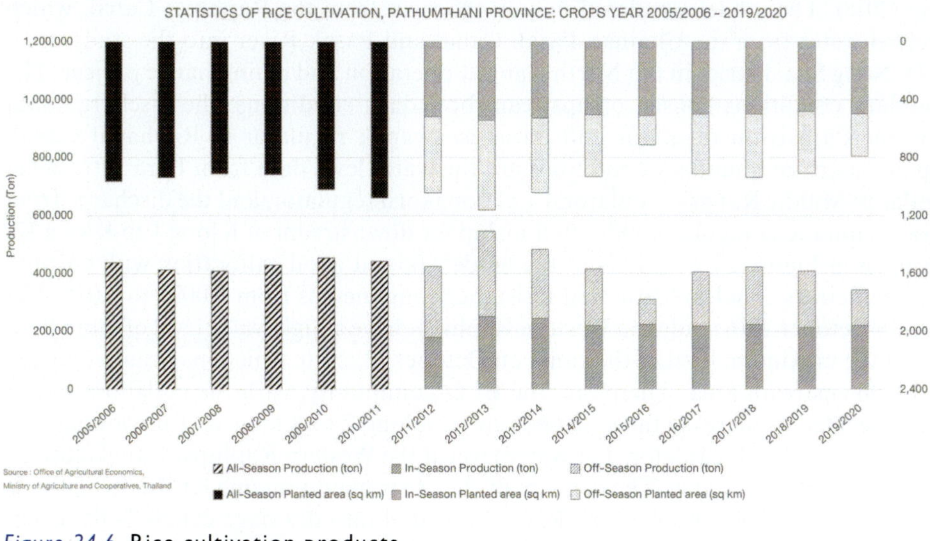

Figure 24.6 Rice cultivation products.

244,275 to 158,024 tons. The area of off-season rice cultivation was also reduced from 523 to 356 km^2. This may be due to less allocation for irrigation water. Farmer started to choose other crops using less water.

24.5 NATURE-BASED SOLUTION FOR PEOPLE AT BUENG CHAM OR NONGSUA DISTRICT

Hydro and Agro Informatics Institute (HAII) introduced science and technology knowledge for survey water allocation and water budgeting. In 2012, the Rangsit community at Bueng Cham Or (BCO), NongSua district, as shown in Figure 24.2, adapted the furrows in their area for Palm oil, which was seriously deteriorated by diseases in orange farms. The community started surveying damaged irrigation infrastructures, deposited canals, and then improved and fixed them.

It has been noted by Utokapat and HAII (2016) that people's income increased along with the improved quality of life. This case verified the concept of water resources management by self-reliance community (CWRM), which followed the water management concept of King Rama IX. Zeeshan et al. (2017) showed that BCO has a high level of satisfaction with income and agricultural yield. In total, 71% of farmers adapt to climate change and can operate regulators for irrigated water management.

During the flood in 2016, flooded water was diverted to the furrows and irrigation canals by the BCO community to save the downstream area from inundation. This practical self-reliance community flood management using nature-based solution as retention and high channel flow type was successful, and it is to be proved by a mathematical model. Furrows systems in the study area are shown in Figure 24.7. Its connection to subcanals, drainage canals, and the Western Raphipat Canal is shown in Figure 24.8. Their function is similar to retention storage. The furrow volume is estimated by its average cross-section and total length of the furrow in the study area and is equal to 470,000 m^3/km^3 or 1.3 MCM.

One dimensional model was developed to simulate diverting water in the Western Raphipat Canal into the provided area using the MIKE 11 model by Ditthabumrung (2018). The model simulated flow in the river from the Raphipat Canal, which received water from the Chainat–Pasak Canal and Pasak River into the study area, BCO, Nong Sua district in the North Rangsit operation and maintenance project. The boundary conditions consist of upstream boundary conditions: the discharge data from the Phrasrison regulator and Phrasrisoawapak regulator in Raphapat Canal. The downstream boundary conditions are the water level data from Phra Intraracha regulator, Middle Rangsit regulator, Chulalongkorn regulator, and the discharge from Phra Thamaracha regulator Q/h relationship for downstream of Klong 1 to Klong 12, as shown in Figure 24.9. Results of the model showed good calibration with correlation coefficients equal to 0.910. Bed resistance roughness is from 0.0025 to 0.035. The crisis in October 2016 made the Western Raphipat have a high water level of more than 90% of the maximum level of the canal on October 13, 2016, which may cause damage to the downstream area. Therefore, the BCO community, with the collaboration of HAII, decided to increase the gate opening of Klong 7 to Klong 10, functioning as a drainage canal head regulator. The water level at the Western Raphipat Canal reduced by 34 cm within half a day. The concept of the crisis operation on October 13, 2016, is to receive water from the Western Raphipat Canal into drainage canals 7–10. Then, the water level in the drainage canal will rise and flow into the furrow by opening the farm turnout. The model results indicated that after opening the gate on October 13,

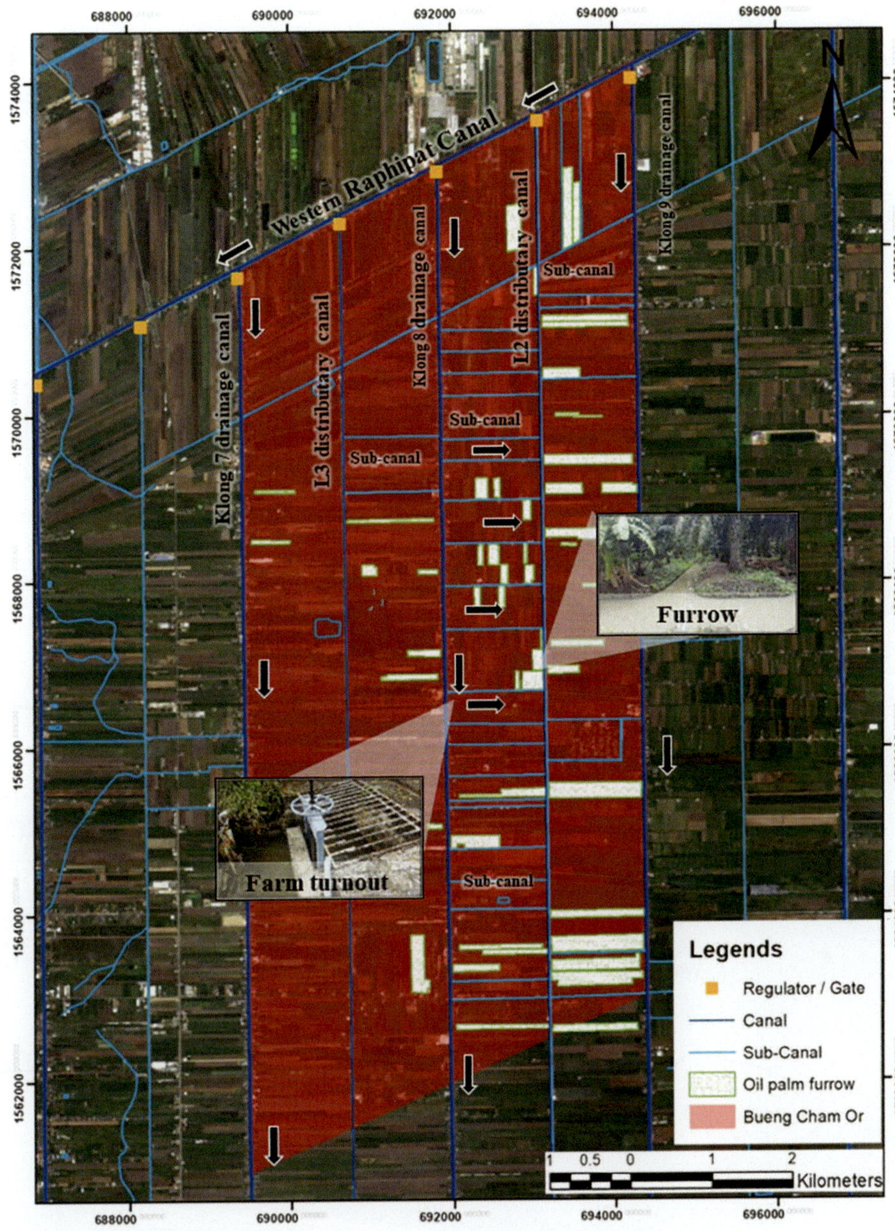

Figure 24.7 Furrows and diversion direction on October 13, 2016.

the water level in the Western Raphipat Canal dramatically decreased from 2.61 m. MSL to 2.27 m. MSL on October 14, as shown in Figure 24.10. Moreover, the Klong 8 drainage canal inflow and outflow showed that the inflow that comes into Klong 8 is higher than the outflow. Thus, the water was stored in the drainage canal and furrows around 5.6 MCM for 19 days, from 13th to 31st October, due to unbalanced inflow and outflow of water volume.

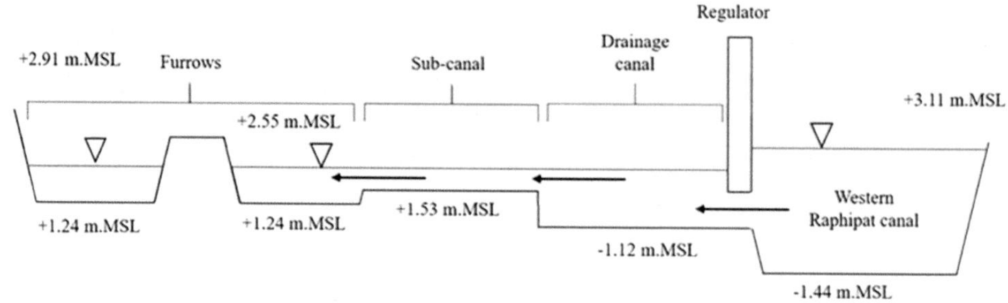

Figure 24.8 Typical cross-section of the connection between furrows and canal.

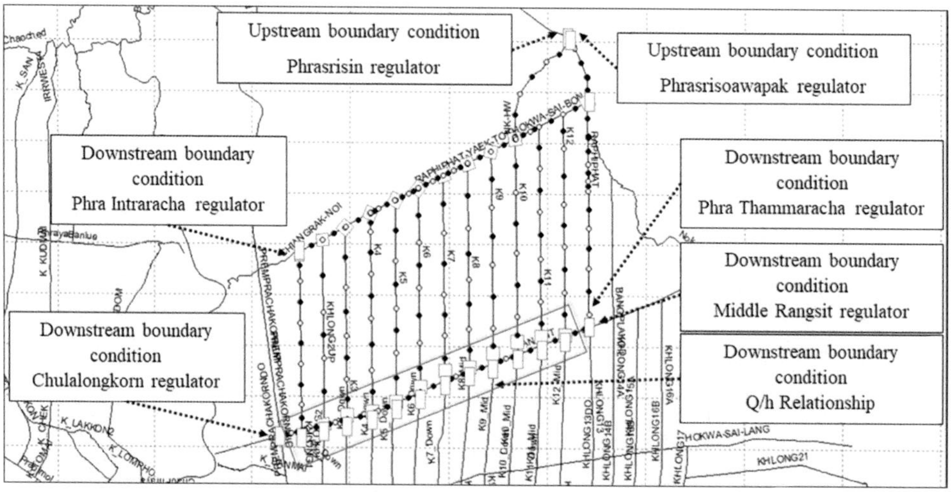

Figure 24.9 Model domain and boundary conditions.

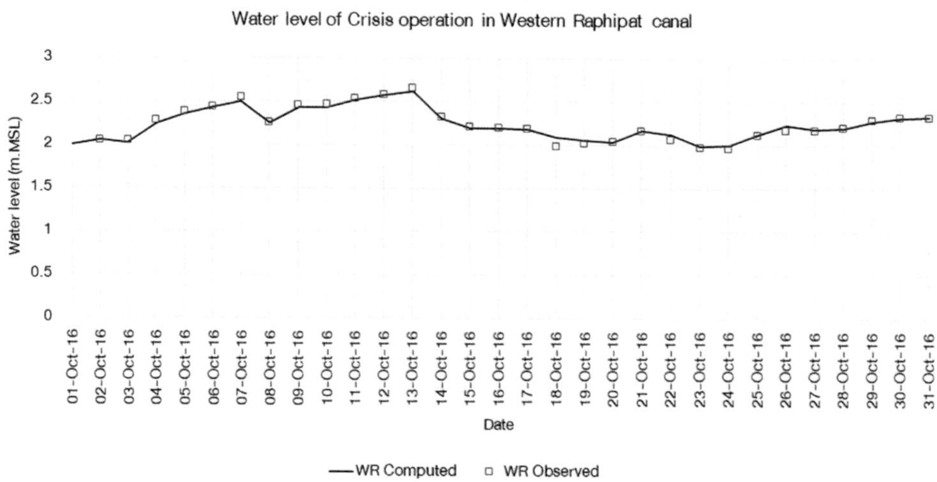

Figure 24.10 Water level in Western Raphipat cabal at crisis operation.

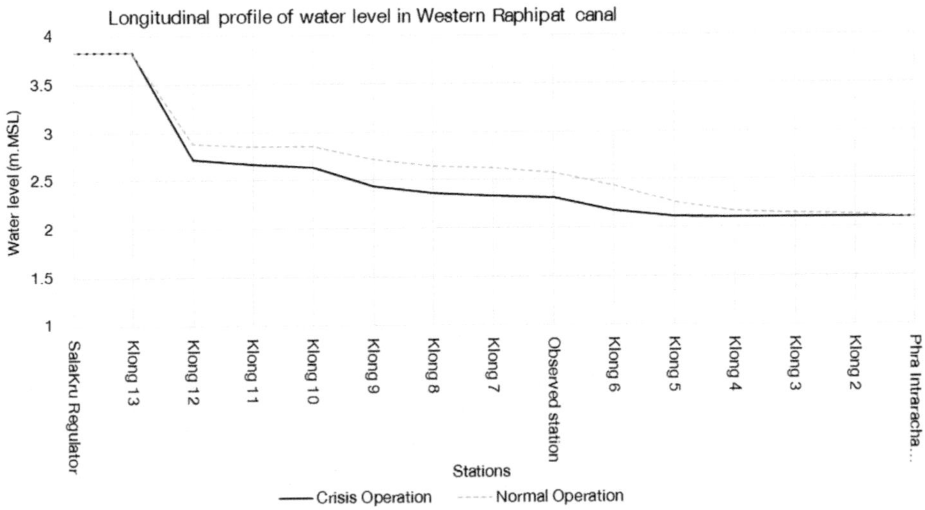

Figure 24.11 Longitudinal profile of water level in Western Raphiphat Canal.

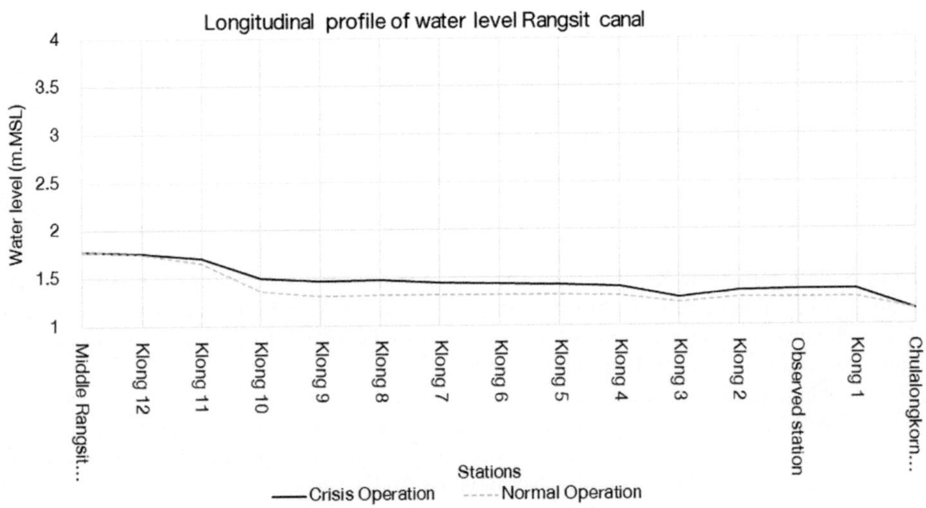

Figure 24.12 Longitudinal profile of water level in Rangsit Canal.

The computed longitudinal profile of the water level presented in Figures 24.11 and 24.12 illustrates the water level difference between normal and crisis operations. The water level in the canal can be reduced for a distance of about 21.86 km from Klong 12 to Klong 4 for the Western Raphipat Canal. In comparison, the water level in the Rangsit Canal increased from Klong 11 to the end of the canal at Chulalongkorn regulator for a distance of about 28.11 km. The results indicated that the water diversion during crisis operation could help reduce the Klong 7–10 area's water level and some distance from the Western Raphipat Canal. Rangsit Canal functioned as receiving water infrastructure. This section shows that the BCO community is a key factor in applying sciences and technology to the historical irrigation project. It can work successfully in normal and crisis operations as a nature-based solution.

24.6 CONCLUSIONS

It has been 117 years already since the Rangsit Canal was completed dredged. It shows that the canal can serve as a multifunction infrastructure. The canal was used for river transport for rice export in the old day. It worked as an irrigation canal to supply water from local and river basins in the central and north of the country to agricultural areas for farmers. This is to maintain Thai culture's original tradition for rice production in the Rangsit area. It is a part of drainage in a flood control system to alleviate floods in the Chao Phraya River basin. Lastly, the community can use it as a part of the local flood diversion system as a nature-based solution for high water levels. The Rangsit Canal has already been proven for multipurpose use in the past. It can continue working as a resilience structure, particularly with more variation and extreme climates in the future.

ACKNOWLEDGMENT

The European Commission partially supports this work under Program HORIZON 2020, the funding scheme "Innovation Action." Project Number is 776866, for the research project RECONNECT (Regenerating ECO systems with nature-based solutions for hydro-meteorological risk rEduCTion).

REFERENCES

Asawai, S. & Limpanart, R. (1997). Chapter 2: One-hundred-year Rangsit: historical, economic, and social study. In C. Suntravanit, T. Tapsawad, W. Chiewchaisak & S. Wankaew (Eds.), *Rangsit canal: 100 years pilot research project for celebrating the golden jubilee of the King*, 8–64, Chulalongkorn University (in Thai).

Ditthabumrung, S. (2018). *Nature-based solution for the water management: a case study in the Rangsit area, Thailand*, Master Thesis, Asian Institute of Technology, May.

Homan Van Der Heide, J. (1903). *General report on irrigation and drainage in the lower Menam valley*, Engineer of the Waterstaat of Netherlands, India, temporarily placed at the disposal of Siam Government, Bangkok, January.

IDRC, CRDI &TDRI. (2016). Analysis of flood management in the Chao Phraya River basin, Policy brief, Thailand Development Research Institute.

JICA. (2013). Project for the comprehensive flood management plan for the Chao Phraya River Basin, September, Research Report.

National Statistical Office of Thailand. (1989). Survey of changing agricultural area in the year 1988 (in Thai).

Office of Agricultural Economics. (2007). Annual agricultural product report in year 2006 (in Thai).

Office of Agricultural Economics. (2021). Annual agricultural product report in year 2020 (in Thai).

Suksri, C. & Prasongsom, K. (1997). Chapter 5: Water resources and irrigation in the Rangsit area. In C. Suntravanit, T. Tapsawad, W. Chiewchaisak & S. Wankaew (Eds.), *Rangsit Canal: 100 years pilot research project for celebrating the golden jubilee of the King*, 138–182, Chulalongkorn University (in Thai).

Utokapat Foundation & Hydro and Agro Informatics Institute (HAII). (2016). *Application of science and technology for community water-related disaster risk reduction*, Utokapat Foundation.

Zeeshan, A. S., Weesakul, S. Kalampabut, N. & Tiewtoy, S. (2017). *Analysis of farmers' choice of adaptation to climate change at NongSua district, Rangsit Canal, Thailand*, THA 2017 International Conference.

Chapter 25

Agriculture irrigation development in Kedah, Malaysia

Strengthen the linkage between national food and water security

Nik Kun Nik Man and Noor Effarizan Ismail
Muda Agricultural Development Authority (MADA), MADA Headquarters

Nor Azazi Zakaria and Chun Kiat Chang
River Engineering and Urban Drainage Research Centre (REDAC), Universiti Sains Malaysia

CONTENTS

25.1 INTRODUCTION

Food and water security are the key challenges highly vulnerable to the continuously changing population growth, managing policies, economic and urbanization, environmental pressures, and societal interactions (Allouche, 2011). Today, agriculture accounts for approximately 70% of worldwide human freshwater use, and the concept of the food–water nexus has dominated the top rank on the global food policy agenda (Gbegbelegbe et al., 2014). The agriculture sector plays an important role in Malaysia's economic development, providing rural employment, uplifting rural incomes, and ensuring national food security. It is about 22% of the land in Malaysia utilized for agriculture. Townships, mining activities, and other land uses take up another 10%, leaving 68% under forest cover. Approximately half the agricultural lands are cultivated with perennial crops. The overall agriculture sector in Malaysia is broad, encompassing industrial crops such as oil palm and rubber, food and cash crops such as paddy, livestock, and specialty products such as edible bird's nests and herbs.

DOI: 10.1201/9781003222736-30

The linkages between agricultural sustainability, food production, water supply, and natural resources in the country have contributed to the country's social, economic, and environmental stability. In line with the National Key Economic Areas (NKEAs) and National Agro-Food Policy (2011–2020), this transformation is anchored on an integrated management model contributing toward better economic growth, social sustainability, and higher agriculture productivity, which complies with the food safety standards to ensure sufficient domestic rice production supply to the country.

Although there is currently limited policy enforcement on integrated food–water security management, this novel concept has been embedded into the theme of the 10th Malaysian Plan to strategically address the availability, accessibility, and affordability of food all over the country.

Despite the paddy and rice industry's small contribution to the nation's Gross Domestic Product (GDP), it has garnered much interest from policymakers, given its complex relationship with food security, culture, and socioeconomic factors. It can be regarded as a strategic intervention to buttress the paddy and rice sectors' development, and to protect the nation's food security. Paddy cultivation in Malaysia comprises the designated granary areas and non-granary areas. Typically, the average yields in the granary areas are higher than in the non-granary areas (4.9 metric tons per ha as compared to 4.0 metric tons per ha in 2016). There are eight main paddy granary areas in Malaysia; however, the yield varies between granary areas being the top performers with yields above 5.0 metric tons per ha. These differences can be attributed to many factors, such as environmental conditions, cultivation areas, soil conditions, weather, farm management, irrigation, and use of technology (Khazanah Research Institute, 2019).

Rice production is one of the largest water consumers in the world and requires large areas to irrigate the paddy field (Gheewala et al., 2014). About 70% of the available water resources in Malaysia are consumed for rice production. In some countries, irrigation accounts for as much as 90% of the available water resources. Such conventional resources consist of rainfall and snowmelt water from rivers, streams, lakes, reservoirs, and aquifers or renewable through the hydrological cycle (Brooks, 2014). Thus, acquiring optimal yield requires a water management practice that is a formidable challenge for food, economic, social, and water security. In this sense, understanding the complex interconnection between food production and water management has become critically important to developing sustainable food and water security management.

25.2 RICE IN MALAYSIA

Malaysia is a humid tropical country. Its climate is characterized by northeast (from November or December until March) and the southwest monsoon (late May to September), which are subject to interference by mountains in Peninsular Malaysia. The annual average rainfall is 2,420 mm for Peninsular Malaysia, the monthly mean air temperature is between 25°C and 28°C in the coastal lowlands, and monthly relative humidity is high and ranges between 75% and 90%. April and October are the wettest months in Peninsular Malaysia, except for the southwest coast, owing to transitional periods during monsoon seasons. Peak rainfall for the southwest coastline occurs in October and November, with minimal rain in February. About 88% of the annual rainfall occurs from May to November—the traditional rice-growing season.

There is insufficient rainfall in December–April to grow a rice crop without supplementary irrigation.

Rice is a strategic commodity for the country, as it is the staple food of most of the population in Malaysia. Now a day, Malaysians consume the grain daily either as cooked rice or indirectly in the form of rice flour. At the same time, commodities are important to the stability and growth of the population, as consumers spend most of their income buying rice. In the early 1900s, there were initial efforts by the government and private enterprises to develop irrigation infrastructure. There were two successes of large-scale irrigation systems in the country: the Wan Mat Saman Scheme in Central Kedah (28,700 ha) and the Kerian Irrigation Scheme in North Perak (23,400 ha). Nowadays, Kedah's rice plains located in the northern part of the West Coast of Peninsular Malaysia produce more than half of Malaysia's homegrown rice supply.

25.3 IRRIGATION AGRICULTURE: PRE-INDEPENDENCE

The northern Kedah plain is low and flat, it extends northwards from the foot of Mount Jerai (Kedah Peak), bordered by the sea to the west, and it extends inland for an average of 20 km to the low foothills of the inner mountain ranges of the peninsular. Most of this plain is less than 3 m above the sea level. A low embankment protects it, and only bunding around the fields prevents regular marine inundation (Dobby, 1951). Besides that, it has faced many water supply problems that have taken place over the years (Hill, 2012). The first major canal in the Kedah plain is believed to have been built during the reign of Sultan Thiauddin Mukarram Shah, who ruled Kedah from 1661 until 1687 (Ahmat, 1984). The opening of the Alor Setar town by Sultan Muhammad Jiwa Zainal Azilin Muazzam Shah II, who ruled Kedah from 1710 until 1778, then brought changes in the domestic water supply and economic activity. Several canals were built to meet the daily needs, especially for paddy fields. Indeed, constructing canals and channelizing natural waterways was uncommon, beginning when rice cultivation was first practiced in Kedah. These undertakings, however, were never large-scale and just expedient measures to meet the local needs. Several others were dug between 1870 and 1890; Wan Mat Saman Canal is one of three major canals constructed (Mahmud, 1972). These canals had become the baselines from which pioneering peasants moved out to reclaim the swamps and the baseline from which subsidiary canals were constructed, making more swamp areas accessible.

Wan Mat Saman Canal (Figure 25.1) was constructed in 1885 by order of Kedah's first 'Prime Minister' at the time, Dato' Wan Muhamad Saman Wan Ismail. (Note: Dato' is the highest state title conferred by the Ruler (Sultan) on the most deserving recipients who have contributed greatly to the state. It is honorary and non-hereditary.) Thailand's agriculture and transportation development inspired the idea of building the canal by Dato' Wan Muhamad Saman during his visit to the country with the Sultan of Kedah, Sultan Ahmad Tajudin Mukarram Shah. In 1885, the Sultan granted him a concession to build the canal (Short & Jackson, 1971). Therefore, Dato' Wan Muhamad Saman invited surveyors from Thailand to help construct the canal after a visit to Bangkok. The canal construction was a "transfer of technology" between Thailand and Kedah. Finally, the canal was cut northwards to the Kedah River at Alor Setar, 35 km from the foothills of Mount Jerai (Kedah Peak) in the south throughout the centre of the Kedah plain. The linear canal was built possible by villagers'

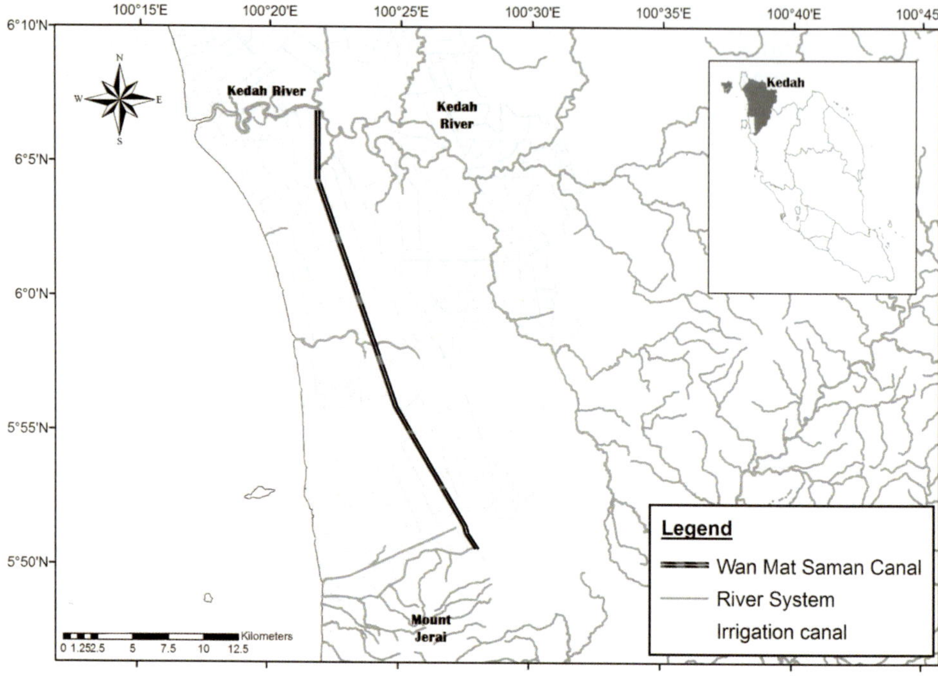

Figure 25.1 Wan Mat Saman Canal, Kedah.

voluntary participation, including the locals' forced labor at that time by digging the canal using traditional tools, such as cam, cop, and machete; the idea of building the canal was strongly rejected. To ensure that the canal ran straight, the initial diggings were done at night, in which rows of "jamung" or traditional torches were lit in a straight line (Figure 25.2). There are three stages of the canal construction as follows.

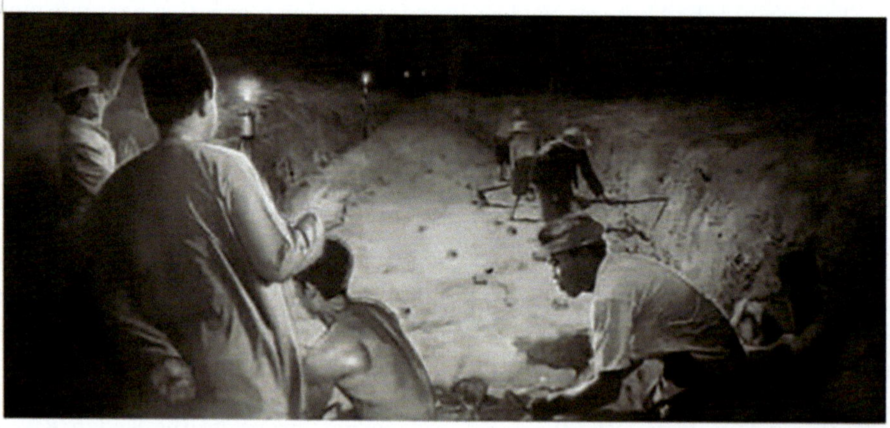

Figure 25.2 Illustration of Wan Mat Saman Canal construction. (Source: Muda Agricultural Development Authority.)

- Stage 1: from Lubuk Katong (known as Tandop) to Kota Sarang Semut
- Stage 2: from Kota Sarang Semut to foothills of Mount Jerai (Kedah Peak)
- Stage 3: from Lubuk Katong to Kedah River

Technically, Lubok Katong was found in an area where a reservoir has been established to control the water level so that the level will not change during monsoon or drought. Therefore, the irrigation system was rudimentary with a dual-function channel, i.e., irrigation during the growing months and drainage system toward the end of the planting seasons (Abdullah, 1974). When the canal was completed in 1895, it was 36 km in length, 7.5 m in width, and had a depth of approximately 2 m. It is the longest canal in Malaysia and continues to be in service today. Abdullah (1974) summarized that the main purpose of the Wan Mat Saman Canal construction was related to the need for water resources for paddy cultivation, which is to irrigate the paddy fields during the dry season or to drain water out during heavy rain to avoid water ponding at the paddy field. Before the canal was built, paddy fields were generally located in small pockets in the floodplains adjacent to the banks of major rivers. Farmers that traditionally planned their crop production primarily relied on uncertain rainfall (Department of Agriculture, 1931). In years of good rainfall, farmers needed no irrigation. However, flooding was often prevalent with the need to provide adequate drainage. After this canal was completed, many other small canals were built connected to the main canal to irrigate the rural areas. As a result, in addition to relying heavily on rivers for water resources, residents had the advantage of taking water from this canal. At the same time, the canal also serves as a route for flood mitigation during heavy rain. Therefore, by building a canal, the excess water could be irrigated during rainy seasons, and the water could be irrigated into the paddy fields during the dry season.

The canal's construction was a monumental achievement during the second-half of the 19th century and turned Kedah into a very successful state for paddy cultivation (Figure 25.3). The canal's construction plays a very significant role in settlement development. It turns all the useless swampy areas into a vast area suitable for rice cultivation, providing a substantial amount of paddy to feed the population. In addition, it is found that the area straddling the Wan Mat Saman Canal had accumulated a higher density of settlement compared with those in other parts of the district, some of which had been settled much earlier (Mahmud, 1972). It was noted that the development of domestic water supply was in line with the development of the agricultural-based

Figure 25.3 Paddy cultivation in Kedah Plain.

economic pattern of the state at that time. By the end of the 19th century, deforestation and clearance works were carried out to construct irrigation canals to irrigate the paddy field (Sadka, 1968). After 1898, building a canal reverted to or was purchased by government. When a British advisory administration emerged in Kedah after 1909, it inherited the basis for more comprehensive development of water-control facilities.

It is indicated that 88,491 residents in the State of Kedah were engaged in agriculture during the 1900s, of which 88,121 residents, or 99.59%, were farmers (Cavendish, 1911). The Wan Mat Saman Canal was a renaissance to the state's agriculture, resulting in widespread paddy cultivation activities and many new settlements along the canal; therefore, a quite extensive network of such irrigation channels in Kedah has been developed. The canal was a modern irrigation system in the state not only to improve the drainage system but also to facilitate journeys of sampan through the canal for the transportation of paddy and other goods. The canal, which helped irrigate the paddy fields in Kedah's plains, enabled the state to boost its rice production (Kedah State Annual Report, 1909). In 1932, the Drainage and Irrigation Department (DID) was established and, together with the Department of Agriculture (DOA), formed the prime movers of organized and systematic irrigation development in the country. These include the continued development of new areas and upgrading of existing schemes.

25.4 IRRIGATION AGRICULTURE: POST-INDEPENDENCE

Kedah's agriculture is often equated with rice cultivation, and for a good reason. Although paddy is not the state's only major crop, it has always been its most important one. Historically, Malaysia has always had production-driven agricultural targets. Starting from 1955 and continuing until a few years after independence, rice sufficiency through local production has become the government's main policy. In the early 1960s, rice production in Malaysia had begun to lag seriously behind consumption, and imports were imposing a serious strain on the balance of payments. Therefore, the government began to plan the large-scale investments needed to approach national self-sufficiency in rice. Hence, the government launched a number of agricultural projects under the First Malaysia Plan (1965–1970) to ensure paddy cultivation activities could be carried out twice a year; This breakthrough came during the late 1960s and early 1970s with Muda Irrigation Scheme, which entire project cost about RM 245 million (USD 840 per ha).

The Muda Irrigation Scheme is situated between latitudes of 5° 40′ N and 6° 30′ N and between longitudes of 100° 5′ E and 100° 35′ E on the Kedah–Perlis coastal plain in northwest Peninsular Malaysia (Figure 25.4). The Muda Irrigation Scheme encompasses an area of 96,000 ha of the Kedah Plain and is cultivated by nearly 60,000 peasant families (De Koninck, 1988). The construction work was started in April 1966. In early 1970, the first stage of irrigation was started in an area of approximately 33,600 ha of off-season crops. Irrigated areas were implemented in four stages and completed in 1974. The project area occupies a flat alluvial plain, about 20 km wide and 65 km long, between the foothills of the Central Range and the Straits of Malacca. As a result, Wan Mat Saman Canal dismissed its use as an irrigation canal as a bigger and mightier irrigation canal, also known as the Muda Irrigation Canal, and was built to replace the Wan Mat Saman Canal. The Irrigation Scheme includes the enlargement, realignment, and improvement works of the 37 km Northern Canal and also the construction of 65 km long Central and Southern Canal (Figure 25.5). DID then handed over the

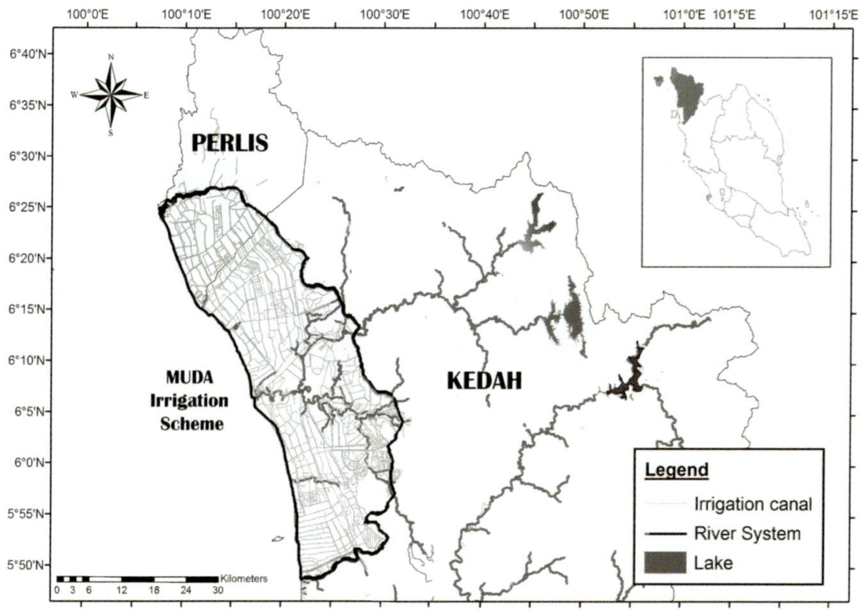

Figure 25.4 Muda irrigation scheme.

Figure 25.5 The Northern, Central, and Southern Canals.

Figure 25.6 The current condition of Wan Muhammad Saman Canal.

Wan Mat Saman Canal (Figure 25.6) to Muda Agricultural Development Authority (MADA) for operation and maintenance.

The construction works for the Muda Irrigation Scheme were started in 1965 and completed in 1970, which included the Muda and Pedu Dams. The construction of two dams on the upper reaches of the Muda River and Pedu River, which is about 80 km to the east of the Muda area, has established two reservoirs, and water from the Muda Reservoir flows to Pedu Reservoir through a tunnel. These reservoirs store irrigation water for the approximately 230,868 acres of the rice-growing area of Muda Valley (Zulkifli, 1980). Under the Muda Irrigation Scheme, a system of irrigation and drainage is provided in parts of western mainland Kedah and Perlis. The irrigation water in the Muda area comes from direct rainfall (52%), the Muda–Pedu reservoir (30%), uncontrolled flow (10%), and recycled drainage water (8%) to cater to both the crop season (Tukimat et al., 2017). Water for irrigation is released into Pedu River, flowing into Padang Terap River, and bifurcates at Pelubang. The flows at Pelubang were diverted into two main channels: the Northern Canal and Southern Canal. Then, the water from the main canals was regulated into the primary, secondary, and tertiary canals before entering the paddy fields.

A breakthrough came during the late 1960s and early 1970s with the establishment of the Muda Irrigation Scheme, covering nearly 100,000 ha. Monitored by the MADA, the scheme, whose major goal was the generalization of the double cropping of paddy, has met with considerable success. Overall, the paddy cultivation activities for Season 1 and Season 2 are according to the MADA's paddy cultivation schedule that has been set. As a result, double cropping is possible in nearly 90% of the scheme's area. Indeed, under pre-project conditions, farmers could grow only a main-season rice crop each year. By the 1980s, the overall output from the scheme had increased by 50% over 1970.

Currently, Muda Irrigation Scheme remains a traditional paddy cultivation area in northern Peninsular Malaysia that has been cultivated for 50 years through paddy cultivation twice a year. In 2016, MADA had an average yield above 5 metric tons per ha, which is higher than the national average yield per hectare. The total paddy cultivation areas under MADA are 130,282 ha, of which 100,685 are paddy fields. The area of paddy cultivation area of the Muda Irrigation Scheme represents 35.13% of the country's paddy area, which is the largest in Malaysia. Through the concept of area development, the Muda area has been divided into 27 development localities, whose supervision is controlled by the four MADA regional offices (Figure 25.7). To ensure

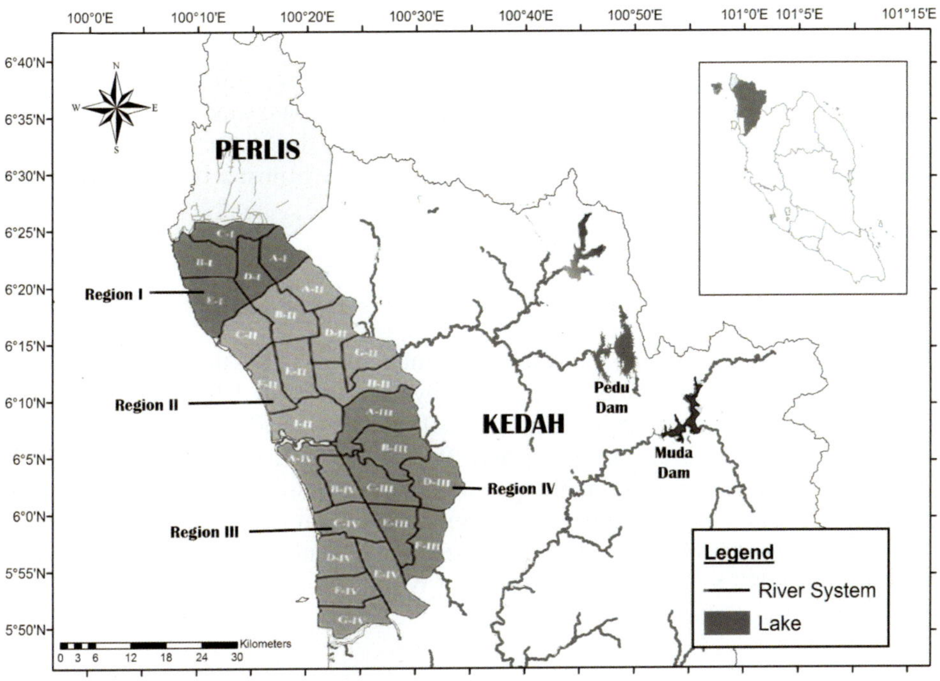

Figure 25.7 Development localities in Muda area.

Table 25.1 Total paddy production and productivity of the Granary areas in 2017

Region	*Area (ha)*
Region I (Perlis)	20,073
Region II (Jitra, Kedah)	32,595
Region III (Pendang, Kedah)	22,682
Region IV (Kota Sarang Semut, Kedah)	25,335
Total	**100,685**

the smooth running of management and administration, MADA is divided into four (4) regions, namely, Region I (Perlis), Region II (Jitra), Region III (Pendang), and Region IV (Kota Sarang Semut) (Table 25.1).

25.5 CONCLUSION

Irrigation development in Malaysia has undergone several distinctive phases, directly linked to the nation's changing agriculture and food industry trends. Since the implementation of the Muda Irrigation Scheme in early 1970, MADA has been established to provide all engineering services to implement socioeconomic development in Muda areas. Under the project, an irrigation and drainage system is provided in parts of western mainland Kedah and Perlis to plant two rice crops a year. In developing paddy cultivation activities in the state of Kedah and Perlis, Muda Irrigation Scheme has since produced nearly 40% of the paddy production in the nation. It generates an industry worth more than a billion ringgit annually, elevating farmers living in the Muda area. The increased total rice and paddy production has allowed Malaysia's self-sufficiency level to hover between 60% and 70%. In addition, recognizing that Wan Mat Saman Canal symbolizes the transition from the conventional way of life to the modern system, MADA has initiated a conservation plan for the construction of a gallery depicting the history of Wan Mat Saman Canal from the Minister of Agriculture and Agro-based Industry Malaysia in 2012.

ACKNOWLEDGMENTS

This article is based mainly on the Wan Mat Saman Canal—Linking Water Resources And Crop Production in Malaysia, authored by us (MADA & USM), and presented and published in Proceedings of International Association for Hydro-Environment Engineering and Research (IAHR)—Asia Pacific Division (APD) Congress: Multi-Perspective Water for Sustainable Development (IAHR-APD 2018) and has been rewritten fitting this monograph, focusing on irrigation agriculture and rice cultivation in Northern of Peninsular Malaysia. The authors would also like to acknowledge IAHR for the 2017 IAHR Hydro-Environment World Heritage Award as international professional recognition of this infrastructure.

REFERENCES

Abdullah, O. (1974). *Terusan Wan Mat Saman*. Mastika, November. Utusan Melayu, 103–104.

Ahmat, S. (1984). *Tradition and change in a Malay State. A study of the economic and political development of Kedah (1878–1923)*. Kuala Lumpur, MBRAS, monograph no. 12.

Allouche, J. (2011). The sustainability and resilience of global water and food systems: political analysis of the interplay between security, resource scarcity, political systems and global trade. *Food Policy*, 36, Supplement 1, S3–S8.

Brooks, J. (2014). Policy coherence and food security: the effects of OECD countries' agricultural policies. *Food Policy*, 44, 88–94.

Cavendish, A. (1911). *Report on the census of Kedah and Perlis* A.H.1329, Penang.

De Koninck, R. (1988). Alor Setar, the capital of Kedah: a city to govern agriculture. *Archipel*, 36(1), 147–164.

Department of Agriculture. (1931). *Straight settlements and federated Malay States*. Guide to the Experimental Station, Serdang. Government Printer, Serdang.

Dobby, E. H. G. (1951). The North Kedah Plain. *Economie Geography*, 27, 287–315.

Gbegbelegbe, S., Chung, U., Shiferaw, B., Msangi, S., and Tesfaye, K. (2014). Quantifying the impact of weather extremes on global food security: a spatial bio-economic approach. *Weather and Climate Extremes*, 4, 96–108.

Gheewala, S. H., Silalertruksa, T., Nilsalab, P., Mungkung, R., Perret, S. R., and Chaiyawannakarn, N. (2014). Water footprint and impact of water consumption for food, feed, fuel crops production in Thailand. *Water*, 6, 1698–1718.

Hill, R. D. (2012). *Rice in Malaya, a study in Malaysia, a study in historical*. Singapore: NUS Press Singapore, pp. 55–70.

Khazanah Research Institute. (2019). *The status of the paddy and rice industry in Malaysia*. Kuala Lumpur, Malaysia: Khazanah Research Institute.

Mahmud, Z. (1972). The population of Kedah in the nineteenth century. *Journal of Southeast Asian Studies*, 3(2), 193–209.

Sadka, E. (1968). *The protected Malay states, 1874–1895*. University of Malaya Press. Malaysia.

Short, D. E. and Jackson, J. C. (1971). The origin of an irrigation policy in Malaya: a review of development prior to establishment of the Drainage and Irrigation Department. *Journal of the Malaysian Branch of the Royal Asiatic Society*, 44(1), 78–103.

Tukimat, N. N. A., Harun, S. and Shahid, S. (2017). Modeling irrigation water demand in a tropical paddy cultivated area in the context of climate change. *Journal of Water Resources Planning and Management*, 143(7), 05017003.

Zulkifli, M. A. (1980). *Biological productivity in Muda-Pedu Reservoir and Canal System*. Penang: Universiti Sains Malaysia.

Chapter 26

Subak irrigation system

A heritage of a sustainable hydro-environment

Pujianiki N. Nyoman
Udayana University

Radianta Triatmadja, Djoko Legono, and Fatchan Nurrochmad
Universitas Gadjah Mada

CONTENTS

26.1 INTRODUCTION

Since more than a 1,000 years ago, Subak has contributed to the community welfare of the island. Subak is a name of a social–religious communal organization of farmers. Subak is also a name of irrigation system in Bali, which is responsible for distributing water for irrigation. It can be said that Subak has been a well-preserved system and organization for more than nine centuries (Sutawan, 1987). The exact time of the beginning of existence, however, is unknown. Archaeological remains like worked stones, some displayed at the Bali Museum, testify that farming in Bali has been developed since the late Neolithic era (AD 300–400). Evans (2012) wrote that the technique has been well-preserved and almost unchanged for 11 centuries in Jatiluwih village, central Bali.

The hydraulic structures as the backbone of the irrigation network have undergone so many destructions due, for example, to floods. However, people rebuilt or restored the structures to keep the irrigation system working. Sutawan (1987) mentioned that before the government-constructed permanent weirs, each active member of Subak had to repair the weir 30–70 times a year due to floods. The combination of

DOI: 10.1201/9781003222736-31

society-culture-religion in the Balinese community has played a central role in driving the community to honor the regulation of Subak as part of their responsibility to other human being, the environment, and God. The three components of responsibility are called *Tri Hita Karana*, meaning three reasons for prosperity (Fox, 2011; Luchman et al., 2015). *Tri Hita Karana*, the operational foundation of the Subak organization, has also encouraged people to sustain the irrigation network.

Today, when Subak has been declared a World Heritage, more tourists (both national and international) are attracted to witness the beautiful natural and man-crafted scenes of paddy fields managed within the Subak system. Tourism may not give much direct benefit for the Subak itself, but it indirectly benefits Bali and supports Bali as a primary tourist destination as an attraction. Subak as a tourist destination has attracted many professionals to start businesses nearby the Subak by building resorts, hotels, and restaurants. The beautiful paddy terrace is the main attraction, so such buildings and restaurants should be as close as possible to the Subak and even within the Subak itself. Such development may change the cultural landscape of Bali and may affect Bali as the foremost Indonesian tourist destination. For this reason, hotel and tourism facilities have to leave at least 70% of the land as a natural paddy field in Jatiluwih village (Jatiluwih Village, 2021) to support the sustainability of the Subak system. Hence, Luchman et al. (2015) suggested that ethnoecologists should take the role of tourist planning.

The future of the Subak system is also challenged by other opportunities to work in a "cleaner environment" for the young generations, as Lorenzen and Lorenzen (2005) mentioned. They also said that the relatively little appreciation compared to the benefit of the Subak may also encourage the young generations to stop working in the Subak system.

26.2 METHOD

The study reviewed available references regarding the Subak, especially those that discussed the hydraulics and environment aspects. A survey was also conducted to enrich and update the current condition data. The sources of the information are relatively old; hence a visit to government offices and the Subak museum were also conducted. A field survey was also conducted to enrich the data. The field survey focused on hydraulic structures. Some of the Subak that we visited were Subak Jatiluwih, Subak Tengitis, Subak Buahan, Subak Selat, and Subak Susut. The leader of the Subak organization, or *Pekaseh*, provided additional information. The information includes the name of the Subak, the service provided by hydraulic structures, and the history of the hydraulic structures, since no written information was available. The data was analyzed using simple statistics and hydraulics to explore the current situation and challenges of Subak as a Hydro-Environment heritage.

26.3 RESULTS AND DISCUSSION

26.3.1 Irrigation tunnel

At the upper part of the river reaches, the landscape of the topography is mountainous, and tunnels are required to distribute water within the Subak irrigation system. Among other structures, tunnels are artificial structures that remain and operate. One example is the tunnel that provides water for four Subaks: Subak Tengitis, Subak Buahan, Subak Selat, and Subak Susut. This tunnel supplies water for approximately 370 ha of paddy fields, and signifies that Subak relies on the available topography and civil engineering.

The water flows through a relatively long upstream tunnel from the intake at River Ayung. Upon exiting the upstream tunnel, the water is divided into the four Subaks. About 30 m downstream of the division structure, the channel irrigating three Subaks (Subak Buahan, Subak Selat, and Subak Susut) has to go through another tunnel. The location of the division structure is approximately 650 m (based on Google Earth) above the sea level and is depicted in Figure 26.1.

An example of the traditional irrigation network of Subak Tedung is presented in Figure 26.2. This irrigation network is laid in Gianyar Districts and was built in 1930 with an area of 48.89 Ha (Dinas Pekerjaan Umum Propinsi Bali, 1997). The total area of the traditional irrigation network of Subaks in Bali Province is approximately 20,000 ha, consisting of several Subaks located in five districts, namely Badung, Bangli, Buleleng, Gianyar, and Tabanan districts. Figure 26.2a shows the position of River Pakerisan on the eastern side of River Ayung and the location of Subak Tedung. The Subak Tedung takes the water from the tributary of River Pakerisan through the various hydraulic structures, including the *empalan* (weir), *awungan* (tunnel), and *abangan* (aqueduct), then drained back to River Pakerisan (see Figure 26.2b).

The typical tunnel width is 100 cm, while the height is 180 cm. This tunnel is probably the minimum size for the tunnel engineers to walk through and work inside it. The tunnel is reported to supply the whole area of the four Subaks when water from River Ayung is sufficient. During the survey, the water depth was approximately 60 cm (Figures 26.3 and 26.4).

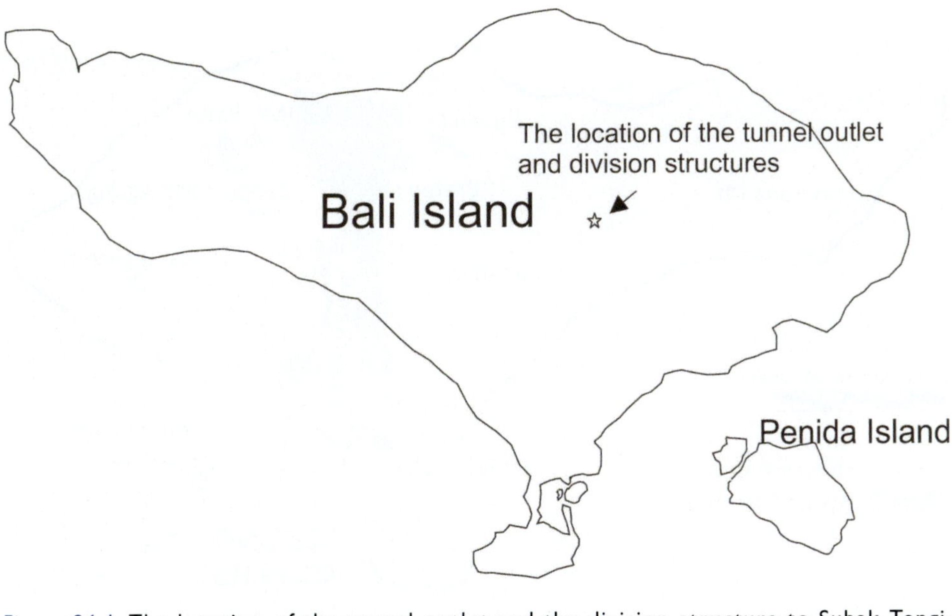

Figure 26.1 The location of the tunnel outlet and the division structure to Subak Tengitis, Subak Buahan, Subak Selat, and Subak Susut. (Map source: Google Earth, attribution in the figure.)

There was no official information on the exact century when the tunnel was constructed. However, the *Pekaseh* mentioned that the tunnel was built at least seven generations ago. Seven generally represents a large number, even though it can be many more than seven. If an average generation is considered as 30 years, this means that the tunnel could have been built much earlier than 200 years ago. Another way to approximate the tunnel's age is that the tunnel is the primary water source for Subak Tengitis, Subak Buahan, Subak Selat, and Subak Susut. The tunnel's age should be approximately the same as those four Subaks. There is no reliable information about the four Subaks; however, Lansing and Fox (2011) mentioned that Subaks in Gianjar near Gunung Kawi Temple are among the oldest. It may be inferred that Subak Tengitis, Subak Buahan, Subak Selat, and Subak Susut near Gunung Kawi in Gianjar, are among the oldest, much earlier than two centuries ago.

The word tunnel engineer is mentioned in an inscription from AD 896 as *Undagi.* The reference to tunnelers is found in a list of professional artisans who worked for pay (Ministry of Culture and Tourism of the Republic of Indonesia, The Government of Bali Province, 2011). The inscription indicated that many Balinese people already dedicated themselves as tunnelers more than a 1,000 years ago. The tunneling engineering for water distribution could have been well understood and practiced. This information supports that the Subak tunnel (not necessarily the above discussed) could have been built even more than a 1,000 years ago.

The tunnel is one of the essential hydraulic irrigation structures sustained for centuries, although they may have undergone many reparations. Unlike other irrigation structures, a tunnel needs a longer time, and a more complicated design and construction process, especially for relatively long and deeply underground ones.

(a)

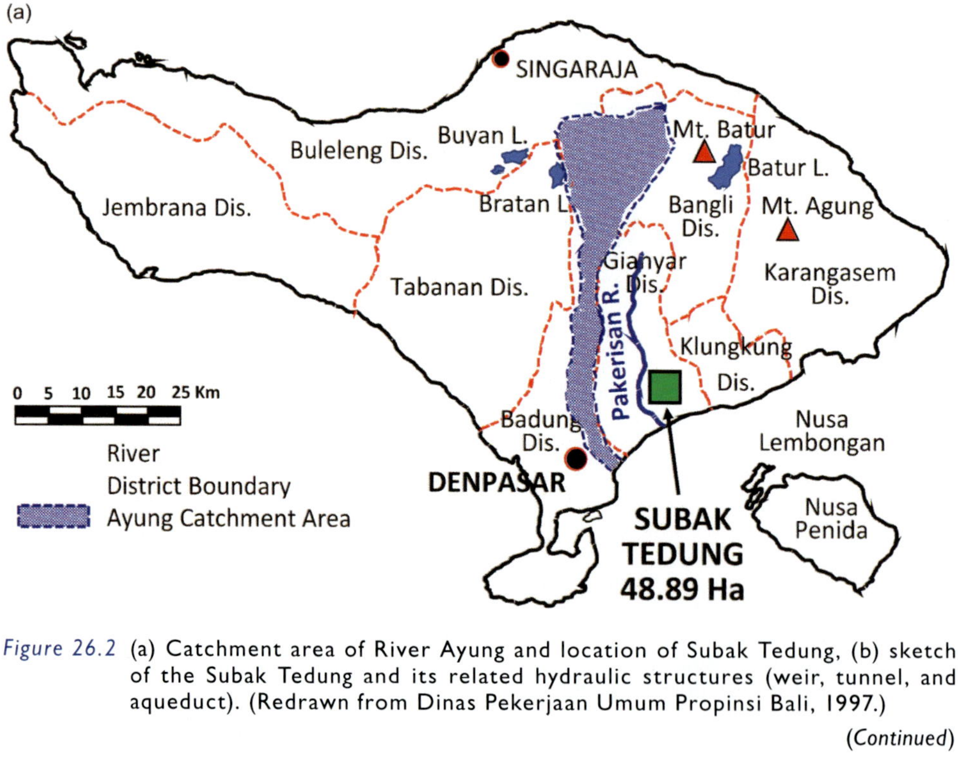

Figure 26.2 (a) Catchment area of River Ayung and location of Subak Tedung, (b) sketch of the Subak Tedung and its related hydraulic structures (weir, tunnel, and aqueduct). (Redrawn from Dinas Pekerjaan Umum Propinsi Bali, 1997.)

(Continued)

(b)

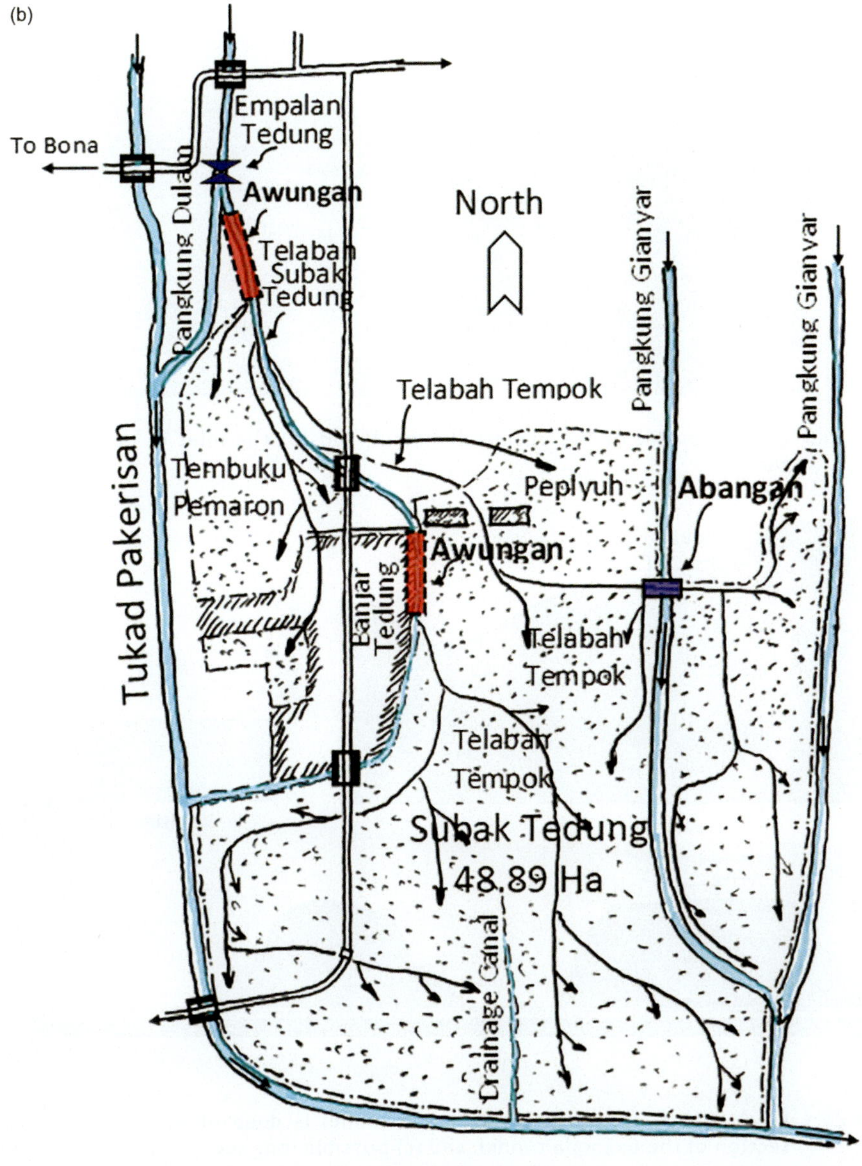

Figure 26.2 (Continued) (a) Catchment area of River Ayung and location of Subak Tedung, (b) sketch of the Subak Tedung and its related hydraulic structures (weir, tunnel, and aqueduct). (Redrawn from Dinas Pekerjaan Umum Propinsi Bali, 1997.)

The complexity includes finding the proper water intake and aligning the tunnel in both vertical and horizontal axis, which was not a simple problem back then and even in some cases. The complexity is not merely related to tunnel construction but also to hydraulics engineering. Not all of the Subaks system is listed as World Heritage, but the ancient tunnels (which may not belong to those Subaks listed as World Heritage) are ancient hydraulic structures worth conserving. In December 2020, Bali Post reported

Figure 26.3 (a) The outlet of the upstream tunnel, and (b) the inlet of the downstream tunnel.

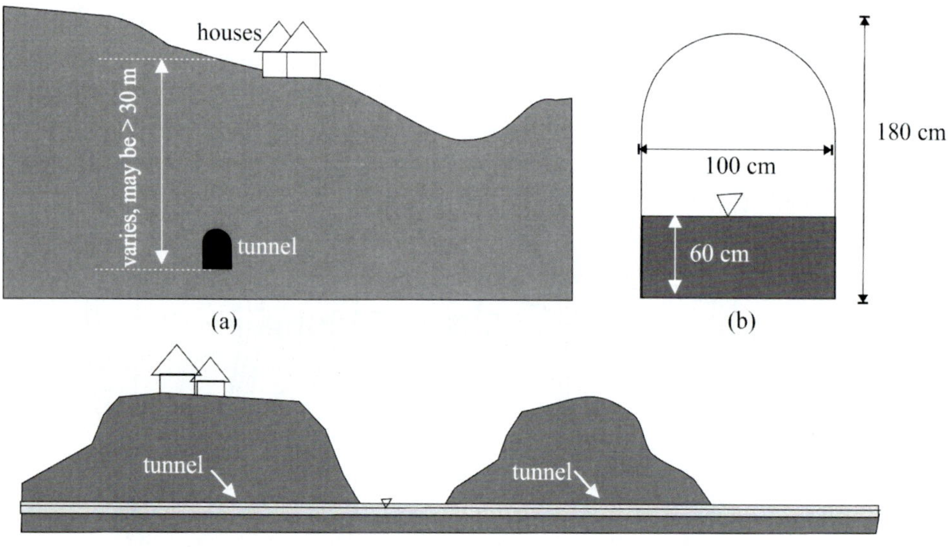

Figure 26.4 (a) A typical cross-section where a tunnel is deep underground, (b) cross-section of the example tunnel, and (c) possible long section of tunnels in series.

that an old tunnel of 480 m long was incidentally found during the construction of the Tamblang Dam. The tunnel was 40 m below the ground, crossing the centerline of the dam foundation. It was said that the tunnel was built during the Dutch colonization era (which ended in 1942) and was probably used for the Subak system. Not all tunnel lines were removed, but some parts of the tunnel were preserved for tourism. The founding also indicates that other tunnels may still exist, which can add to the richness of the heritage.

Figure 26.5 A traditional flow division structure made of wood.

26.3.2 The traditional division structures

The irrigation network of Subak relies on gravitation where no pump is used. The land contour enables the water to be distributed through channels quite efficiently. Such landscape consists of many paddy fields terraces, which present a beautiful view that attracts tourism. Theoretically, the terraces reduce surface water flow velocity by destroying potential energy at the steps between the terraces, reducing soil erosion. However, the hillside channels are frequently used as conveyance and drainage systems due to topographical conditions and lack of space. Additionally, the reuse of the excess irrigation water from the upstream portion of a Subak by the downstream part of the same Subak is common (Dinas Pekerjaan Umum Propinsi Bali, 1997).

Although the network is ancient, the design is similar to the present engineered irrigation systems. It may consist of weirs, intake, primary, secondary, tertiary, quarterly channels, drainage, divider boxes, siphons, tunnels, and overflow structures (Norken et al., 2016). There is a need to divide the water discharge accurately for two reasons. First, the land should get the right amount of water for irrigation, and, second, the farmers must compensate for the water they get. Hence, a good understanding between farmers is required to distribute the water fairly and proportionally using flow division structures. The traditional flow division structure is made of wood. Such a structure can still be found within the irrigation networks (Figure 26.5). As shown in Figure 26.5, the flow is divided into two directions. The minor discharge goes to the PVC pipe, while the rest flows downstream of the channel.

The traditional division structure is usually made of wood. Although the farmers do not know the hydraulics formulae behind the division structure, they understand the flow principles. They know that the width and the water depth at the weir play essential roles in the discharge through the weir. They even understand that the flow upstream of the weir should more or less be perpendicular to the weirs that divide the flow. They underline the weakness of the traditional division structures: the sensitiveness of the installation related to the water depth at the weir. If the height of one part of the weir is not the same as the other part, the flow may be significantly different, although the width is the same. Theoretically, the water depth above the weir is more dominant, as the flow discharge is a function of $h^{3/2}$, where h is the water depth slightly upstream above the weir. Although the farmers do not understand the formula precisely, they can sense that the depth affects the flow significantly. Therefore, the depth above the weir at a division structure should be the same for each Subak as standard.

Another type of traditional broad-crested weir that provides a certain amount of water to a Subak is given in Figure 26.6 after Norken et al. (2015). Figure 26.6 shows a measure of one *tektek,* which provides enough water for approximately 0.3–0.5 ha (Norken et al., 2015). There are many forms of equations to compute the flow over the weir. The equation can be written as Eq. (26.1) (see, for example, Tracy, 1957)

$$Q = C_d b H^{1.5} \quad (26.1)$$

where Q is the discharge across the weir, C_d is the discharge coefficient, b is the width of the weir, and H is the total head (including the velocity head) above the weir (Figure 26.6). The coefficient C_d is a function of h/L, where h is the head above the weir crest away from the weir, and L is the length of the weir. The difference between H and h usually is small upstream of the weir. Based on the available data, Tracy (1957) indicated that C_d ranges from 2.65 to 3.5 $\mathrm{ft}^{1/2}$/s or 1.46 to 1.93 $\mathrm{m}^{1/2}$/s. Taking the minimum and the maximum value of C_d and applying them to Eq. (26.1) yields flow discharges of 0.33 and 0.44 l/s, respectively.

Such a quantity of continuous water flow is enough to irrigate 0.3 to 0.5 ha of paddy field, similar to that of Norken et al. (2015).

Rao and Muralidhar (1963) proposed Eq. (26.2) to avoid iteration for practical reasons.

$$Q = C_d b h^{1.5} \quad (26.2)$$

Based on Eq. (26.2), they found that C_d ranges from 2.65 to 3.65 $\mathrm{ft}^{1/2}$/s, or 1.46 to 2.02 $\mathrm{m}^{1/2}$/s, which confirms that the effect of velocity head is small.

When the water distribution structure is a sharp-crested weir, Eq. (26.1) still applies. Rao and Muralidhar classified the crest as long crested for $h/l \leq 0.2$, broad crested

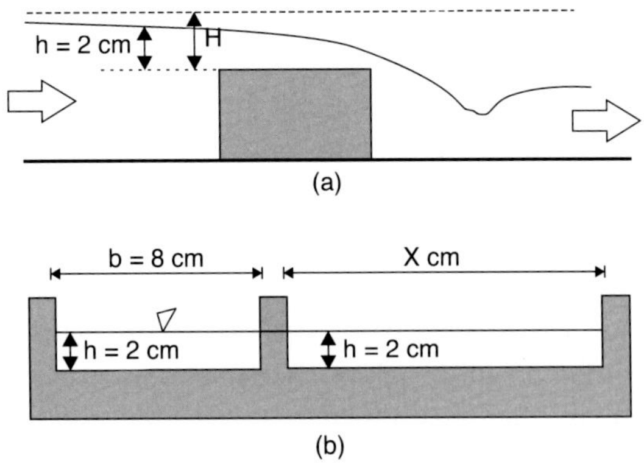

Figure 26.6 Traditional control devices providing one *tektek* (Norken et al., 2015), (a) long section, and (b) cross-section.

for 0.1 ≤ h/l ≤ 0.4, narrow crested for 0.4 ≤ h/l ≤ 1.5, and sharp crested for 1.5 ≤ h/l ≤ 1.9. The only difference between the weir formulas is the coefficient C_d, which changes slightly. Hence, the flow discharge through different weir crest widths slightly changes, but the distribution's proportion remains acceptable. The simple rules in the Subak system seem accurate, and fair water distribution can be attained.

26.3.3 The challenge of heritage sustainability

Based on the statistic (Badan Pusat Statistik Bali, 2014; Badan Pusat Statistik, 2016), the ratio of rice production to the island's population is 0.21 tons/person/year. This ratio is comparable to 0.28 tons/person/year, the ratio of the total rice production of Indonesia to the total number of Indonesian populations. The average rice consumption of Indonesian is 0.114 tons/person/year (Waspada Online, 2016). Therefore, it seems that Bali has been able to supply its rice demand since the old days, where the irrigation network has also been the backbone to sustain the environment.

Based on Table 26.1, from 2009 to 2013, the average conversion of the Subak (an irrigated paddy field in Bali) into other business lands such as offices, hotels, restaurants, and housing is much less than other rapidly growing cities in Indonesia. Table 26.1 shows that the Special Capital Region of Jakarta's agricultural land is vastly converted into other types of land use. As the capital city, Jakarta has probably attracted many investors to provide more housing and offices that lead Jakarta as the fasted converted land. The trend increased within the last four years (2013–2017) when irrigated land conversion reached more than 11%. Although Bali developed rapidly to be the primary tourist destination of Indonesia, it managed to slow down the conversion until 2013. As indicated in Table 26.1, the average yearly conversion of the Subak from 2009 to 2013 is 0.17%, significantly less than the average of Java Island (0.46%).

Table 26.1 Conversion of agricultural land in Java and Bali in Indonesia

Province	*Area of irrigated agricultural land (ha)*			*Reduction of irrigated agricultural land (ha)*		*Average reduction of irrigated agricultural land (% year^{-1})*	
	Year			*Year*		*Year*	
	2009	2013	2017	2009–2013	2013–2017	2009–2013	2013–2017
Jakarta Special Capital Region	1,184	870	477	314	393	6.63	11.29
Banten	111,084	104,385	102,862	6,699	1,523	1.51	0.36
West Java	759,499	744,090	723,636	15,409	20,454	0.51	0.69
Central Java	689,383	683,735	675,694	5,648	8,041	0.20	0.29
East Java	879,958	862,590	834,902	17,368	27,688	0.49	0.80
Yogya Special Region	46,547	46,165	42,081	382	4,084	0.21	2.21
Java Island (total)	2,487,655	2,441,835	2,379,652	45,820	62,183	0.46	0.64
Bali	78,683	78,163	73,739	520	2,424	0.17	0.80

Source: Center for Agriculture Data and Information System, Secretariat General—Ministry of Agriculture, 2014 and 2020.

Bali's irrigation land conversion is even less than the average Sumatra agricultural land conversion from 2009 to 2013, which amounts to 3.5%. The lower percentage of paddy field conversion indicates that the *Tri Hita Karana* (the operational foundation of Subak) plays a vital role in sustaining the land for irrigation and its environment. However, within the last four years (2013–2017), there was a leap in irrigated land conversion in Bali up to 0.80% per year, which is slightly higher than the average of Java Island.

According to Badan Pusat Statistik (2016), Bali's average reduction of paddy fields is 900 ha/year. However, between 2018 and 2020, there was a significant reduction in harvested area, production, and productivity of paddy in Bali (Badan Pusat Statistik, 2021). Within the last two years, the harvested area reduced by 18%, from 110,978 to 90,982 ha or nearly 10,000 ha/year. The paddy production reduced by 20% from 667,069 tons to 532,168, and the productivity reduced by 2.6% from 6.011 to 5.849 ton/ha. The reduction signifies the threat of recent development in Bali.

Such development indicates that the problem of sustaining the heritage is now threatened more by financial opportunities of other businesses that lead to reducing the number of farmers and agricultural land, as suggested by Lorenzen and Lorenzen (2005). In addition to involving ethnoecologists' role in tourist planning, as suggested by Luchman et al. (2015), other benefits for the farmer within *Tri Hita Karana* should be introduced to balance the opportunities in the paddy field workforce. Norken et al. (2016) have identified eight challenges regarding the existence of Subak. The most severe challenge is probably the yearly decrease of paddy fields due to land conversions into tourism-related housing and commercial buildings, especially near the town. Such conversions may be encouraged by opportunities for a more financially appreciated working field for the young generation, as Lorenzen and Lorenzen (2005) also mentioned. When less or no more young generation is willing to work in the field, they tend to turn the paddy fields into other forms of businesses when possible. Such conversion is relevant to concerns about how the *Tri Hita Karana* survives under the dynamic changes in Bali's human civilization (Yekti et al., 2017).

An additional challenge to the sustainability of the heritage is Bali's population growth, which is relatively high due to a significantly high number of migrations from other islands in Indonesia to Bali. Between 2000 and 2010, Bali's population grew by 2.15%, adding approximately 0.74 million residences to the island (Erviani, 2011) and consequently encouraging the conversion of agricultural land. Other than that, Pujianiki (2016) pointed out that water availability is decreasing due to land-use conversion and climate change. Water availability may eventually affect the sustainability of the paddy field.

Within the last five years, farmers competed to sell their Subak's land due to the increased taxes. Such conversions could have been caused, because the tourism business is more promising than managing paddy fields, and the new generation of Balinese may not be interested in becoming farmers (Lorenzen and Lorenzen, 2005). Figure 26.7 shows some increasing scenes where the business industry, hotels, and restaurants are built in the middle of the paddy fields.

Norken et al. (2016) also mentioned poor irrigation networks' physical conditions. The average paddy field owned by farmers is relatively small, making it less profitable. Pollution along the rivers may also threaten the sustainability of Subak.

Sutawan (1987) indicated that the government involvement in the Subak management and infrastructures might be classified as either a less successful project, a somewhat successful project, or the successful one. The less successful project has caused

Figure 26.7 (a) Converted paddy fields into business and workshop areas, (b) converted paddy fields into buildings and restaurants.

some damage to the original network performance, while the somewhat successful projects have created new problems while solving the other. The successful projects have improved the Subak performances and helped the farmers.

These lessons learned so far are invaluable for both the community and the government of Indonesia to sustain the heritage. The government should control the conversion of paddy fields into other more benefiting assets, especially as Subak has been listed as one of the World Heritage Sites.

26.4 CONCLUDING REMARKS

Subak system is recognized as a world cultural and natural heritage by UNESCO, and a hydraulics environment heritage. Its hydraulics tunnels and networks system prove that sophisticated hydraulic environment engineering was practiced more than a 1,000 years ago in Bali, Indonesia. Such heritage should be well maintained and conserved. The threat to the sustainability of the heritage is increasing and becoming more complex. The government must take more significant measures to help sustain the heritage.

ACKNOWLEDGMENTS

The authors are in debt to the Ministry of Public Works and Housing of Indonesia for providing the information related to the study. Our sincere thanks also go to Subak Jatiluwih, Subak Buahan, and Subak Payangan for their help during the survey.

REFERENCES

Badan Pusat Statistik. (2016). *Proyeksi Penduduk menurut Provinsi*, 2010–2035, http://www.bps.go.id/linkTabelStatis/view/id/1274, Accessed January 2016

Badan Pusat Statistik Provinsi Bali. (2014). *Bali dalam Angka 2014,* http://bali.bps.go.id/tabel_detail.php?ed = 607002&od = 7&id = 7

Badan Pusat Statistics Provinsi Bali. (2021). *Luas Panen, Produksi, dan Produktivitas Padi Menurut Kabupaten/Kota di Provinsi Bali*, 2018–2020, https://bali.bps.go.id/statictable/2019/10/17/169/

luas-panen-produksi-dan-produktivitas-padi-menurut-kabupaten-kota-di-provinsi-bali-2018-2020.html
Center for Agriculture Data and Information System. (2014). *Statistic of Agricultural Land 2009–2013*, Secretariat General - Ministry of Agriculture. Available at http://www.pertanian.go.id/file/Statistik_Lahan_2014.pdf
Center for Agriculture Data and Information System. (2020). *Statistic of Agricultural Land 2015–2019*, Secretariat General- Ministry of Agriculture. Available at http://epublikasi.setjen.pertanian.go.id/download/file/559-buku-statistik-data-lahan-tahun-2015-2019
Dinas Pekerjaan Umum Propinsi Bali. (1997). Subak *in Bali*.
Erviani, N. K. (2011). Bali records highest population growth in 50 years, *The Jakarta Post*, Thu, December 15.
Evans, K. (2012). World heritage listing for Bali's 'Subak' tradition, Newsline, Accessed January 19, 2016, http://www.abc.net.au/news/2012-06-27/an-balis-Subak-world-heritage-listed/4096016
Fox, R. (2011). *Critical reflections on religion and media in contemporary Bali*, pp 118, Brill.
Jatiluwih Village Website. (2021). General Information, https://jatiluwih.id/about/ retrieved 2021.
Lansing, J. S., and Fox, K. M. (2011). Niche construction on Bali: the gods of the countryside, *Philosophical Transactions of the Royal Society B* 366, 927–934.
Luchman, H., Kim, J. E., and Hong, S. K. (2015). Cultural landscape and ecotourism in Bali Island, Indonesia, *Journal of Ecology and Environment* 32(1), 1–8, available from http://koreascience.or.kr/article/JAKO200911440577766.page
Lorenzen, R. P., and Lorenzen, S. (2005). Changing realities - perspectives on Balinese rice cultivation, *Human Ecology*, 39(1), 29–42.
Ministry of Culture and Tourism of the Republic of Indonesia, The Government of Bali Province. (2011). Cultural landscape of Bali province, *Nomination for Inscription on The UNESCO World Heritage List*, 2011.
Norken, I. N., Suputra, I. K., and Arsana, I. G. N. K. (2015). Water resources management of Subak Irrigation System in Bali, *Applied Mechanics and Materials* 776, 139–144.
Norken, I. N., Suputra, I. K., and Arsana, I. G. N. K. (2016). *Tantangan Dalam Melestarikan Sistem Subak Sebagai Warisan Budaya Dunia Di Bali*, National Seminar INACID Werdhapura, Sanur – Bali.
Pujianiki, N. P. (2016). *Upaya Mitigasi Kekeringan Daerah Irigasi Jatiluwih*, National Seminar INACID Werdhapura, Sanur – Bali
Rao, N.S.G., and Muralidhar, D. (1963). Discharge characteristics of weirs of finite-crest width, *La Houille Blanche* 49(5), 537–545, https://doi.org/10.1051/lhb/1963036
Sutawan, N. (1987). Farmer manages irrigation systems and the impact of government assistance: a note from Bali, Indonesia, International Irrigation Management Institute (IIMI) and Water and Energy Commission Secretariat (WECS) of the Ministry of Water Resources, Government of Nepal, *Public Intervention in Farmer-Managed Irrigation Systems*. Digana Village via Kandy, Sri Lanka: IIMI Pub. iv, 323p.
Tracy, H. J. (1957). *Discharges characteristics on broad-crested weirs*, Geological Surveys Circular 397, Washington, DC, USA.
Waspada Online. (2016). *Konsumsi beras orang Indonesia, 114 kg per tahun per kapita*, http://waspada.co.id/warta/indonesia-hari-ini/konsumsi-beras-orang-indonesia-114-kg-per-tahun-per-kapita/, Accessed January 2016.
Yekti, M., Schultz, D, Norken, I. N., Gany, A. H. A., and Hayde, L. (2017). Learning from experiences of ancient Subak schemes for participatory irrigation system management in Bali, *Irrigation and Drainage* 66(4), 567–576.

Chapter 27 (Special contribution)

Water culture of the people in Uzbekistan

Ancient traditions, structures, and modern global problems

Yulduz A. Ergasheva and Alisher N. Khazratov
Karshi Engineering Economics Institute, Karshi, Uzbekistan

CONTENTS

27.1 INTRODUCTION

Central Asia, located at the crossroads of the ancient Great Silk Road, has always been one of the world's most important regions until global trade started flowing through ocean routes (Frankopan, 2016). It is no coincidence that, from the time of the primitive community system of history, the Great Silk Road cities like Bukhara, Samarkand, and Tashkent are located near rivers, lakes, streams, and springs. The living conditions of humans have been directly related to water and water management. That is why issues related to the water problem have been in the spotlight of the ruling circles throughout history.

27.2 WATER CULTURE IN ANCIENT UZBEKISTAN

In Central Asia, especially in Uzbekistan, every drop of water has long been considered gold, precious, and sacred. Therefore, when we think about the value and importance of water and our people's traditions and modern views on the culture of use, our ancestors deeply appreciated it through various wise sayings and proverbs that respect the water.

DOI: 10.1201/9781003222736-32

They emphasized the importance of saving water, not polluting it, and keeping it clean and pure: "Water is dearer than gold," "Water is gold, irrigator is jeweler," "Water has come—light has come," "Watered land is a flower garden, waterless land is a grave," A bird does not land on the ground without water, "Water is the blood of a farmer," "A saint is the one who brings water," "A person who digs a well knows the value of water," "Where there is water, there is human," "Pollution of water is a sin," etc. (Ashirov, 2020).

In Central Asia, where social development and agriculture have relied on irrigated agriculture for thousands of years, land and water conservation have been a constant focus of people, and water has been valued as the foundation of life. As a source of life in our region, water has positively impacted the formation of noble, benevolent human qualities, and served as a factor in strengthening mutual understanding, interaction, and good neighborly relations between different peoples, nations, and states. In our arid country, the culture of high-level irrigated agriculture through the rational use of land and water has been formed since the ancient times. Although the formation of new and modern values of rational water use is one of the most pressing issues today, the worship of nature was first formed as primitive religious ideas, and later reflected in Zoroastrianism and other religious teachings.

Respect for water and rain is particularly important in all parts of Central Asia, where irrigation is based on agriculture. The sacred book of Zoroastrianism—the Avesta pays special attention to the reverence and appreciation of water, the source of life. In the Ardvisura (Amu Darya) image, a special dedication is made to it. The waters of the Amu Darya were described as "waters above the waters," the greatest of all waters. In Zoroastrianism, certain rules have been developed for water, water was considered a sacred element, and people were concerned about its purity. Therefore, water sources, rivers, lakes, and wells are protected from impurities. If it is necessary to wash something that is not ritually clean, the water is collected in a special place. As stated in the Avesta, "The punishment of the one who testifies before the pure water and the burning fire is the most terrible punishment in the world." In Avesta, Anahita is interpreted as the god of water and fertility (G'ulomov, 1959).

Islam also calls for water conservation, especially not wastage. Waste and pollution of water is a grave sin. According to our people's tradition, a traveler who drank a sip of water from a house became a close person to that house. A person who fasts also breaks his fast with water in the mouth. Even if a person is in deep sorrow, he can alleviate his grief by looking at the flowing water. The man who had a bad dream also went to the water's edge. That is why our people have the phrase "Say your dream to the water." Buckets, jugs, and other utensils are always filled with water and sealed, so the house is always blessed.

Digging new canals and ditches has been considered one of the most rewarding jobs. Traditionally, wasting, polluting, and spitting water is strictly forbidden. Irrigation networks are cleaned yearly in the early spring by "Hashar" (traditional voluntary help for a community).

In Uzbekistan, many ditches and canals that flow in early spring have been dug using the "Hashar" method. When the water was opened at the end of a Hashar, the population collected money, wheat, and food, and bought and sacrificed animals (calves, goats, white horses, rams, goats). In some parts of Central Asia, the blood of a slaughtered animal is poured into the water, and the carcass is thrown into the water.

When a new ditch was dug in Khorezm and connected to the Amu Darya, 8–10 cows and bulls were sacrificed according to the Khan's order. Slaughtered animals were removed from the water and cooked. A sacrificial ceremony was also held in the Fergana Valley. When it rained, a sacrifice was made in the cemetery. Frogs were hung at the edge of the fields to call rain (Ashirov, 2007).

From ancient times, the issue of water and its supply has always been to the attention of scientists. One of the great thinkers, Ahmad al-Farghani, originally from the Fergana region of Uzbekistan, built a water meter—"Miqyos an-Nil" on the island of Rada for the Nile River in Egypt, which is proof of our opinion. This indicates that a method of quantitative control was developed to make the right decisions about water in the 9th century. The practical tasks related to creating and managing water resources were skillfully solved. This structure has not lost its significance for centuries. A statue of the scientist was unveiled in 2007 on the island of Rada.

In his book *Monuments left from ancient people*, the great Uzbek scholar Abu Rayhan Beruni (XI century) describes the customs and festivals dedicated to water. In this book, the scientist informs that one of the traditions of Navruz is that people sprinkle water on each other so that this year would be wet and productive: "That is why people used to sprinkle water on Navruz" (Beruni, 1968).

Munis and Agahi—the great representatives of classical literature also served the people wholeheartedly in their duties as "Chief mirabs" (Chief hydraulic engineers) in the Khiva khanate. Because, from time immemorial, being "Chief mirab" has always been one of the most responsible tasks. Well-known, reputable, and honest people were elected to this post. The famous scholar of the Muslim world Burhaniddin Marginoni (XII century), in his famous work about Islamic jurisdiction – "Hidoya," detailed the rights to use water. This is proof of the ancient development of water science in our country.

27.3 ANCIENT HYDRAULIC STRUCTURES

It is known that irrigation networks and dams were built in Khorezm in the VI–VII centuries BC. In the VIII century, irrigation canals were built in the Zarafshan valley. On the present "First May" water intake site, a dam was built according to the works of Arab historian Ibn-Khaukal. The Fergana canal supplied water to Samarkand, and the Shohrud canal supplied water to Bukhara. Greek historians have noted the existence of about 50 large canals in the Shosh land (around Tashkent). Large irrigation canals were built in the Fergana Valley in the XVI–XVII centuries. At the beginning of the 19th century, there were more than 60,000 "Chig'irs" (mechanical water lifters) in the lower reaches of the Amu Darya alone that carried water to a higher altitude.

The presence of Zakh, Iskandar, Bozsuv, Salar, Khan, Polvon, Shavat, Gazavot, Dargam, Norpay, and Shumanay canals, Khan and Abdullakhan dams, which have survived to our time, as well as 1.6–1.8 million ha of irrigated arable land in Uzbekistan at the beginning of the last century, is a vivid example of the large-scale construction of hydraulic engineering in our history (Bakiyev et al., 2008a).

The Persian word "Koriz"—an underground construction has transported water through underground tunnels. Koriz was widely spread in mountain and foothill regions of Central Asia. Until the beginning of the 20th century, Koriz water was used to irrigate crops in Central Asia. The art of excavating Koriz was required to know the changes in the underground water levels and determine the point of the terrain ramp. It was built with particular tools. After signing routes of the Koriz along those highways, wells were dug in every 5–40 m. The wells' depths ranged from 14–15 m, sometimes to 60–70 m. Water wells were united into a one-layer circuit through the tunnels. The tunnel's height was 1.5 m, the width was 1 m, its walls were made from wood and stone, and the length extended to a few meters depending on the terrain ramp. For instance, Koriz, called Maston, had 280 wells with a depth of 15 m and a length of 3 km in the Nurata district, Navoi region (Khasanov, 2016).

The results of archeological excavations in our country show that 3,000–3,200 years ago, in the Bronze Age, irrigation works were developed, and waterworks were built. In excavating the Dargam canal that took water from the Zarafshan River and was built only by manual labor two-and-a-half thousand years ago, hydraulic engineers took into account the land's topography with great precision (Vatandosh, 2021).

From the beginning of the epoch, water-lifting devices—"Chig'irs" were built to irrigate the lands above the river water level. Our ancestors built tunnels under the high banks of the Amudarya and Zarafshan to get water for irrigation. In the Fergana Valley, the river water level was controlled by simple but robust "Sepoya," "Chorpoya," and "Labigardon" structures that control the direction of the water against the rapid flow of the valley (streams) and a device that directs it to the general flow of the river. In addition, the discovery of the wheelchair dates back to the time of the Kushan kingdom. As a result of discovering the lifting mechanism of water from the hills, many lands were developed, and settlement in Central Asia accelerated (Mukhammadzhanov, 1975).

In the Nurata region, farmers and irrigators have discovered some of the most remarkable irrigation techniques of their time to find and use additional water resources. They built small reservoirs—"Locked Pools" to collect and use a small amount of spring water. These reservoirs were locked (The water supply pipes were closed with plugs.), and opened from it depending on the necessity and used for irrigation.

Our people used sardobas, pools, and wells in water storage, and, in everyday life, they used containers made from ceramics such as hum, jug, and tumcha. The water stored in that container evaporated from the outer side of the container and was kept

Figure 27.1 Karshi sardoba, built in the XVI century, located in Kashkadarya region, Uzbekistan.

cold even in the summer heat without being wasted. Pools have been built in the Mahalla–Guzars (local neighborhoods) and used wisely by the population for water consumption. Drinking water is often stored in ceramic containers that are hygienically convenient and do not harm human health. In the Karmana, Karshi, and other deserts of Central Asia in the late Middle Ages, more than 40 sardobas and the remains of ancient koriz systems and wells are still found (Figure 27.1) (Mavlonov, 2008).

In the Middle Ages, hydraulic structures were important for water collection and storage, and were divided into naturally nourished sardobas (from snow, rain, and spring water) and artificially nourished sardobas (from wells, canals, etc., rivers, and streams). Various sardobas have been built around the Great Silk Road to meet the water needs of travelers and tourists. Raboti Malik, Qaravulbozor, Eshan Imlo, Khalifai Khudoydod, Jizzakh-orda, Yogochli, Murza Rabot, Jikka, Kaltepa, Jizzakh, Raboti hadith, Shohruhiya, Kaynarbulak, Muborak, Surkhi, Fatta, Mirhaydar Sultan, Maymanak, and others are among those sardobas (Figure 27.2).

Medieval sardobas are complex irrigation-engineering structures built to suit each place's geographical climate. The water in the pool of this structure has been kept clean for up to a year. During the Mongol invasion, the waterways in Central Asia were destroyed. For example, a large city in Khorezm, such as Old Urgench, was flooded, and the rare "Jui Arzis" lead canal in Samarkand was demolished. It should be noted that the Afrosiyab fortress is supplied with water through a special pipe, the bottom of the ditch is mounted on arched pillars, and the sides are reinforced with lead. There are speculations that Jui Arzis flowed at the height of 10m. From the same unique ancient hydraulic structures, the pyramid-type aqueducts of the upper reaches of the Yakkabogdarya around the ancient fortress and the aqueducts over the Bandikhansoy supply water in the Kashkadarya region have been preserved until today (Bakiyev et al., 2008b).

During the reign of Amir Temur and the Timurids, irrigation was given extraordinary importance and was considered the basis of agriculture. Canals were dug in Khorasan and elsewhere. During this period, irrigation canals were rehabilitated, koriz systems, countless gardens, and waterworks were built, and more attention was paid to agriculture. In particular, about the development of agriculture, various hydraulic structures have been built in the country, irrigation systems have been developed, and the Mugan desert has been turned into irrigated land.

Amir Temur, in his book *The Code of Amir Temur*, expressed the state's attitude to irrigated lands as follows: "Let them take into account the lands irrigated continuously by koriz systems, springs, and rivers, and take two-thirds of the harvest from such lands for the people, and one-third for the royal treasury" (Temur, 2011). It is known from historical sources that Amir Temur built many canals in his homeland and Khorasan, Iran, and the Caucasus.

According to written sources about the Oqsaroy palace in Shahrisabz, which was built by Amir Temur and had a total height of about 70m, a pool was built on the palace's roof (Figure 27.3). The spring water from the Takhtikoracha pass flowed into the pool through lead pipes. One can only imagine architects' knowledge in the Middle Ages when there were no pumping stations to bring water to such heights. Two huge pillars of the entrance roof from the Oqsaroy palace have been preserved; its height is 38m (G'afforov, 2011; Meros, 2017).

Several large and small dams were built to construct reservoirs in the rivers of Forish and the neighboring district of the Nurata mountain system. Examples are the

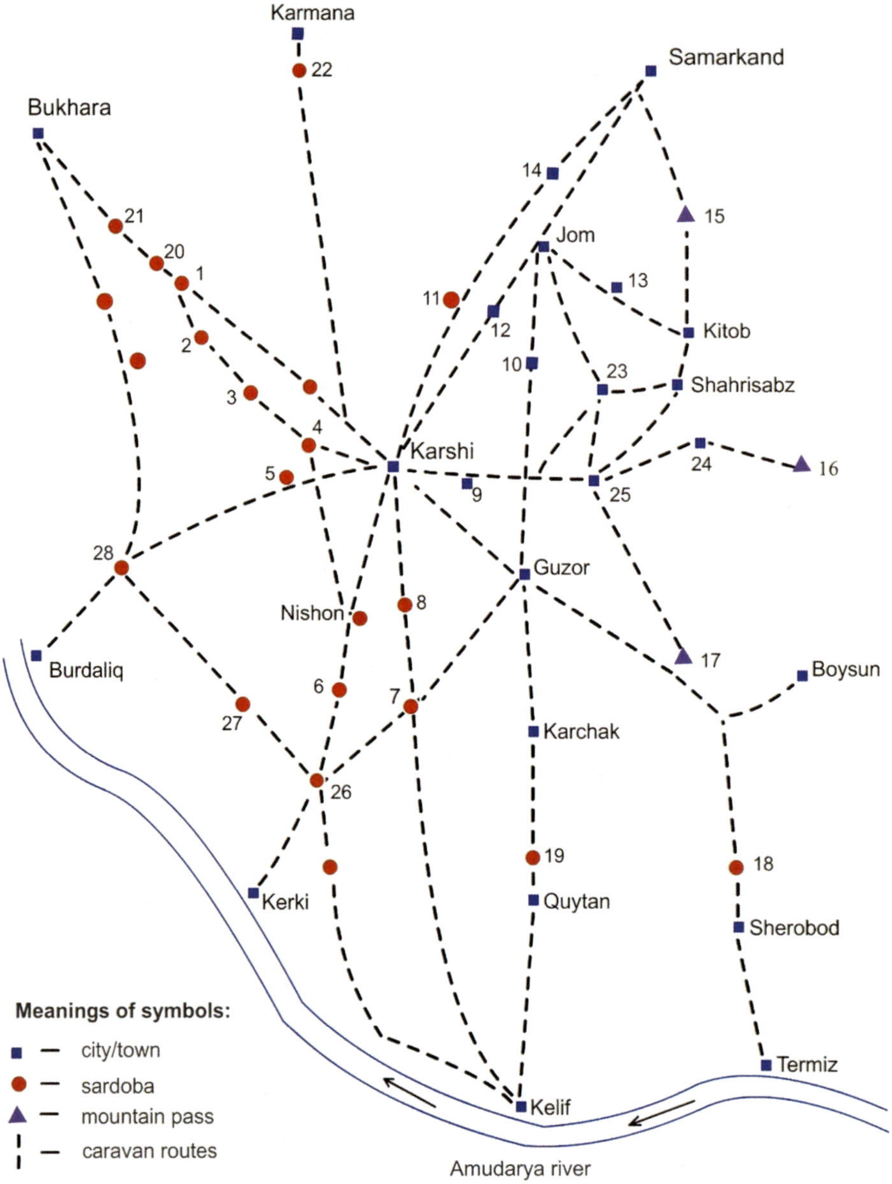

Figure 27.2 Ancient trade routes and sardobas in southern part of Uzbekistan (1-Buzachi sardoba, 2-Begichak sardoba, 3-Yakka sardoba, 4-Buzurg sardoba, 5-Qamashi sardoba, 6-Talimarjon sardoba, 7-Urasi sardoba, 8-Yusuf sardoba, 9-Yangikent village, 10-Chiyalquduq village, 11-Shurbozor sardoba, 12-Shurquduq village, 13-Makrid village, 14-Ajrim village, 15-Takhtikoracha pass, 16-Toshkurgan pass, 17-Oqrabod pass, 18-Sherobod sardoba, and 19-Abdullakhan sardoba). (Adapted from Mavlonov (2008), Ancient Roads of Central Asia, Akademiya: Tashkent.)

Figure 27.3 Remains of Oqsaroy palace pillars located in Kashkadarya region, Uzbekistan.

Khan Dam, built in the 10th century on the Osmon River in Forish district, the brick dam built in the 12th century in Kattakurgan district, and the Abdullah Khan Dam on the Akhchop River east of Nurata.

The Abdullah Khan Dam was completed in the 16th century. The dam is 15m high, 73m long at the top, 4.5m wide, 73m long at the bottom, and 15m wide. When the water depth in front of the dam was 15m, the reservoir length was extended to a distance of 1,250m. The Khan Dam was built on a narrow ridge of the Osmon River. Hard rocks (stones) were joined together in the dam using alabaster. The water surface length in the reservoir is 700m long, 50m wide in front of the dam, and 200m wide at the endpoint of the reservoir. This allowed irrigating two-thirds of 1,000ha of land in the Kaltepa desert, which at that time needed to be developed. To drain the collected water during the flood from the reservoir and irrigate crops from it, on the right side of the dam, close to the shore, a pipe in the form of a porch with a width of 50cm and a height of 7m was installed. The dam has survived until today, but the reservoir has lost its volume over the centuries due to the rock and mud that flowed from the Osmon River (Bakiyev et al., 2008a).

As we analyze the water culture history in Uzbekistan, we can observe that the human–nature interaction changes from nature's dominance to human dominance as humankind evolves. According to historians and the Khan of Khiva khanate, between 1644 and 1664, Abulg'oziy Bahodirxon, the Amudarya River, has flowed into the Caspian Sea since ancient times. In 1573, however, the Amudarya changed its course toward the Aral Sea and began to flow there. Previously, the Aral Sea was called the Syr Sea, and only the Syrdarya River flowed into it (Bahodirxon, 1990). Consequently, Urgench city became desert, and the farmers dependent on water from the Amudarya River started suffering from water shortage.

27.4 RECENT DEVELOPMENTS IN THE 20TH CENTURY

However, the 20th century has seen an abrupt interference with nature by humans. In the 1920s and 1930s, the construction of hydraulic structures, mainly based on

Figure 27.4 Construction of Great Fergana Canal. (In August 1939, the 270 km long Great Fergana Canal was constructed in 45 days in a very short time with the people's Hashar (voluntary help) in 1939 by 160,000 Uzbek and Tajik workers.)

manual power, was carried out on a large scale by mechanization after the 1950–1960s. Because of big Soviet irrigation projects, water from the Amudarya and Syrdarya rivers started diverting to deserts for agricultural development in the 1960s. One of the big irrigation projects before the Second World War was the 270 km long Great Fergana Canal, which was constructed in 45 days in a very short time with the people's manual labor in 1939 by 160,000 Uzbek and Tajik workers (Meros, 2015) (Figure 27.4).

During this period, much experience was gained due to the irrigation of the Mirzachul, Karshi, and Surkhan–Sherabad deserts. Of course, during the 20th century, the rapid development of water resources in the country, the development of many lands, and the construction of new reservoirs and hydraulic structures have changed the nature and image of our country. As a result, new cities and villages have been built in our country, and a completely new water management system has emerged (Figure 27.5).

Following this, the science in water management has developed, and the training of personnel required for the needs of this field has reached a new level. Research in water management and reclamation in Central Asia began in the late 19th–early 20th centuries, and the field reached its peak in the 60s of the last century.

However, in the former Soviet Union, large-scale desertification and the dominance of cotton cultivation led to a sharp decline in the flow of the Amudarya and Syrdarya rivers into the Aral Sea, which had a negative effect. Water used in various industries and utilities was dumped into rivers and streams, and the ecological situation deteriorated sharply. Especially in the 70s and 80s of the 20th century, the river flows of the region were used wastefully. The ecological degradation and drying up of the Aral Sea, which has occurred due to the violation of centuries-old traditions of rational use of water in our region and our country, has become a global problem. The traditions of our people, such as the efficient use of water and its purity, have been neglected.

In addition to the locally caused environmental problems, global climate change has also contributed to water shortages in Central Asia. More frequent low-water periods are observed in the region because of global climate change two or three years in a row. Before the 1970s, water-scarce years appeared every 10–12 years; during the last

Figure 27.5 Mirishkor canal built in 1973 (one of the big irrigation projects to divert water from Amudarya to agricultural development of Karshi steppe).

40 years, they are observed every four to six years. The degradation of permanent glaciers caused a short-term increase in large river flows in Central Asia. The water flows of the basins of small rivers like Surkhandarya and Kashkadarya have noticeably decreased. The water content variability on these rivers increased with the transition from snow-feeding to rain-feeding (Salokhiddinov, 2021).

27.5 AFTER GAINING INDEPENDENCE OF UZBEKISTAN

Currently, there is 15.9 million ha of irrigated land in the country. Irrigated area is 4.3 million ha or 9.6% of the 44.9 million ha total area. Irrigated lands account for 95% of agricultural production. Water is essential for everything in Uzbekistan, located in an arid region, far from the oceans and major seas. The average annual volume of water resources used in Uzbekistan has been 51–53 km^3. This is a sign that the total volume of water taken from rivers has dropped significantly by 20% compared to the 1980s. This is because river water flow has also been declining under the influence of climate change and due to the problems of transboundary water use. From 1980 to the present, the population of Uzbekistan rose from 15.0 to 34.0 million. As a result, the per capita water consumption has decreased (from 4,403 m^3 per capita in 1,980 to 1,600 m^3 in 2019). About 80% of the water resources used in our country (about 41.5 km^3/year) are formed in neighboring republics. The main part of water resources used in the country is used for irrigation (90%), utilities (2.3%), energy (1.4%), industry (1.2%), fisheries (0.6%), and other purposes (4.5%) (Ministry of Water Resources of the Republic of Uzbekistan, 2020).

It is estimated that a 10%–20% reduction in water supply due to climate change in the future could have serious consequences for the size of irrigated land and employment, resulting in a decline in gross domestic product. Effective water resources management to meet the needs of irrigated agriculture, utilities and industry, the environment, and other sectors is crucial to ensure the country's sustainable economic development. Our country has been gradually implementing the necessary measures using several methods to save water in recent years. Water-intensive crops such as cotton and rice have been reduced and replaced by cereals, vegetables, melons, orchards, and vineyards. In particular, compared to the 1980s, the cotton fields have been reduced by almost 50%, and the rice fields have been reduced by 75%, bringing water savings to 20% (IFAS, 2020).

To reduce water loss, in 2020 alone, water-saving technologies were implemented on 133,600 ha in the country. In this connection, during the past year, the area of water-saving technologies implementation has reached 291,200 ha —about 7% of the total area of irrigated land. The state budget allocated subsidies of 132.8 billion sums for farms and cluster organizations implementing drip irrigation on 17,053 ha of cotton fields. As a result of water-saving irrigation technologies during the vegetation, 280 million m^3 of water was saved, the water supply of more than 300 thousand ha of irrigated land was improved, and the opportunity was created to irrigate additional crops with an area of more than 16000 ha (Ministry of Water Resources of the Republic of Uzbekistan, 2020).

Uzbekistan is planting thousands of trees on the dried Aral seabed with the help of international partners to combat the environmental impacts. Initiatives like "GreenAralSea" by UNDP and "Aral Forest" by the Ministry of Innovation Development of the Republic of Uzbekistan are among them (UNDP, 2020; IICAS, 2020).

The country has 28,440 km of irrigation networks, including 18,701 km of earthen canals, 9,203 km of concrete-lined canals, and 536 km of reinforced concrete canals. Moreover, 54,432 different hydraulic structures have been built. Seventy reservoirs and flood reservoirs with a total volume of 19.4 billion m^3 are in use. To improve the reclamation of irrigated lands, a total length of 142.9 thousand km, including 106.2 thousand km of open and 36.7 thousand km of closed horizontal collector-drainage network, 172 reclamation pumping stations, and 3,897 vertical drainage wells are used.

In the medium and long term, to effectively manage water resources, organize the rational use of water in the country, and reform water management, the "Concept of Water Resources Development of the Republic of Uzbekistan in 2020–2030" was developed. It is planned to achieve the following indicators by the end of 2030:

- increase the efficiency of irrigation systems from 0.63 to 0.73;
- reducing the area of irrigated lands with low water supply from 560 to 190 thousand ha;
- reduction of annual electricity consumption of pumping stations in the system of the Ministry of Water Resources from 8.0 to 6.0 billion kWh;
- increasing the total area covered by water-saving technologies for irrigation of crops to 2.0 million ha, including drip irrigation technology, to 500,000 ha;
- reduction of total irrigated saline areas from 1,935 to 1,603 thousand ha, medium- and high-salinity lands from 607 to 258 thousand ha;
- increasing the number of water facilities with the introduction of automated water metering and control system—"Smart Water" to 1,000;
- automation of water management processes at 100 large water facilities;
- implementation of 50 projects in the water sector based on public–private partnerships (IFAS, 2020).

Development and implementation of the "Concept of water resources development of the Republic of Uzbekistan in 2020–2030" are playing an extremely important role in ensuring the country's water and food security; through the effective management of water resources and the organization of rational use of water in the country, the reform of water management, and the widespread introduction of market principles and mechanisms and information and communication technologies, as well as the effective use of scientific potential.

27.6 CONCLUSIONS

The analysis of water culture, the development of water management and irrigated agriculture, the construction of hydraulic structures and the development of irrigation systems, and global water problems show that the traditional attitude of the Uzbek people to water, restoration of historical values, its preservation, increasing irrigated lands using only water-saving technologies, and the transition to a drip irrigation system remains essential. At the same time, through environmental education in the family, it is important to form the idea that water—a unique gift for the population should be used with care, and all hydraulic structures should be built and operated with environment-friendly technologies.

ACKNOWLEDGMENTS

The authors would like to acknowledge Prof. Robert Ettema of Colorado State University for his encouragement in preparing the paper.

REFERENCES

Ashirov, A. (2007). *Ancient beliefs and ceremonies of the Uzbek people,* Publishing House of the National Library of Uzbekistan named after Alisher Navoi: Tashkent.

Ashirov, A. (2020). *Water in Uzbek culture*, Akademnashr: Tashkent.

Bahodirxon, A. (1990). *Shajarai Turk*, Cho'lpon: Tashkent.

Bakiyev, M., Majidov, I., Nosirov, B., Xo'jaqulov, R., Rahmatov, M. (2008a). *Hydraulic structures textbook*, Yangi asr avlodi: Tashkent.

Bakiyev, M. R., Qodirov, O., Murodov, R. A. (2008b). *Architecture,* Tashkent Institute of Irrigation and Agricultural Mechanization Engineers: Tashkent.

Beruni, A. R. (1968). *Monuments left from ancient people* (Vol.1), Fan: Tashkent.

Frankopan, P. (2016). *The Silk Roads: a new history of the world* (18th edition), Bloomsbury Publishing Plc: Harvard.

G'afforov, Sh. S., Yunusov, M. A., Sharipov, S. I., Saidov, M. M. (2011). *History of architectural monuments of Uzbekistan*, Samarkand State University: Samarkand.

G'ulomov, Ya. G. (1959). *History of irrigation of Khorezm: from ancient times to the present*, Fan: Tashkent.

IFAS. (2020). Decree of the President No. UP-6024, *On approval of the water sector development concept of Uzbekistan for 2020–2030*, https://aral.uz/en/blog/2020/

International Innovation Center for the Aral Sea Basin under the President of the Republic of Uzbekistan, IICAS. (2020). #AralForest, *Let's plant 1,000, 000 trees*, https://aralforest.org/

Great Fergana Canal. (2021). Ministry of Water Resources of the Republic of Uzbekistan, (2020), https://water.gov.uz/public/files/1/5f48e327b7732.pdf. Accessed September 2021.

Khasanov, A. O. (2016, June). About several infrastructure constructions of the Great Silk Road, *International Journal of Innovative Science, Engineering & Technology*, *3*(6), 295–299.

Mavlonov, U. (2008). *Ancient Roads of Central Asia*, Akademiya: Tashkent.
Meros. (2015, October 29). http://world.meros.uz/en/wonder/view?id = 697
Meros. (2017, April 26). https://meros.uz/object/oqsaroy
Mukhammadzhanov, A. R. (1975). *Towards the history of irrigation in the Kushan era. In the book. Central Asia in the Kushan era* (Vol. 3), Nauka: Moscow.
Salokhiddinov, A. T., et al. (2021). Impact of regional climate changes on changes in river water content in Uzbekistan, *Journal of Irrigation and Melioration*, *2*(24), 7–12.
Temur, A. (2011). *The code of Amir Temur*, Gafur Gulam: Tashkent.
United Nations Development Programme (UNDP). (2020). *Green Aral Sea Initiative – Planting a Forest on the Aral Seabed*, https://www.uz.undp.org/content/uzbekistan/en/home/presscenter/pressreleases/2020/020/green-aral-sea-initiative--planting-a-forest-on-the-aral-seabed.html. Accessed June 2020.
Vatandosh. (2021). http://vatandosh.uz/2021/06/dargom-kanalining-tarixi/

Index

For Product Safety Concerns and Information please contact our EU representative GPSR@taylorandfrancis.com Taylor & Francis Verlag GmbH, Kaufingerstraße 24, 80331 München, Germany